Advanced Technologies in Physical and Mechanical Wood Modification

Advanced Technologies in Physical and Mechanical Wood Modification

Guest Editors

Radoslaw Mirski
Dorota Dziurka

MDPI

Basel • Beijing • Wuhan • Barcelona • Belgrade • Novi Sad • Cluj • Manchester

Guest Editors

Radoslaw Mirski
Department of Mechanical
Wood Technology
Poznan University of Life
Sciences
Poznań
Poland

Dorota Dziurka
Department of Wood-Based
Materials
Poznań University of Life
Sciences
Poznań
Poland

Editorial Office
MDPI AG
Grosspeteranlage 5
4052 Basel, Switzerland

This is a reprint of the Special Issue, published open access by the journal *Forests* (ISSN 1999-4907), freely accessible at: https://www.mdpi.com/journal/forests/special_issues/wood_techlologies.

For citation purposes, cite each article independently as indicated on the article page online and as indicated below:

Lastname, A.A.; Lastname, B.B. Article Title. *Journal Name* **Year**, *Volume Number*, Page Range.

ISBN 978-3-7258-5201-7 (Hbk)
ISBN 978-3-7258-5202-4 (PDF)
https://doi.org/10.3390/books978-3-7258-5202-4

Contents

About the Editors vii

Preface ix

Dorota Dziurka and Radosław Mirski
Advanced Technologies in Physical and Mechanical Wood Modification
Reprinted from: *Forests* **2025**, *16*, 1388, https://doi.org/10.3390/f16091388 1

He Sun, Xun Chang, Changqing Fu, Yuntian Yan, Chunlei Dong and Taian Chen
Effect of Impregnation with a Low-Concentration Furfuryl Alcohol Aqueous Solution on Hygroscopic Properties of Chinese Fir and Poplar Wood
Reprinted from: *Forests* **2022**, *13*, 1176, https://doi.org/10.3390/f13081176 5

Waldemar Perdoch, Ewelina Depczyńska, Karolina Tomkowiak, Monika Furgał, Mariola Kurczak and Bartłomiej Mazela
The Impact of Vinylotrimethoxysilane-modified Linseed Oil on Selected Properties of Impregnated Wood
Reprinted from: *Forests* **2022**, *13*, 1265, https://doi.org/10.3390/f13081265 19

Aleksandra Jeżo, Anita Wronka, Aleksander Dębiński, Lubos Kristak, Roman Reh, Janis Rizhikovs and Grzegorz Kowaluk
Influence of Upcycled Post-Treatment Bark Biomass Addition to the Binder on Produced Plywood Properties
Reprinted from: *Forests* **2023**, *14*, 110, https://doi.org/10.3390/f14010110 31

Radosław Auriga, Piotr Borysiuk, Maciej Latos, Alicja Auriga, Łukasz Kwaśny and Joanna Walkiewicz
Impact of Sugar Beet Pulp Share on Selected Physical and Mechanical Properties of Particleboards
Reprinted from: *Forests* **2023**, *14*, 40, https://doi.org/10.3390/f14010040 45

Izabela Betlej, Bogusław Andres, Krzysztof Krajewski, Anna Kiełtyka-Dadasiewicz, Piotr Boruszewski, Dominika Szadkowska, et al.
The Effect of Ethanol Extracts and Essential Oils Obtained from Different Varieties of Mint on Wood Molding
Reprinted from: *Forests* **2023**, *14*, 1522, https://doi.org/10.3390/f14081522 53

Qian Lang, Vladimirs Biziks and Holger Militz
Influence of Phenol–Formaldehyde Resin Oligomer Molecular Weight on the Strength Properties of Beech Wood
Reprinted from: *Forests* **2022**, *13*, 1980, https://doi.org/10.3390/f13121980 70

Qian Lang, Vladimirs Biziks and Holger Militz
Water Vapor Sorption Kinetics of Beech Wood Modified with Phenol Formaldehyde Resin Oligomers
Reprinted from: *Forests* **2023**, *14*, 2015, https://doi.org/10.3390/f14102015 81

Andi Detti Yunianti, S Suhasman, A Agussalim, Musrizal Muin and Heru Arisandi
Timber Properties and Cellulose Crystallites Size in Pine Wood Cut in Different Sawing Patterns after Pretreatment with CH_3COOH and H_2O_2 and Densification
Reprinted from: *Forests* **2021**, *12*, 1607, https://doi.org/10.3390/f12111607 91

Marek Grześkiewicz, Sławomir Krzosek, Izabela Burawska, Piotr Borysiuk and Piotr Mańkowski
Influence of Thermo-Mechanical Densification (TMD) on the Properties of Structural Sawn Timber (*Pinus sylvestris* L.)
Reprinted from: *Forests* **2023**, *14*, 231, https://doi.org/10.3390/f14020231 . . . . **100**

Shibin Yu, Wen Pan, Hexian Su, Liaoyuan Ye and Daohang Wang
Experimental Study on Tenon and Mortise Joints of Wood-Structure Houses Reinforced by Innovative Metal Dampers
Reprinted from: *Forests* **2022**, *13*, 1177, https://doi.org/10.3390/f13081177 . . . . **112**

Yuanyuan Guo, Wei Wang and Xuewei Jiang
Molecular Dynamics Study on Mechanical Properties of Cellulose with Water Molecules Diffusion Behavior at Different Oxygen Concentrations
Reprinted from: *Forests* **2023**, *14*, 371, https://doi.org/10.3390/f14020371 . . . . **129**

Aneta Gumowska and Grzegorz Kowaluk
Physical and Mechanical Properties of High-Density Fiberboard Bonded with Bio-Based Adhesives
Reprinted from: *Forests* **2023**, *14*, 84, https://doi.org/10.3390/f14010084 . . . . **142**

Istie Rahayu, Fitria Cita Dirna, Akhiruddin Maddu, Wayan Darmawan, Dodi Nandika and Esti Prihatini
Dimensional Stability of Treated Sengon Wood by Nano-Silica of Betung Bamboo Leaves
Reprinted from: *Forests* **2021**, *12*, 1581, https://doi.org/10.3390/f12111581 . . . . **158**

Aleš Straže, Jure Žigon, Stjepan Pervan, Mislav Mikšik and Silvana Prekrat
The Influence of Processing Conditions on the Quality of Bent Solid Wood from European Oak
Reprinted from: *Forests* **2023**, *14*, 1047, https://doi.org/10.3390/f14051047 . . . . **167**

Piotr Borysiuk, Radosław Auriga, Jacek Wilkowski, Alicja Auriga, Adrian Trociński and Lee Seng Hua
A Study on the Susceptibility of PLA Biocomposites to Drilling
Reprinted from: *Forests* **2022**, *13*, 1950, https://doi.org/10.3390/f13111950 . . . . **178**

Zbigniew Malinowski, Jakub Kawalerczyk, Joanna Walkiewicz, Dorota Dziurka and Radosław Mirski
The Effect of the Tree Dieback Process on the Mechanical Properties of Pine (*Pinus sylvestris* L.) Wood
Reprinted from: *Forests* **2023**, *14*, 274, https://doi.org/10.3390/f14020274 . . . . **188**

About the Editors

Radoslaw Mirski

Radoslaw Mirski is a full professor at the Faculty of Forestry and Wood Technology, Poznań University of Life Sciences (Poland). Prof. Mirski's research is focused on enhancing the efficiency and sustainability of wood and non-wood raw materials in the production of composite and prefabricated materials for the construction sector. His major research interests include the durability and environmental performance of OSB panels, the use of agricultural residues in insulation materials, WPC technology, and the modification of phenolic resins. He also investigates energy and resource efficiency in technological processes and recycling potential of wood-based products.

He is the author or co-author of over 220 scientific publications, including more than 100 papers indexed in JCR journals, and co-inventor of 14 patents. He has participated in numerous R&D projects, including national strategic initiatives like BIOSTRATEG. Prof. Mirski has supervised several doctoral theses and is an experienced academic teacher in the areas of wood technology, electrotechnics, automation, and structural engineering.

As a scholar with interdisciplinary expertise combining forestry, material science, and civil engineering, Prof. Mirski is actively engaged in the development of innovative, bio-based materials and solutions for sustainable construction.

Dorota Dziurka

Dorota Dziurka is a full professor at the Faculty of Forestry and Wood Technology, Poznań University of Life Sciences (Poland). Her research interests focus on wood-based technologies, adhesive resin modification, and the potential use of renewable and alternative lignocellulosic raw materials to produce composite materials. She has over 200 scientific publications, including over 80 in JCR-listed journals. She is also the co-author of 9 patents. She has participated in many research and development projects funded by, among others, the National Centre for Research and Development and the National Science Centre, including those concerning low-emission materials for the furniture and construction industries. Prof. Dziurka is actively involved in teaching, conducting classes on wood-based materials while also supervising diploma and doctoral theses. Her work combines technological experience with a pro-environmental approach and the practical application of research results.

Preface

The dynamic development of the wood industry and growing demands for wood-based materials are forcing the search for new methods to improve their performance properties. As a renewable and low-emission raw material, wood is becoming a key material in sustainable construction and the circular economy. However, the natural limitations of wood, such as its sensitivity to moisture, fungi, deformation, and structural variability, require implementing strategies for its permanent modification.

This Special Issue of Advanced Technologies in Physical and Mechanical Wood Modification brings together sixteen original scientific articles addressing the topic of wood modification from various perspectives: chemical, biological, physical, and molecular and material modelling. The articles cover, among other things, the influence of phenol-formaldehyde resin oligomers on the sorption kinetics and strength of beech wood, the effect of natural antifungal agents obtained from mint, the process conditions for bending oak wood, and the behaviour of cellulose in environments with varying oxygen concentrations. Among the experimental works, research on the impact of heat and chemical treatment on Scots pine wood, the use of food and wood industry waste in composites, and the properties of wood-based panels bonded with bio-based adhesives occupy a special place. Another important novelty is the reports on the impact of nanoparticles and structural modifications at the molecular level, which bridge the gap between materials science and wood engineering.

The collected publications highlight the interdisciplinary nature of contemporary wood research, combining natural sciences, engineering, and technology, and show that effective wood modification requires both an understanding of mechanisms at the micro level and their transfer to the industrial scale. This collection is valuable to developing knowledge about wood modification and may inspire further research into sustainable and environmentally friendly new-generation wood materials.

The Guest Editors would like to thank all authors from Poland, China, Slovakia, Slovenia, Croatia, Germany, Lithuania and Indonesia who contributed to this Special Issue. We want to express our special thanks to Nikola Jovanović Editor of this Special Issue, for his professionalism, kindness, and assistance at every stage of our cooperation.

Radoslaw Mirski and Dorota Dziurka
Guest Editors

forests

MDPI

Editorial

Advanced Technologies in Physical and Mechanical Wood Modification

Dorota Dziurka * and Radosław Mirski

Department of Mechanical Wood Technology, Faculty of Forestry and Wood Technology, Poznań University of Life Sciences, 60-627 Poznań, Poland; radoslaw.mirski@up.poznan.pl
* Correspondence: dorota.dziurka@up.poznan.pl

In recent years, research on wood modification, wood composites, and the use of renewable raw materials and plant industry by-products in materials engineering has grown dynamically. Sustainable development, the need to reduce chemical emissions, and the improvement of the functional properties of wood and wood-based materials are key challenges facing modern science and the wood industry.

The publications collected in this "Advanced Technologies in Physical and Mechanical Wood Modification" Special Issue address these challenges comprehensively and in an interdisciplinary manner. They contain the results of experimental and modeling studies on the impact of chemical, physical, and thermomechanical treatment on the properties of solid wood, composite materials, and biopolymers. Particular attention is paid to issues related to the following:

- Modification of wood using organic compounds such as furfuryl alcohol, which at low concentrations (10%) reduced total water vapour sorption by approx. 28% in poplar wood and 35% in Chinese fir [1]. In turn, the use of silane-modified linseed oil contributed to a 38.6% increase in dimensional stability and a reduction in water swelling of over 30% [2], which, according to the authors, should translate into increased resistance of wood to colonising and decomposing fungi.
- Waste biomass, such as tree bark or beet pulp, used to produce wood-based materials. It has been shown that in the case of plywood with 10% bark biomass added, an increase in shear strength at the glue line of approximately 18% was observed [3]. In the case of particleboard, beet pulp content of up to 25% may be acceptable for producing lower-density boards that meet the requirements of EN 312 for general-purpose P2 boards (used in dry conditions) [4].
- Improving biological durability and resistance to fungi. Betlej et al. [5] researched the effectiveness of ethanol extracts and essential oils obtained from various varieties of mint (*Mentha* sp.) in limiting the growth of mould fungi and wood discolouration. Ethanol extracts used at a dose of $\geq$40 g/m^2 showed vigorous fungistatic activity—mycelium growth inhibition reached up to 90% on agar medium. Although essential oils were more active (especially against Chaetomium globosum), they did not show a full biocidal effect on wood. The authors emphasise that the chemical composition of the extracts, mainly the presence of oxygenated monoterpenoids and monoterpenes, determines the strength of the biological effect, and the use of natural plant-based agents can be an ecological supplement to traditional wood protection methods, especially in the early stages of their use.
- Analysis of wood's mechanical, sorption, and structural properties after modification, e.g., with phenol–formaldehyde resins [6–8]. Lang et al. [6] investigated the effect of

the molecular weight of phenol–formaldehyde (PF) resin oligomers on the mechanical properties and dimensional stability of beech wood impregnated with these resins. The results showed that resins with a lower molecular weight (approx. 237–305 g/mol) provide better penetration and more uniform filling of cell walls. This translates into significantly increased dimensional stability (less shrinkage and swelling) and greater weight gain (WPG), reaching 24.7% at a 20% solution concentration. The authors showed that the observed decrease in modulus of elasticity (MOE) and impact resistance (IBS)—by as much as 60–64% compared to unmodified wood—is a consequence of the formation of a rigid, crystalline PF structure in the cell wall, which limits the plasticity of the material. Microscopic studies confirmed that low-molecular-weight resins penetrate cell walls better, while those with a higher molecular weight are mainly deposited in the lumen of the vessels. In another publication [7], the same authors showed that thanks to PF wood modification, water vapour sorption kinetics decreased by 25–30%.

- The use of thermomechanical modification. It has been shown [9] that thermomechanical treatment of pine wood significantly increases its density (35%), stiffness, bending strength (47%), and resistance to cracking, making it a more valuable construction material. In addition, long-term deformation (creep) is significantly reduced, improving structural applications' durability and reliability. Furthermore, by controlling the process parameters (pressure, temperature, and degree of compression), the characteristics of the final product can be adjusted. In summary, the authors state that TMD is an effective way to modify low-grade wood, making it suitable for more demanding engineering applications.
- Innovative structural solutions, such as the use of metal dampers in tenon and mortise joints, which increased the shear strength of the joints by 20–25% and improved the behaviour of the structure under cyclic dynamic loads [10]. The authors researched using these innovative dampers in traditional tenon and mortise joints in wooden structures. The results of quasi-static experiments showed that using dampers made of Q235 steel resulted in a significant increase in stiffness, load-bearing capacity, and energy absorption by the joints. The reinforced joints showed more stable and symmetrical hysteresis curves and reduced residual displacements, which means less play and better behaviour under cyclic loads. As the researchers point out, this technology can be used to reinforce new wooden structures and protect historic buildings without compromising their aesthetic and structural integrity. The dampers are discreetly mounted on the side surfaces.
- Using numerical and spectroscopic methods. Molecular dynamics simulations have shown that the oxygen content in the cellulose environment has a significant impact on hydrogen bonding with water, which may affect the hygroscopicity and stability of cellulose materials [11]. Guo et al. [11] conducted molecular dynamics (MD) studies to evaluate the effect of oxygen concentration on the mechanical properties and structure of cellulose. The study analysed models with 0–10% oxygen content, reflecting the conditions of steam heat treatment of wood. It was shown that at an oxygen concentration of about 2%, the highest values of Young's and shear modulus are obtained, indicating greater stiffness of cellulose. Higher concentrations led to the degradation of internal hydrogen bonds, increased water diffusion, and greater mobility of cellulose chains, resulting in reduced mechanical properties. The study results provide theoretical support for the design of thermal–steam treatment processes at the molecular level, with the possibility of optimising conditions to achieve better wood stiffness and material stability.

One of the essential aspects of the research is its environmental benefits. Modifying wood using natural agents—such as vegetable oils, bioadhesives, or natural compounds—reduces the use of toxic chemicals, including traditional preservatives and formaldehyde-based adhesives [2,12]. Research on the use of bio-based adhesives, based on natural raw materials, especially starch, in the production of HDF boards has shown that they can provide parameters comparable to conventional synthetic adhesives [12]. In addition, using agricultural and food industry by-products as additives to wood-based materials supports the circular economy by reducing waste and improving resource efficiency [3,4]. Using such solutions promotes the creation of materials with a smaller carbon footprint and greater durability, which in the long term leads to a reduction in the exploitation of forest resources.

A key result of the research published in this SI is a significant increase in the durability of wood and wood-based materials. The use of appropriate modification methods—such as chemical impregnation, heat treatment, or the use of bioactive additives—contributes to improving the resistance of wood to biological factors (fungi, insects), moisture, and changing weather conditions [1,5,13]. The increases the service life of the materials, affecting their suitability for structural and outdoor applications and reducing the need for frequent replacement or maintenance, thereby reducing costs and the environmental footprint of use.

The practical applications of the developed solutions cover a wide range of industries. In construction, modified wood is used as a structural element, as a facade element, or in window and door joinery [9,14], characterised by increased durability and resistance to external conditions. In the furniture and finishing industries, improving wood's mechanical and aesthetic properties allows for the production of durable and environmentally friendly products [12]. In the packaging and transport sector, biodegradable wood composites and materials reinforced with natural fibres and bioadhesives are being developed [15]. Finally, using agricultural waste as additives to wood-based materials opens up new opportunities for agriculture and the food industry, promoting local processing and regional innovation [4].

It is also worth noting that wood used in the timber industry increasingly comes from trees weakened or dying due to environmental stress, disease, or climate change. As shown by research on Scots pine (*Pinus sylvestris* L.) wood originating from deadwood [16], the process of biological dieback significantly affects its mechanical properties. The reduction in the strength parameters of such raw material may hinder its direct use, creating the need for appropriate physical, chemical, or structural modification before such wood is approved for structural or composite use. Wood modification may therefore prove to be not only a way to improve the quality of the material in the future, but also a requirement for the effective use of raw material with reduced technical value.

Modern wood modification methods also include advanced artificial intelligence algorithms that allow for precise modeling of changes in wood properties, e.g., colour due to heat treatment [17]. The high accuracy of such models indicates their potential in predicting the effects of technological processes and optimising production parameters. In turn, biological approaches using fungal inoculation and controlled mechanical damage allow for the activation of trees' natural defence mechanisms and the stimulation of the production of desired secondary metabolites, such as sesquiterpenes [18]. Such approaches are critical in producing high-value aromatic raw materials like agarwood. These types of approaches are focused on the intentional design of wood properties.

The research presented in this SI deepens knowledge in the field of wood technology and materials science and responds to the real needs of the wood, furniture, and construction industries. The results can be applied both in developing new material production technologies and in designing modern, durable, and environmentally friendly wood products.

Author Contributions: Conceptualisation, D.D. and R.M.; formal analysis, D.D. and R.M.; writing and editing, D.D. All authors have read and agreed to the published version of the manuscript.

Funding: This research received no external funding.

Conflicts of Interest: The authors declare no conflicts of interest.

References

1. Sun, H.; Chang, X.; Fu, C.; Yan, Y.; Dong, C.; Chen, T. Effect of Impregnation with a Low-Concentration Furfuryl Alcohol Aqueous Solution on Hygroscopic Properties of Chinese Fir and Poplar Wood. *Forests* **2022**, *13*, 1176. [CrossRef]
2. Perdoch, W.; Depczyńska, E.; Tomkowiak, K.; Furgał, M.; Kurczak, M.; Mazela, B. The Impact of Vinylotrimethoxysilane-Modified Linseed Oil on Selected Properties of Impregnated Wood. *Forests* **2022**, *13*, 1265. [CrossRef]
3. Jeżo, A.; Wronka, A.; Dębiński, A.; Kristak, L.; Reh, R.; Rizhikovs, J.; Kowaluk, G. Influence of Upcycled Post-Treatment Bark Biomass Addition to the Binder on Produced Plywood Properties. *Forests* **2023**, *14*, 110. [CrossRef]
4. Auriga, R.; Borysiuk, P.; Latos, M.; Auriga, A.; Kwaśny, Ł.; Walkiewicz, J. Impact of Sugar Beet Pulp Share on Selected Physical and Mechanical Properties of Particleboards. *Forests* **2023**, *14*, 40. [CrossRef]
5. Betlej, I.; Andres, B.; Krajewski, K.; Kiełtyka-Dadasiewicz, A.; Boruszewski, P.; Szadkowska, D.; Zawadzki, J.; Radomski, A.; Borysiuk, P. The Effect of Ethanol Extracts and Essential Oils Obtained from Different Varieties of Mint on Wood Molding. *Forests* **2023**, *14*, 1522. [CrossRef]
6. Lang, Q.; Biziks, V.; Militz, H. Influence of Phenol-Formaldehyde Resin Oligomer Molecular Weight on the Strength Properties of Beech Wood. *Forests* **2022**, *13*, 1980. [CrossRef]
7. Lang, Q.; Biziks, V.; Militz, H. Water Vapor Sorption Kinetics of Beech Wood Modified with Phenol Formaldehyde Resin Oligomers. *Forests* **2023**, *14*, 2015. [CrossRef]
8. Yunianti, A.D.; Suhasman, S.; Agussalim, A.; Muin, M.; Arisandi, H. Timber Properties and Cellulose Crystallites Size in Pine Wood Cut in Different Sawing Patterns after Pretreatment with CH3COOH and H2O2 and Densification. *Forests* **2021**, *12*, 1607. [CrossRef]
9. Grześkiewicz, M.; Krzosek, S.; Burawska, I.; Borysiuk, P.; Mańkowski, P. Influence of Thermo-Mechanical Densification (TMD) on the Properties of Structural Sawn Timber (*Pinus sylvestris* L.). *Forests* **2023**, *14*, 231. [CrossRef]
10. Yu, S.; Pan, W.; Su, H.; Ye, L.; Wang, D. Experimental Study on Tenon and Mortise Joints of Wood-Structure Houses Reinforced by Innovative Metal Dampers. *Forests* **2022**, *13*, 1177. [CrossRef]
11. Guo, Y.; Wang, W.; Jiang, X. Molecular Dynamics Study on Mechanical Properties of Cellulose with Water Molecules Diffusion Behavior at Different Oxygen Concentrations. *Forests* **2023**, *14*, 371. [CrossRef]
12. Gumowska, A.; Kowaluk, G. Physical and Mechanical Properties of High-Density Fiberboard Bonded with Bio-Based Adhesives. *Forests* **2023**, *14*, 84. [CrossRef]
13. Rahayu, I.; Dirna, F.C.; Maddu, A.; Darmawan, W.; Nandika, D.; Prihatini, E. Dimensional Stability of Treated Sengon Wood by Nano-Silica of Betung Bamboo Leaves. *Forests* **2021**, *12*, 1581. [CrossRef]
14. Straže, A.; Žigon, J.; Pervan, S.; Mikšik, M.; Prekrat, S. The Influence of Processing Conditions on the Quality of Bent Solid Wood from European Oak. *Forests* **2023**, *14*, 1047. [CrossRef]
15. Borysiuk, P.; Auriga, R.; Wilkowski, J.; Auriga, A.; Trociński, A.; Li, S.H. A Study on the Susceptibility of PLA Biocomposites to Drilling. *Forests* **2022**, *13*, 1950. [CrossRef]
16. Malinowski, Z.; Kawalerczyk, J.; Walkiewicz, J.; Dziurka, D.; Mirski, R. The Effect of the Tree Dieback Process on the Mechanical Properties of Pine (*Pinus sylvestris* L.) Wood. *Forests* **2023**, *14*, 274. [CrossRef]
17. Liang, J.; Wang, W.; Qu, Z.; Cao, Y.; Gong, J. Prediction of Color Change in Heat-Treated Wood Based on Improved Zebra Algorithm Optimized Deep Hybrid Kernel Extreme Learning Machine Model (IZOA-DHKELM). *Forests* **2025**, *16*, 253. [CrossRef]
18. Zhao, W.; Song, X.; Zhou, Z.; Liu, G.; Zhang, Q.; Pang, S. Effects of Different Levels of Physical Damage Combined with Fungal Induction on Agarwood Formation. *Forests* **2024**, *15*, 168. [CrossRef]

forests

MDPI

Article

Effect of Impregnation with a Low-Concentration Furfuryl Alcohol Aqueous Solution on Hygroscopic Properties of Chinese Fir and Poplar Wood

He Sun, Xun Chang, Changqing Fu, Yuntian Yan, Chunlei Dong and Taian Chen *

College of Materials Science and Engineering, Southwest Forestry University, Kunming 650224, China; hesun1993@163.com (H.S.); 18811358536@163.com (X.C.); cq292560@163.com (C.F.); 15862231716@163.com (Y.Y.); dchunlei@swfu.edu.cn (C.D.)

* Correspondence: nfucta@sina.com

Abstract: Furfurylation with a low concentration of furfuryl alcohol (FA) promotes the improvement of the properties and the effectiveness of FA on cell–wall action without darkening the furfurylated wood to the point that it affects its applications. In this paper, the effects of furfurylation on the hygroscopicity and water uptake dimensional stability of poplar (*Populus* sp.) and Chinese fir (*Cunninghamia lanceolata*) were analyzed. Meanwhile, the distribution of FA resin, the relationship between wood and water, the change in pore size distribution, and the weight percentage gain and cell wall bulking coefficient of wood were also investigated. The results were as follows: (1) A low concentration of FA could better enter the cell walls of the Chinese fir than the poplar, as FA resin was almost cured in the secondary walls, cell corners, and compound middle lamellae when a 10% concentration of FA was applied to the Chinese fir and poplar. When the FA concentration was increased to 30%, there were no significant increases in the amount of FA entering the cell walls and the amounts of FA cured in the cell lumen of the poplar were greater than those of the Chinese fir. Meanwhile, the modification of cell walls was more suitable in poplar than in Chinese fir. (2) The pointed ends of the pit chambers and the pit apertures (800–1000 nm) in the poplar and the small pores of the pit membranes and the pit apertures (1–6 μm) in the Chinese fir were partially deposited by the FA resin, which formed new pores in the size ranges of 80–600 nm and 15–100 nm, respectively. The porosity of the poplar was greater than that of the Chinese fir, and the bulk density of the poplar was less than that of the Chinese fir before and after modification. (3) Furfurylation with a low concentration of FA was able to better reduce the equilibrium moisture content, improve the anti-swelling efficiency, and enhance the dimensional stability of the poplar wood compared to the Chinese fir. Furfurylation effectively reduced water uptake due to the hydrophobic property of the FA resin. The water uptake of the Chinese fir increased by 17%–19% in second cyclic water soaking when treated with FA with various concentrations, which indicated the loss and leaching of FA resin during the test. Low-field NMR was used to demonstrate that the furfurylation not only reduced the amount of water but also affected the combination state of bound and free water with wood. Thus, furfurylation at a low concentration is a feasible method by which to extend applications of furfurylated wood.

Keywords: furfurylated wood; moisture absorption; water uptake; dimensional stability; LF-NMR; pore size distribution; Chinese fir and poplar

1. Introduction

Chinese fir (*Cunninghamia lanceolata*) and poplar (*Populus* sp.) are the main plantation-grown species with great importance in the timber supply system in China. However, both the Chinese fir and poplar have significant disadvantages in terms of their poor mechanical properties and dimensional instability, limiting their high-value utilization in the wood-processing industry. Numerous studies have been conducted to improve their fast-growing wood density and application performance by using various methods such as UF and PF resin impregnation modification [1,2]. Furfurylation is an impregnation modification method [3,4], and it consists of impregnation and curing processes. Furfuryl alcohol (FA) can be impregnated into wood and then cured into resin under the action of heat and a catalyst, which can effectively increase wood density, enhance the mechanical properties, and improve dimensional stability [5–11]. It is of utmost importance for FA to be derived from biomass materials such as corn cobs and non-fossil resources [12]. Therefore, furfurylation is considered to be an environmentally friendly modification method [13–16].

The development of furfurylated wood originated in North America during the 1950s from I.S. Goldstein [17], and the furfurylation method of commercial wood represented by Kebony® in Norway is widely used in outdoor places such as wooden buildings and trestles [18,19]. However, the treatment process has always employed high levels of weight percentage gain, and large amounts of FA tend to accumulate and deposit in the cell lumen. The effectiveness of FA-modified cell walls also diminishes with increasing concentrations of FA. Additionally, furfurylated wood of a high concentration darkens the color of modified wood and does not significantly improve dimensional stability compared to furfurylated wood of a low concentration [19,20]; neither of them are compatible with the interior furniture and wood product requirements for light colors and high dimensional stability [21]. There are also differences in the wood properties between Chinese domestic, fast-growing trees and European temperate species, and previous studies indicated that the concentration of FA-modified solutions determines the distribution area in wood; FA is mainly deposited in wood's cell walls and is almost absent in the cell lumen when the concentration of FA is low [22]. Furthermore, compared to its greater improvement of the mechanical and hygroscopic properties of furfurylated wood with high concentrations of FA, the relationship between the modified wood with a low concentration of FA and hygroscopic properties for the main purpose of cell wall modification has yet to be scientifically evaluated.

This paper focused on the hygroscopicity and dimensional stability properties of furfurylated wood with a low FA concentration (below 30%) treatment. The effectiveness of the modification on the cell wall was analyzed by correlating the weight percentage gain and cell wall bulking coefficient. The changes in physical aspects such as the furfuryl alcohol resin distribution and pore size distribution were also analyzed. Finally, the effect of the modification treatment on the wood–water relationship of the Chinese fir and poplar was analyzed with low-field NMR.

2. Materials and Methods

2.1. Materials

Poplar and Chinese fir were harvested from Jiangsu Province and Zhejiang Province, China, respectively. The soil type of both the poplar and Chinese fir was clayey, the height of the poplar from sea level was 70 m, the height of the fir from sea level was 1500 m, and the diameters of the poplar and fir were 30 cm or more. The trees were over 25 years old. The air-dry density of the poplar was found to be about 480 kg/m^3, and that of the Chinese fir was about 310 kg/m^3. Specimens with straight grain and no visible defects were selected and processed into 20 × 20 × 20 mm (tangential × radical × longitudinal) when the moisture content was below 12%. Samples were selected from the sapwood of the Chinese fir and poplar. Furfuryl alcohol (FA), maleic anhydride, and borax were purchased from Tianjin Fengchuan Chemical Reagent Technologies Co., Ltd. (Tianjin, China).

2.2. Methods

2.2.1. Preparation of Furfurylated Wood

Distilled water was used as the solvent to prepare the modified solution with FA concentrations of 10%, 20% and 30% in which furfuryl alcohol, maleic anhydride, and borax were used in the mass ratio of 1:0.06:0.01, respectively. The impregnation processes were as follows.

(1) Drying: The specimens were placed in an oven at 103 °C for a 24 h drying process, and then the oven-dries mass and tangential, radical, and longitudinal directions were measured after cooling in a desiccator with an accuracy of 0.0001 g and 0.01 mm, respectively.
(2) Impregnation (full cell process): The specimens were first treated under a negative pressure of 0.06 MPa for 0.5 h, then kept under a pressure of 0.4 MPa for 2 h, and finally treated under a negative pressure of 0.06 MPa for 0.5 h [23]. After the impregnation was completed, the excess impregnation solution on the surface was wiped away. Meanwhile, in order to more uniformly diffuse the furfuryl alcohol, each specimen was wrapped in aluminum foil and kept at 30 °C for 72 h.
(3) Curing and drying: The samples were cured in an oven at 103 °C for 6 h, and then the aluminum foil was removed. The furfurylated wood mass M_1 and dimensions were obtained after drying; the drying process was carried out at 50 °C for 12 h, followed by 70 °C for 12 h, and finally 103 °C for 24 h. According to the weight percentage gain, the furfurylated poplar and Chinese fir wood impregnated with 0%, 10%, 20%, and 30% concentrations were labeled as PW0, PW17, PW35, PW51, SW0, SW23, SW45 and SW75, respectively.

The weight percentage gain (WPG) and bulking coefficient (BC) were calculated using the following equation:

$$\mathrm{WPG}(\%) = \frac{M_1 - M_0}{M_0} \times 100 \tag{1}$$

where M_0 and M_1 are the oven-dried weights (g) before and after FA treatment, respectively.

$$\mathrm{BC}(\%) = \frac{V_1 - V_0}{V_0} \times 100 \tag{2}$$

Here, V_0 and V_1 are the oven-dried volume (mm^3) before and after treatment, respectively.

2.2.2. Morphology and Microstructure

Each sample was placed in a beaker filled with distilled water; then, the beaker was placed in a desiccator with a pressure of −0.06 MPa until the sample was submerged; afterwards, the sample was softened in a water bath at 90 °C for 2 h. Transverse and longitudinal sections with thicknesses of 10–12 μm were sectioned with a microtome (SM2000R, Leica company, Weitzlar, Germany).

Transverse sections of the samples were observed with confocal laser scanning microscopy (SP8, Leica company, Weitzlar, Germany) using a 63× oil microscope, with a 633 nm excitation state and a detector range of 650–700 nm. Meanwhile, the distribution position of FA in the cell wall was observed with a scanning electron microscope (Quanta 200, FEI, Portland, OR, USA).

2.2.3. EMC and Dimensional Stability of Adsorption

Samples were conditioned at 20 °C and relative humidity (RH) of 33%, 65%, and 95% through moisture absorption with 12 replicates [24]. The humidity and temperature in the moisture absorption test were controlled using a climate chamber (KMF720, Binder, Tuttlingen, Germany) with a temperature accuracy of ±0.1 °C and a humidity accuracy of ±2.5%. The measurement accuracy of the balance was 0.0001 g [25]. The moisture adsorption tests were continuously conducted from the lowest to higher RH values after the samples were treated in the oven at 103 °C for 24 h. The weight and dimensions were measured when mass variations were lower than 0.1% after daily mass readings at each RH. The reduced equilibrium moisture content (EMC_R) and anti-swelling efficiency (ASE*) were calculated using the following equation [26,27]:

$$\mathrm{EMC}(\%) = \frac{M_2 - M_1}{M_1} \times 100 \tag{3}$$

where M_2 is the mass of absorption equilibrium.

$$\mathrm{EMC_R} = \mathrm{EMC}(1 + \mathrm{WPG}) \quad (4)$$

$$\mathrm{ASE}(\%) = \frac{\alpha_1 - \alpha_0}{\alpha_0} \quad (5)$$

Here, α_1 and α_0 are the volume swelling efficiency of untreated and treated wood in absorption, respectively.

$$\alpha(\%) = \frac{V_a - V_u}{V_u} \times 100 \quad (6)$$

$$\mathrm{ASE}^* = \mathrm{ASE}(1 + \mathrm{BC}) - \mathrm{BC} \quad (7)$$

Here, V_a is the swollen volume and V_u is the oven-dried volume.

2.2.4. Mercury Intrusion Porosimetry (MIP)

Mercury intrusion porosimetry (AutoPore IV, Micromeritics, Norcross, GA, USA) was used to analyze the pore size distribution, cumulative pore volume, bulk density, and samples porosity with dimensions of 6 × 6 × 5 mm^3 (R × T × L) under oven-dried conditions. Hg intrusion was performed using a pressure between 0 and 414 MPa with an equilibrium time of 10 s at the initial and the final pressure levels during the MIP test process.

2.2.5. LF-NMR Analysis

For the analysis of the effect of FA modification on bound and free water, low-field NMR (MicroMR-60H, Niumag, Shanghai, China) was used to analyze the control and furfurylated wood with 30% FA and to condition samples to become fully water-saturated via the full-cell vacuum-pressure process. The Carr–Purcell–Meiboom–Gill (CPMG) pulse sequence was used to calculate the T_2 of the samples, and 18,000 echoes with 64 scans were acquired. The echo time was 0.2 ms, LF-NMR data were inverted with a multi-exponential function, and the algorithm used SIRT. The test temperature was room temperature (25 °C).

2.2.6. Cyclic Water Soaking and Dimensional Stability

In order to evaluate the effect of FA treatment on the water uptake and dimensional stability, the samples were first placed in an oven at 103 °C for 24 h. Second, we conditioned samples to become fully water-saturated using a full-cell vacuum-pressure process and then soaked them in water at 30 °C for 24 h. Finally, we oven-dried them again. This process was defined as a drying–soaking–drying cycle, which was repeated three times [28]. We tested 12 replicates for each group. The reduced water uptake ($\mathrm{MC_{WAR}}$) and reduced anti-swelling efficiency ($\mathrm{ASE_{ws}}$) with water saturation were calculated with each cycle repetition and using the following equations:

$$\mathrm{MC_{WA}}(\%) = \frac{G_i - G_{i0}}{G_{i0}} \times 100 \quad (8)$$

$$\mathrm{MC_{WAR}} = \mathrm{MC_{WAR}}(1 + \mathrm{WPG}) \quad (9)$$

where G_i is the mass of a fully water-saturated sample and G_{i0} is the mass of an oven-dried sample in the same number of cycles.

Reduced anti-swelling efficiency ($\mathrm{ASE_{ws}}$) values with water saturation were calculated with Equations (5)–(7) described above [25,26] during cyclic water soaking.

2.2.7. Statistical Analysis

IBM SPSS Statistics (V.23) software was employed to assess the effect of the furfuryl alcohol treatment concentration on weight percentage gain, bulking coefficient, equilibrium moisture content, water uptake, and volume swelling efficiency, according to the methods of a previous study [29]

3. Results and Discussion

3.1. WPG and BC of Furfurylation, Hygroscopicity and Dimensional Stability

The WPG and the FA concentration showed a significant positive relationship, as seen in Table 1. However, there were considerable differences in the WPG of the Chinese fir and poplar. At a 10% furfurylation concentration, the WPG values of the Chinese fir and poplar were about 23% and 17%, respectively. When the concentration increased to 30%, the WPG values of the Chinese fir and poplar were 75% and about 51%, respectively, indicating that the WPG of the Chinese fir was generally higher than that of the poplar. However, the air-dry densities of the poplar and fir were about 480 and 310 kg/m^3, respectively, which led to differences in the permeability and porosity of the Chinese fir and poplar and may have resulted in different weight percentage gains in furfurylation. Additionally, there was a large number of alternate pits in the vessel that may have resulted in a lower retention of furfuryl alcohol in the cell walls of the poplar. It was also suggested that the comparatively lower WPG of poplar wood is caused by its easier evaporation during the curing and drying stages [30].

Table 1. WPG and BC and variations in EMC_R and ASE* at 20 °C for Chinese fir and poplar wood with furfurylation (%).

Sample	FA wt%	WPG	BC	RH/EMC$_R$			RH/ASE*		
				33%	65%	95%	33%	65%	95%
SW0	-	-	-	5.91 ^{a} (0.42)	11.23 ^{a} (0.02)	25.36 ^{a} (0.69)	-	-	-
SW23	10	23.69 ^{c} (1.25)	6.22 ^{b} (0.98)	4.18 ^{b} (0.11)	9.25 ^{b} (0.12)	23.55 ^{b} (0.52)	33.23	28.83	31.42
SW45	20	45.80 ^{b} (2.52)	9.31 ^{a} (0.93)	4.43 ^{c} (0.14)	9.14 ^{c} (0.29)	21.89 ^{b} (0.79)	30.91	41.77	37.61
SW75	30	75.02 ^{a} (3.19)	9.85 ^{a} (1.24)	3.83 ^{d} (0.12)	8.68 ^{d} (0.14)	22.36 ^{b} (0.96)	36.42	58.24	42.80
PW0	-	-	-	5.15 ^{a} (0.04)	11.04 ^{a} (0.37)	25.22 ^{a} (0.92)	-	-	-
PW17	10	17.66 ^{c} (1.31)	11.70 ^{b} (0.66)	3.83 ^{b} (0.11)	8.82 ^{b} (0.07)	22.73 ^{b} (1.18)	48.45	34.94	44.98
PW35	20	35.96 ^{b} (3.96)	13.73 ^{a} (0.67)	3.77 ^{c} (0.07)	8.63 ^{c} (0.29)	21.49 ^{c} (0.65)	42.51	40.79	36.78
PW51	30	51.07 ^{a} (6.07)	14.02 ^{a} (1.16)	3.79 ^{d} (0.07)	8.53 ^{d} (0.62)	22.66 ^{c} (1.10)	77.57	51.29	55.82

The standard deviation is given in parentheses; the same letter after a mean indicates no significance at $p \leq 0.05$.

As can be seen from Table 1, the BC increased accordingly alongside the WPG. The increases in the WPG of the Chinese fir were similar to those of the BC from 10% to 20% and from 20% to 30% of FA concentration, though the increases in BC lowered in the second stage. Statistical analysis showed significant differences between various concentrations of FA and the WPG, but there were no significant differences between furfurylated wood and BC at concentrations of 20% and 30%, indicating that the amount of FA entering the cell wall did not significantly increase and the effectiveness of FA on cell wall modification decreased with the increase in WPG. In addition, the BC of the poplar was significantly higher than that of the Chinese fir, indicating that the furfurylation of the poplar cell wall required lower concentrations than that of the Chinese fir.

Dimensional stability is one of the important properties of wood products in usage, and it is related to moisture absorption. As can be seen from Table 1, The EMC_R of the furfurylated wood was lower than that of the control wood. The EMC of SW75 was reduced by 23%–36% compared to the control wood of below 95% RH, and the EMC was reduced by about 11% at 95% RH while the EMC reduction in poplar was smaller than that of the Chinese fir. These results indicate that furfurylation could effectively reduce the hygroscopicity of the poplar and Chinese fir; this may have been due to the curing of FA resin in the cell wall that covered moisture adsorption sites such as hydroxyl groups in the amorphous areas [11], but the differences were not significant across the various FA concentration treatments. The changes in the hygroscopic process of ASE* after furfurylation were more intuitive than those of EMC_R. It was found that the average ASE* value of SW75 was more than 45% and that the average ASE* value of PW51 was more than 61%. It was shown that a low concentration of furfurylation could better improve the hygroscopic dimensional stability of the Chinese fir and poplar, but it was more effective in improving the dimensional stability of the poplar, which was consistent with higher BC values also observed in the poplar.

3.2. Mercury Intrusion Porosimetry (MIP)

The MIP method is a common means of analyzing the large pores of wood. The cumulative pore volume reflects the total pore volume in a sample, and the log-differential pore size distribution represents the pore volume in different pore size ranges. Figure 1 (left) shows the cumulative volume and log-differential pore size distributions for Chinese fir samples. The 1–6 μm diameter variation of the pore volume of the control wood was caused by the pointed ends of the pits, the voids of the pit membranes, and the pit openings [31,32]. The log-differential distribution of the pore diameters between 1 and 6 μm decreased with increasing concentrations of FA due to the deposition of FA resin at the pointed ends of pits and tracheids. The furfurylated Chinese fir showed new pores in the range of 80–600 nm, and larger concentrations of FA led to the formation of smaller new-pore sizes because the FA resin was cured on the pores of the pit openings, pit membranes, and pointed ends of the tracheids. The diameters of between 6 and 20 μm may have been caused by latewood tracheids, and pore sizes of over 20 μm represented the earlywood tracheids in the control Chinese fir [31]. The pore volume in the 6–20 μm range decreased with increasing impregnation concentrations, but the level of change in pore volume for diameters of over 20 μm was less than 6–20 μm due to furfurylation, indicating that the amount of cured FA in the cell lumina of latewood was greater than that of earlywood in the Chinese fir.

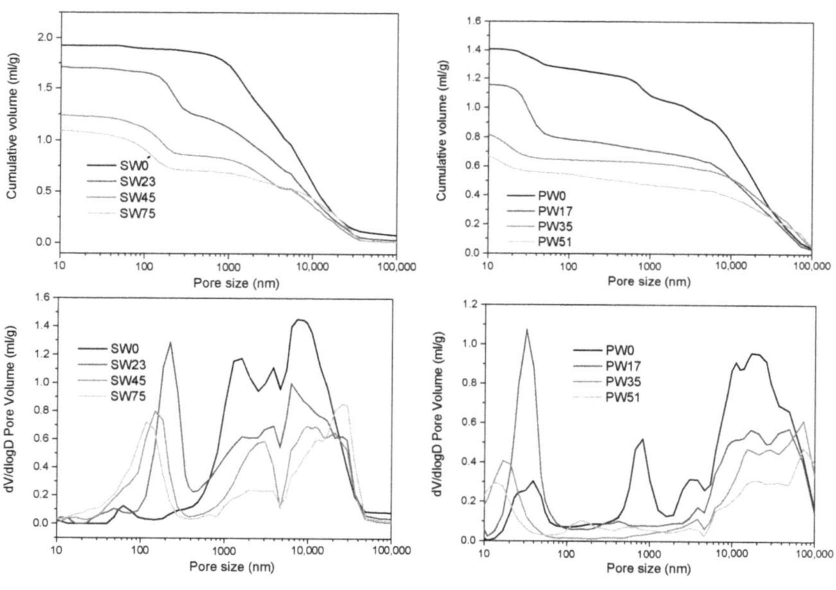

Figure 1. Cumulative pore volume and log-differential intrusion versus pore size of Chinese fir (**left**) and poplar (**right**).

Figure 1 (right) shows the cumulative pore volume and log-differential pore-size distributions of the poplar wood. The pore volume of the control wood significantly changed at 800–1000 nm according to the log-differential pore size due to the point ends of all the cell elements and pit chambers, openings, and fibers [33]. However, the furfurylated poplar wood showed almost no pore size distribution between 800 and 1000 nm, indicating that FA was cured at the point ends of the pit chambers, openings, and wood fibers. The pore diameters between 10 and 15 μm represented the voids of the wood fibers, and the pore diameters over 30 μm were assigned to the vessels in the control wood [33,34]. Pore volume decreased with increasing impregnation concentrations in the range of 10–15 μm, indicating that the amount of FA cured in the wood fibers increased with increased concentrations. The pores with diameters of between 15 and 80 nm reflected the voids of the pit membranes in the control wood. The pore volume of PW17, between 15 and 80 nm, was larger than that of other furfurylated poplar wood, which implied that the new pores were formed on the basis of the pores at the point ends of the pit chambers; on the other hand, smaller pore diameters formed as the WPG of FA increased in PW35 and PW51.

The newly formed pore size distribution of the furfurylated poplar (10–80 nm) was smaller than that of the Chinese fir (80–600 nm), which may have been related to the fact that poplar wood contains a much higher number of pits and smaller pore diameters in pit membranes. The amount of FA resin in the wood fibers and latewood tracheids significantly increased with increasing concentrations, as shown in Figure 1. The FA modification affected the pit structure from low concentrations, which may have been related to the fact that the pit is an important channel by which an impregnating solution can enter wood.

Table 2 shows the results of the MIP measurements. The bulk density of the furfurylated wood increased and the porosity gradually decreased with increasing concentrations of FA. The WPG of the poplar was less than that of the fir, the bulk density of the poplar was greater than that of the fir, and the porosity of the poplar was still less than that of the fir before and after modification. On the other hand, according to previous studies and changes in cumulative pore volume and log-differential pore-size distributions after furfurylation, the pore size distributions were grouped into three pore classes (below 800 nm, 800–5000 nm, and greater than 5000 nm) [32]. Additionally, the pore volume proportion was calculated as shown in Figure 2. The percentages of pore volumes below 800 nm for SW0, SW23, SW45, and SW75 were 9.9%, 33.3%, 33.8%, 37.9%, and 41.6%, respectively, and those of 800–5000 nm were 27.6%, 23.9%, and 13.6%, respectively. The trends of the pore volume percentages of the poplar and fir below 800 nm and from 800 to 5000 nm were generally consistent. These results indicate that the pore volume of the furfurylated wood decreased in the range of 800–5000 nm and increases below 800 nm, which led to changes in bulk density and porosity.

Table 2. Bulk density and porosity of control wood and furfurylated poplar and Chinese fir wood.

Sample	PW0	PW17	PW35	PW51	SW23	SW45	SW75
Total intrusion volume mL/g	1.41	1.19	0.85	0.72	1.74	1.25	1.11
Bulk density g/cm^3	0.47	0.52	0.64	0.68	0.40	0.47	0.52
Porosity %	66.42	61.81	54.32	49.55	69.52	58.94	58.31

3.3. Characterization of FA Distribution

The CLSM and SEM images of the control and furfurylated poplar wood are shown in Figure 3a–d,e–h respectively. There was fluorescence in the cell walls as a whole in PW17 (Figure 3b), indicating that furfuryl alcohol could effectively penetrate the cell wall and deposit into the cell walls, especially in the lignin-rich areas such as the cell corners (CC), secondary walls (S_2) and compound middle lamellae (CML). As the concentration of FA increased to 20%, the distribution of PW35 (Figure 3c) in the cell walls was consistent with that of PW17 (Figure 3b), but there was brighter fluorescence in the wood fiber lumen. These results could imply that FA was deposited in the inner walls of the wood fibers. It can also be seen from the SEM images in Figure 1 that the wood fibers in PW35 (Figure 3g) changed from clean and tidy in the lumen of PW17 (Figure 1f) cells after FA was attached. PW51 cell walls and cell lumen showed greater changes than PW17 and PW35. The cell walls and lumen of PW51 showed brighter fluorescence, and changes in the cell walls occurred due to a large number of furfuryl alcohol resin curing and bulking processes. Changes in cell lumen were caused by differences in FA conjugation lengths between the cell walls and cell lumen, resulting in stronger fluorescence in cell lumen [22]. Furthermore, the wood fiber cell morphology of PW51 (oval shaped) became crowded. The microstructure of a poplar longitudinal section can be observed in Figure 4a–d, which shows that furfurylation led to the deposition of FA resin in the alternate pits and wood fiber tracheids of the poplar that became more obvious with increases in furfuryl alcohol concentration.

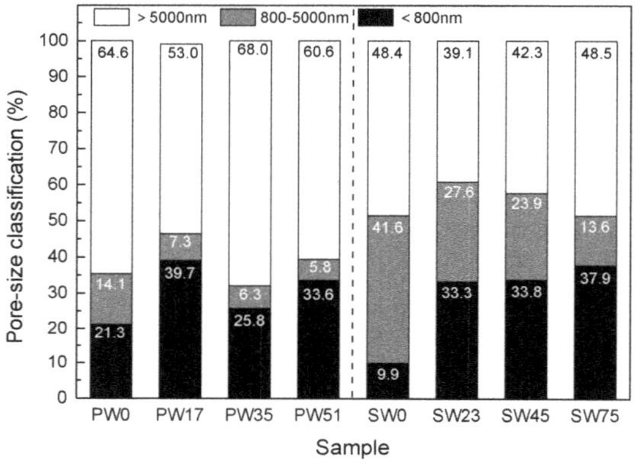

Figure 2. Comparison of the average pore size classification of untreated and furfurylated Chinese fir and poplar wood.

The CLSM and SEM images of the Chinese fir shown in Figure 3i–p also show that the FA deposited in the Chinese fir cell walls was consistent with that of the poplar, with both mainly located in the lignin-rich area. The WPG of the Chinese fir was generally higher than that of the poplar, but the fluorescent brightness seemed to be weaker than that of the poplar at each modification level, which may have been related to the fact that the cell wall BC of the Chinese fir was smaller than that of the poplar. Meanwhile, compared to that of SW23 and SW45, the cell morphology of SW75 was deformed and apparent fluorescence in the tracheid cell cavities was observed. In addition, the amount of FA resin in the cell lumen of the 30%-concentration furfurylated Chinese fir was smaller than that of the poplar, probably due to the fact that Chinese fir cell walls could accommodate more furfuryl alcohol resin. According to the microstructure of the Chinese fir longitudinal section in Figure 4e–h, the filling of the Chinese fir pits and tracheids became obvious with increasing FA concentrations, and the pit aperture was almost completely filled at the 30% concentration.

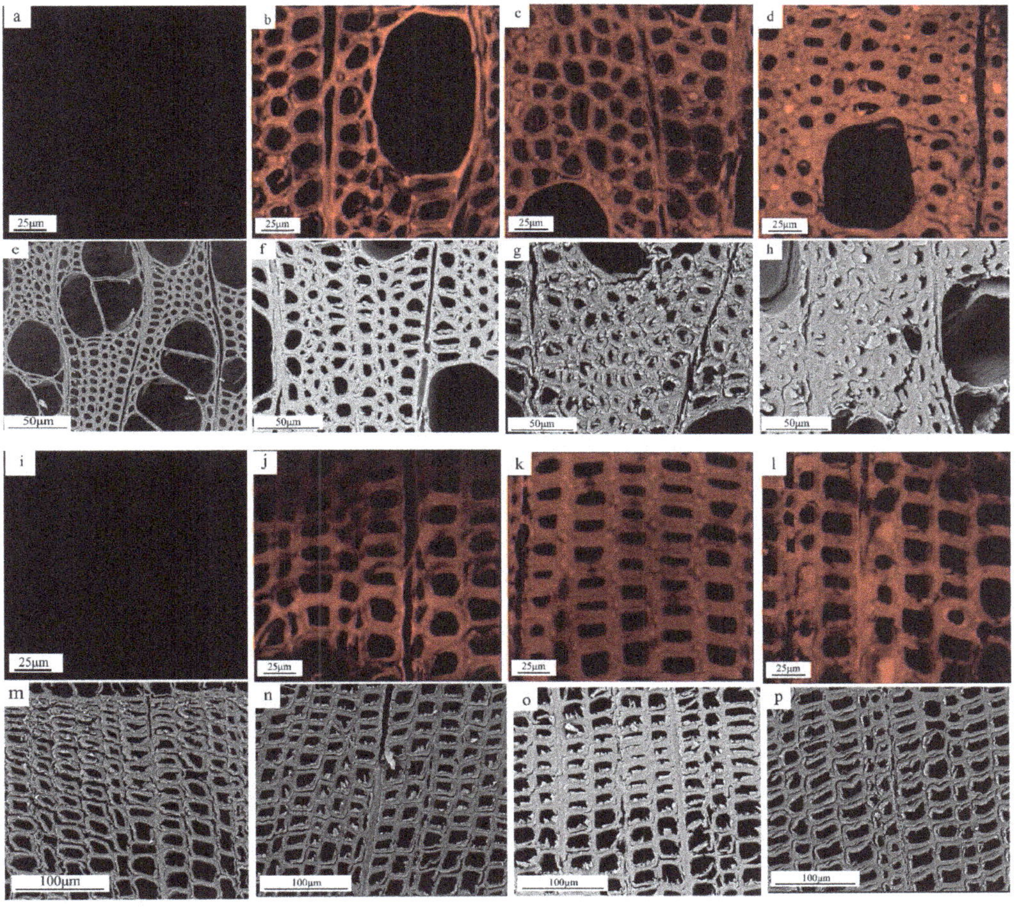

Figure 3. (**a**–**d**) CLSM images of control poplar wood (**a**), SW23 (**b**), SW45 (**c**), and SW75 (**d**); (**i**–**l**) CLMS images of control Chinese fir wood (**i**) PW17 (**j**), PW35 (**k**), and PW51 (**l**). (**e**–**h**) SEM images of control poplar wood (**e**), PW17 (**f**), PW35 (**g**), and PW51 (**h**); (**m**–**p**) SEM images of control Chinese fir wood (**m**), SW23 (**n**), SW45 (**o**), and SW75 (**p**).

3.4. LF-NMR

Figure 5 and Table 3 show the T_2 relaxation time distribution, mean value, EMC (water saturation), and proportion of furfurylated and untreated wood, which indicate the microscopic state of water-bound hydrogen in wood and the difference in the degree of water binding to the wood according to the work of previous studies [35–39]. The relaxation times of the three main peaks of the T_2 spectrum of SW0 were 1.05 ms, 12.75 ms, and 109.70 ms, and the proportions of their water populations were 9.9%, 3.0%, and 86.7%, respectively, which could be attributed to bound water in the cell walls and free water in the tracheids of latewood and earlywood. The results were generally consistent with those of previous studies [40]. The T_2 values of SW75 were 0.56 ms, 219.64 ms, and 1245 ms, and the proportions of their water populations were 14.57%, 67.57%, and 17.22%, respectively, which could be attributed to cell walls, tracheids, and incompletely cured FA resin, respectively. The T_2 values of water-saturated PW0 were 2.25 ms, 54.79 ms, and 471.38 ms, and their water populations accounted for 12.2%, 44.7%, and 41.0%, respectively; the shortest relaxation time was caused by bound water in the cell walls, and the others were caused by free water in wood fibers and vessels [41]. The T_2 values of PW51 were 0.74 ms, 102.34 ms, and 880.49 ms, and the proportions of the their water populations were 15.7%, 5.2%, and 78.3%, respectively, caused by cell walls, wood fibers, vessels, and incompletely cured furfuryl alcohol resin, respectively. Compared to control wood, the relaxation times of bound water were reduced by 46% for SW75 and 67% for PW51, though the relaxation time of free water was prolonged due to furfurylation, indicating that furfurylation tightened the bound water and freed the free water. The change in the T_2 value of the bound water may have been due to the increase in density and the decrease in moisture content after FA treatment. On the other hand, the effects of FA on free and bound water were different because FA resin is a hydrophobic substance that enlarged the cell wall contact angle more than that of the control wood, resulting in increasing T_2 values between the wood and free water [42,43]. In addition, the proportions of the different water populations of furfurylated and control wood showed that although furfurylation reduced the moisture content of the wood, it increased the proportion of bound water and decreased the proportion of free water in the poplar and Chinese fir wood following water-saturation treatment, during which the proportion of free water in the vessel increased and that in the wood fiber decreased.

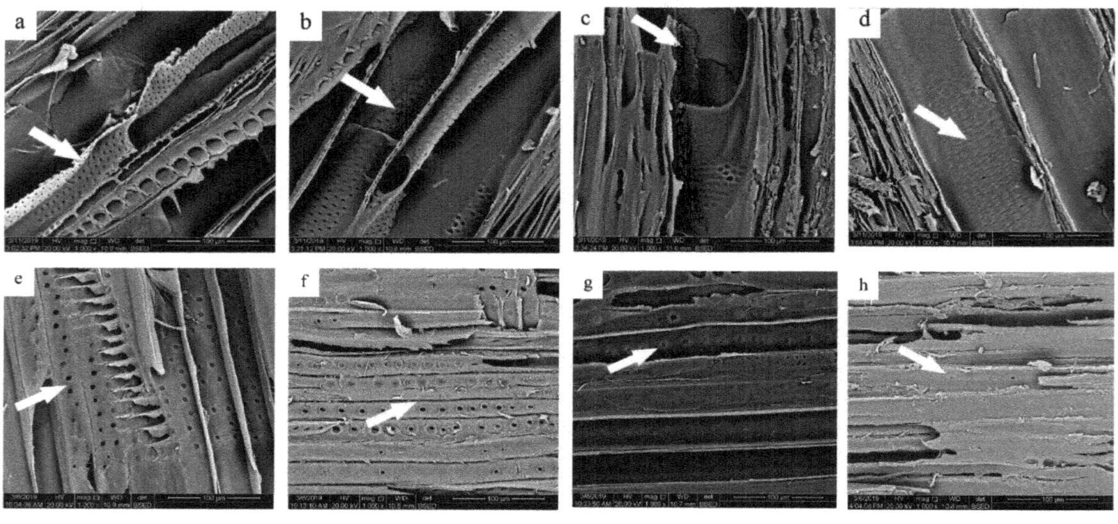

Figure 4. Microstructures of longitudinal sections of control poplar wood (**a**), PW17 (**b**), PW35 (**c**), and PW51 (**d**); microstructures of longitudinal sections of control Chinese fir wood (**e**), SW23 (**f**), SW45 (**g**), and SW75 (**h**).

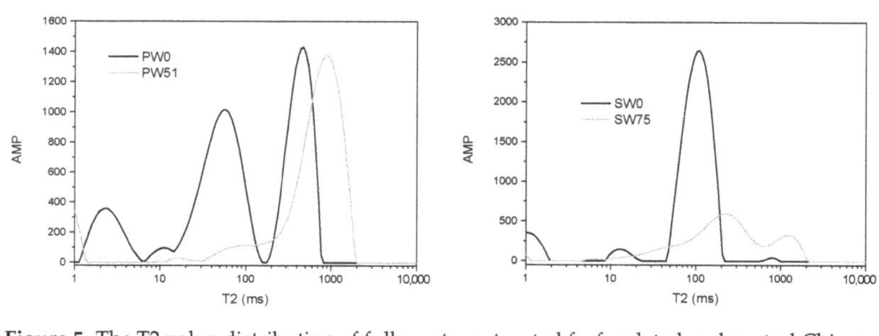

Figure 5. The T2 value distribution of fully water-saturated furfurylated and control Chinese fir and poplar wood.

Table 3. The T_2 mean value, peak area, and proportion of furfurylated and untreated Chinese fir and poplar wood.

Sample	Peak	Fully Water-Saturated (%)	T_2 Value (ms)	Peak Proportion (%)
SW0	1	251.7	1.1	9.9
	2		12.8	3.0
	3		109.7	86.7
	4		821.4	0.5
	Sum			100
SW75	1	70.0	0.6	14.6
	2		6.4	0.6
	3		219.6	67.6
	4		1245.9	17.2
	Sum			100
PW0	1	169.6	2.25	12.2
	2		11.1	2.0
	3		54.8	44.7
	4		471.4	41.0
	Sum			100
PW51	1	79.7	0.7	15.7
	2		15.7	0.8
	3		102.3	5.2
	4		880.5	78.3
	Sum			32,528.3

In summary, the effects of furfurylation on bound and free water included reductions in the amount of moisture content, tighter bonds between bound water and the wood, and a freer relationship between the free water and wood. The first effect can be attributed to the cell wall densification and the reduction in bound water following the furfuryl alcohol impregnation treatment, and the last effect may be attributed to the attachment of the hydrophobic FA resin to the cell walls.

3.5. Cyclic Water Soaking and Dimensional Stability

Water uptake represents the ability of wood to hold an amount of water, and it is closely related to porosity. Accordingly, water uptake also reflects the effect of FA treatment on the pore size distribution of cell walls and cell cavities [25]. Table 4 shows the MC_{WAR} and ASE_{WS} values of furfurylated and untreated wood during the water uptake test. It can be seen that the FA treatment could effectively reduce the MC_{WAR} of the Chinese fir and poplar, as the MC_{WAR} of the Chinese fir and poplar decreased along with increasing FA concentrations, with average reductions of 56%–79% for the Chinese fir and 8%–44% for the poplar compared to the control wood. Furthermore, Tables 2 and 3 show that the MC_{WA} of PW0 was 58% of that of SW0 and that the porosity of PW0 was 95.16% of that of SW0 in the control wood, mainly due to porosity. Moreover, the MC_{WAR} of the furfurylated wood was different to control samples. The porosity of furfurylated Chinese fir was 13.51%–27.54% higher than that of the poplar, but the MC_{WAR} of the furfurylated poplar wood was 22.17%–67.98% higher than that of the Chinese fir following furfurylation with various concentrations, indicating that the higher the WPG, the more significant the reduction in MC_{WAR}. In addition, the ASE_{WS} correspondingly increased with increases in FA concentration. However, changes in the amount of free water demonstrated insignificant effects on the dimensional stability of the wood. Thus, although the ASE_{WS} increased with increasing WPG, it was not as significant as the decrease in MC_{WAR}.

Table 4. Variation in MC_{WAR} (fully water-saturated) and ASE_{WS} of furfurylated and untreated poplar and Chinese fir with water uptake.

Sample	MC_{WAR} (%)			ASE_{WS} (%)		
	1st	2nd	3rd	1st	2nd	3rd
SW0	278.57 [a] (8.85)	276.04 [a] 9.94	277.40 [a] (8.85)			
SW23	95.80 [b] (9.75)	114.15 [b] (11.10)	120.88 [b] (9.67)	56.56	38.42	27.40
SW45	60.24 [c] (1.86)	70.35 [c] 2.69	74.20 [c] (50.89)	70.22	47.07	41.20
SW75	57.19 [c] (1.39)	66.70 [c] 3.29	62.66 [d] (9.21)	70.90	51.16	42.88
PW0	162.22 [a] (9.65)	161.66 [a] (10.31)	160.64 [a] (10.97)			
PW17	145.17 [b] (2.62)	145.92 [b] (3.54)	147.68 [b] (2.36)	50.48	23.32	14.90
PW35	115.24 [c] (8.68)	118.18 [c] (10.49)	123.04 [c] (9.07)	63.79	35.08	21.40
PW51	89.80 [d] (12.08)	99.15 [d] (11.74)	102.91 [d] (14.34)	72.36	45.37	23.47

The standard deviation is given in parentheses; the same letter after the mean indicates no significance at $p \leq 0.05$.

The number of water soaking cycles also had an effect on furfurylated wood, as seen in Table 4. As the cyclic water soaking times increased, the MC_{WAR} of the poplar and fir control wood did not significantly change at around 276% and 161%, respectively. The ASE_{WS} of the control wood decreased as the number of cycles of water soaking increased, but the third cycle of water soaking was less variable compared to the second cycle. Compared to the control wood, the MC_{WAR} and ASE_{WS} of the furfurylated wood slowly increased. The MC_{WAR} of the second cycle of the furfurylated Chinese fir wood increased by 17%–19% compared to the first cycle, which indicated that there may have been a loss of FA impregnated in the wood during the test that resulted in increases in pore space and the water-uptake capacity. Furthermore, the loss of furfuryl alcohol was consistent with the large T2 values produced by the uncured furfuryl alcohol in low-field NMR. On the other hand, the ASE_{WS} of PW51 decreased by more than 37% after each repetition of cyclic water uptake, which could imply that although furfurylation had a significant bulking effect on the poplar cell walls, it may have been unstable due to the large number of pits in the poplar cell walls.

4. Conclusions

(1) Low concentrations of furfuryl alcohol can better enter the cell walls of the Chinese fir than those of poplar. Compared to an FA concentration of 10%–20%, there were no significant increases in the amount of FA entering the cell walls when the FA concentration was increased to 30% in both the Chinese fir and the poplar. Furthermore, the furfurylation of the poplar cell walls was more suitable than those of the Chinese fir cell walls. The FA resin was almost cured in the secondary walls, cell corners, and compound middle lamellae when furfurylated with the 10% concentration. Once the concentration increased to 30%, the amount of FA cured in the cell lumen of the poplar was greater than that of the Chinese fir.

(2) The poplar's pointed ends of pit chambers and pit apertures (800–1000 nm) and the Chinese fir's small pores of pit membranes and pit apertures (1–6 μm) were partially infiltrated and deposited by FA resin at the 10%–30% concentrations and formed new pore sizes in the pore size distribution ranges of 80–600 nm and 15–100 nm, respectively. The amount of FA cured in the tracheids of the Chinese fir latewood and poplar wood rays increased along with increasing FA concentrations. The porosity of

the poplar was greater than that of the Chinese fir, and the bulk density was lower in the poplar than the fir before and after modification.

(3) Wood impregnation with a low concentration of FA was able to reduce the EMC_R and improve the ASE*, but the dimensional stability of the poplar wood was more significantly improved. Additionally, furfurylation effectively reduced water uptake due to the hydrophobicity properties of the furfuryl alcohol resin. Furthermore, with increases in cyclic water soaking times, furfurylation demonstrated significant effects of MC_{WAR} reduction and ASE_{WS} improvement. The MC_{WAR} of the Chinese fir increased by 17%–19% when treated with FA at various concentration in secondary cyclic water soaking, suggesting the loss and leaching of FA resin during the test. In addition, low-field NMR showed that the effects of modification on bound and free water included reductions in the amount of moisture content, tighter bonds between the bound water and wood, and a freer relationship between the free water and wood. It can therefore be concluded that furfurylation is a feasible method by which to extend furfurylated wood applications when using FA at low concentrations.

Author Contributions: Conceptualization, T.C.; experiments and data analysis, T.C., H.S. and Y.Y.; writing—original draft preparation, H.S. and X.C.; resources, T.C.; writing—review and editing, T.C. and C.F.; supervision, T.C. and C.D.; project administration, T.C. All authors have read and agreed to the published version of the manuscript.

Funding: This research was funded by the National key technology research and development project (2017YFD0600202), the Natural National Science Foundation of China (31560190) and the Natural National Science Foundation of China (31960291).

Institutional Review Board Statement: Not applicable.

Informed Consent Statement: Not applicable.

Data Availability Statement: The data presented in this study are available on request by contact with the corresponding author.

Conflicts of Interest: The authors declare no conflict of interest.

References

1. Hosseinpourpia, R.; Adamopoulos, S.; Mai, C. Dynamic vapour sorption of wood and holocellulose modified with thermosetting resins. *Wood Sci. Technol.* **2016**, *50*, 165–178. [CrossRef]
2. Klüppel, A.; Mai, C. The influence of curing conditions on the chemical distribution in wood modified with thermosetting resins. *Wood Sci. Technol.* **2013**, *47*, 643–658. [CrossRef]
3. Lande, S.; Westin, M.; Schneider, M. Properties of furfurylated wood. *Scand. J. For. Res.* **2011**, *19*, 22–30. [CrossRef]
4. Esteves, B.; Nunes, L.; Pereira, H. Properties of furfurylated wood (*Pinus pinaster*). *Eur. J. Wood Wood Prod.* **2011**, *69*, 521–525. [CrossRef]
5. Morozovs, A.; Keke, A.; Fisere, L.; Spulle, U. Wood modification with furfuryl alcohol and furfurylated wood durability in water. In Proceedings of the 17th International Scientific Conference Engineering for Rural Development, Jelgava, Latvia, 23–25 May 2018.
6. Yang, T.; Cao, J.; Ma, E. How does delignification influence the furfurylation of wood? *Ind. Crops Prod.* **2019**, *135*, 91–98. [CrossRef]
7. Herold, N.; Pfriem, A. Impregnation of veneer with furfuryl alcohol for an improved plasticization and moulding. *Eur. J. Wood Wood Prod.* **2013**, *71*, 281–282. [CrossRef]
8. Kong, L.; Guan, H.; Wang, X. In situ polymerization of furfuryl alcohol with ammonium dihydrogen phosphate in poplar wood for improved dimensional stability and flame retardancy. *ACS Sustain. Chem. Eng.* **2018**, *6*, 3349–3357. [CrossRef]
9. Buchelt, B.; Dietrich, T.; Wagenführ, A. Testing of set recovery of unmodified and furfurylated densified wood by means of water storage and alternating climate tests. *Holzforschung* **2014**, *68*, 23–28. [CrossRef]
10. Dong, Y.; Ma, E.; Li, J.; Zhang, S.; Hughes, M. Thermal properties enhancement of poplar wood by substituting poly(furfuryl alcohol) for the matrix. *Polym. Compos.* **2019**, *41*, 1066–1073. [CrossRef]
11. Shen, X.; Guo, D.; Jiang, P.; Yang, S.; Li, G.; Chu, F. Water vapor sorption mechanism of furfurylated wood. *J. Mater. Sci.* **2021**, *56*, 11324–11334. [CrossRef]
12. Deka, H.; Mohanty, A.; Misra, M. Renewable-resource-based green blends from poly(furfuryl alcohol) bioresin and lignin. *Macromol. Mater. Eng.* **2014**, *299*, 552–559. [CrossRef]
13. Lande, S.; Westin, M.; Schneider, M.H. Eco-efficient wood protection: Furfurylated wood as alternative to traditional wood preservation. *Manag. Environ. Qual.* **2004**, *15*, 529–540. [CrossRef]

14. Pilgård, A.; Treu, A.; van Zeeland, A.N.T.; Gosselink, R.J.A.; Westin, M. Toxic hazard and chemical analysis of leachates from furfurylated wood. *Environ. Toxicol. Chem.* **2010**, *29*, 1918–1924. [CrossRef] [PubMed]
15. Pilgård, A.; De Vetter, L.; Van Acker, J.; Westin, M. Toxic hazard of leachates from furfurylated wood: Comparison between two different aquatic organisms. *Environ. Toxicol. Chem.* **2010**, *29*, 1067–1071. [CrossRef] [PubMed]
16. Vetter, L.D.; Pilgård, A.; Treu, A.; Westin, M.; Acker, J.V. Combined evaluation of durability and ecotoxicity: A case study on furfurylated wood. *Wood Mater. Sci. Eng.* **2009**, *4*, 30–36. [CrossRef]
17. Goldstein, I.; Dreher, W. Stable Furfuryl Alcohol Impregnating Solutions. *Ind. Eng. Chem.* **1960**, *52*, 57–58. [CrossRef]
18. Mantanis, G.I. Chemical modification of wood by acetylation or furfurylation: A review of the present scaled-up technologies. *Bioresources* **2017**, *12*, 4478–4489. [CrossRef]
19. Hadi, Y.S.; Nawawi, D.S.; Abdillah, I.B.; Pari, G.; Pari, R. Evaluation of discoloration and subterranean termite resistance of four furfurylated tropical wood species after one-year outdoor exposure. *Forests* **2021**, *12*, 900. [CrossRef]
20. Li, W.; Wang, H.; Ren, D.; Yu, Y. Wood modification with furfuryl alcohol catalysed by a new composite acidic catalyst. *Wood Sci. Technol.* **2015**, *49*, 845–856. [CrossRef]
21. Sedliačiková, M.; Moresová, M.; Aláč, P.; Malá, D. What is the supply and demand for coloured wood products? An empirical study in Slovakian practice. *Forests* **2021**, *12*, 530. [CrossRef]
22. Thygesen, L.G.; Barsberg, S.; Venås, T.M. The fluorescence characteristics of furfurylated wood studied by fluorescence spectroscopy and confocal laser scanning microscopy. *Wood Sci. Technol.* **2010**, *44*, 51–65. [CrossRef]
23. Pu, H.; Chen, T.; Li, Y. Influences of furfurylation modification on the wood properties of rubberwood. *J. For. Eng.* **2014**, *28*, 50–53.
24. Yang, L.; Tian, G.; Yang, S.; Shang, L.; Liu, X.E.; Jiang, Z. Determination of fiber saturation point of rattan (*Calamus simplicifolius*) using the LF-NMR and two conventional methods. *Wood Sci. Technol.* **2020**, *54*, 667–682. [CrossRef]
25. Zelinka, S.L.; Glass, S.V.; Thybring, E.E. Evaluation of previous measurements of water vapor sorption in wood at multiple temperatures. *Wood Sci. Technol.* **2020**, *54*, 769–786. [CrossRef]
26. Thybring, E.E. The decay resistance of modified wood influenced by moisture exclusion and swelling reduction. *Int. Biodeterior. Biodegrad.* **2013**, *82*, 87–95. [CrossRef]
27. Chen, J.; Wang, Y.; Cao, J.; Wang, W. Improved water repellency and dimensional stability of wood via impregnation with an epoxidized linseed oil and carnauba wax complex emulsion. *Forests* **2020**, *11*, 271. [CrossRef]
28. Chen, T.; Luo, P.; Xu, Z.; Luo, X.; Hu, Z. Effects of furfuryl alcohol impregnation on physical and mechanical properties of densified poplar wood. *J. For. Eng.* **2016**, *1*, 21–25.
29. Boneka, A.S.; Uyup, M.K.A.; Hua, L.S.; Siam, N.A.; Salim, S.; Ashaari, Z. Sorption isotherm and physico-mechanical properties of kedondong (*Canarium* spp.) wood treated with phenolic resin. *Constr. Build. Mater.* **2021**, *288*, 123060. [CrossRef]
30. Li, W.; Liu, M.; Wang, H.; Ren, D.; Yu, Y. Fabrication of highly stable and durable furfurylated wood materials. Part II: The multi-scale distribution of furfuryl alcohol (FA) resin in wood. *Holzforschung* **2020**, *74*, 1147–1155. [CrossRef]
31. Zauer, M.; Hempel, S.; Pfriem, A.; Mechtcherine, V.; Wagenführ, A. Investigations of the pore-size distribution of wood in the dry and wet state by means of mercury intrusion porosimetry. *Wood Sci. Technol.* **2014**, *48*, 1229–1240. [CrossRef]
32. Li, L.; Wang, X.; Yan, Y.; Ping, L. Pore analysis of thermally compressed Scots pine (*Pinus sylvestris* L.) By mercury intrusion porosimetry. *Holzforschung* **2017**, *72*, 57–63. [CrossRef]
33. Plötze, M.; Niemz, P. Porosity and pore size distribution of different wood types as determined by mercury intrusion porosimetry. *Eur. J. Wood Wood Prod.* **2011**, *69*, 649–657. [CrossRef]
34. Jansen, S.; Choat, B.; Pletsers, A. Morphological variation of intervessel pit membranes and implications to xylem function in angiosperms. *Am. J. Bot.* **2009**, *96*, 409–419. [CrossRef]
35. Araujo, C.D.; MacKay, A.L.; Whittall, K.P.; Hailey, J.R.T. A diffusion model for spin-spin relaxation of compartmentalized water in wood. *J. Magn. Reson.* **1993**, *101*, 248–261. [CrossRef]
36. Cai, C.; Zhou, F.; Cai, J. Bound water content and pore size distribution of thermally modified wood studied by NMR. *Forests* **2020**, *11*, 1279. [CrossRef]
37. Cox, J.; Mcdonald, P.J.; Gardiner, B.A. A study of water exchange in wood by means of 2D NMR relaxation correlation and exchange. *Holzforschung* **2010**, *64*, 259–266. [CrossRef]
38. Xu, K.; Lu, J.; Gao, Y.; Wu, Y.; Li, X. Determination of moisture content and moisture content profiles in wood during drying by low-field nuclear magnetic resonance. *Dry. Technol.* **2017**, *35*, 1909–1918. [CrossRef]
39. Thybring, E.E.; Digaitis, R.; Nord-Larsen, T.; Beck, G.; Fredriksson, M. How much water can wood cell walls hold? A triangulation approach to determine the maximum cell wall moisture content. *PLoS ONE* **2020**, *15*, e0238319. [CrossRef]
40. Fredriksson, M.; Thygesen, L.G. The states of water in norway spruce (*Picea abies* (L.) Karst.) Studied by low-field nuclear magnetic resonance (LFNMR) relaxometry: Assignment of free-water populations based on quantitative wood anatomy. *Holzforschung* **2017**, *71*, 77–90. [CrossRef]
41. Almeida, G.; Gagné, S.; Hernández, R.E. A NMR study of water distribution in hardwoods at several equilibrium moisture contents. *Wood Sci. Technol.* **2007**, *41*, 293–307. [CrossRef]
42. Thygesen, L.G.; Elder, T. Moisture in untreated, acetylated, and furfurylated Norway spruce studied during drying using time domain NMR. *Wood Fiber Sci.* **2008**, *40*, 309–320.
43. Thygesen, L.G.; Elder, T. Moisture in untreated, acetylated, and furfurylated Norway spruce monitored during drying below fiber saturation using time domain NMR. *Wood Fiber Sci.* **2009**, *41*, 194–200.

forests

MDPI

Article

The Impact of Vinylotrimethoxysilane-modified Linseed Oil on Selected Properties of Impregnated Wood

Waldemar Perdoch [1], Ewelina Depczyńska [1], Karolina Tomkowiak [1], Monika Furgał [2,*], Mariola Kurczak [2] and Bartłomiej Mazela [1]

[1] Faculty of Forestry and Wood Technology, Poznań University of Life Sciences, Wojska Polskiego 28, 60-637 Poznan, Poland
[2] Tikkurila Polska S.A., ul. Ignacego Mościckiego 23, 39-200 Dębica, Poland
* Correspondence: monika.furgal@tikkurila.com; Tel.: +48-6-0080-5828

Abstract: This study aimed to examine the effect of organosilicon compounds-modified linseed oil parameters on wood utility properties. Linseed oil silylation with an organosilicon compound containing a vinyl group (vinyltrimethoxysilane) has made it possible to synthesize products characterized by high stability in storage. The impregnation of wood with these resulting products contributed to increased resistance of the protected wood to water exposure when compared to unprotected wood or wood protected by polymerized oil that was not subject to the silylation reaction. The protected wood was characterized by a lower water absorption rate in liquid as well as vapor forms contained in the humid air. This property immediately translated into an increase in the resistance of the wood exposed to *Basidiomycota*. The loss of mass of *Pinus sylvestris* L. (pine wood) exposed to the *Coniophora puteana* was approximately 3%. This parameter did not improve when the wood was subjected to the washing process (mass loss approx. 3.6%). Thus, the resulting products are those demonstrating the highest potential for use in the preparation of wood protection means.

Keywords: biobased materials; linseed oil; silane; VTMOS; wood

1. Introduction

Despite a broad range of existing wood protection products, the search for solutions that would meet growing market expectations, such as water resistance and microbial attack, UV radiation, or low toxicity relative to the natural environment, still continues. Due to the limitations of synthetic materials, it is also important to remember that the natural origin of raw materials is often defined by their so-called biogenic carbon content. Oils of natural origin, in particular, attract increasing attention [1–5]. It is generally known that the impregnation of wood with natural oils, including linseed oil or tung oil, contributes to the reduction of water absorption and results in the wood demonstrating higher resistance to wood decay fungi (white and brown rot) [6–9]. Despite the aforementioned advantages in terms of the use of natural oils in wood protection methods, these products have numerous disadvantages. The major imperfections are coating hardening, high viscosity, relatively low mechanical parameters of the coating (hardness, flexibility, abrasion), a relatively high water washout, low resistance to UV radiation, and they tend to change color (most often, they are prone to yellowing). Thus, increasing attention is paid to wood modification with oils used for its protection. Due to the unsaturated transtype bonds in its content, linseed oil is characterized by higher reactivity than other commonly used oils, e.g., tung oil [10]. Veigel et al. [11] observed that adding the modified nanocellulose to linseed oil improves its mechanical parameters (resistance to abrasion) and resistance to weather conditions. Several authors described linseed oil modification methods through uretining reactions in the acetone process in which the product formed coatings fit for wood protection [4,8,12–14]. These tests were developed by implementing modified oils with metal-containing microbiocides, i.e., silver, zinc, manganese, lead, or calcium. The authors stated that the most efficient products were achieved by implementing silver compounds. Other tested parameters, i.e., hardness, impact strength, binding strength, adhesion, and resistance to light, have not changed compared to the coatings that do not contain microbiocides]. Linseed and soy oil have also been subject to in situ epoxidation and copolymerization with vinyl acetate. Positive effects of the modification have been observed only for the linseed oil-based product. Pinewood impregnated with the above product has a higher resistance to decay fungi (*Trametes versicolor, Postia placenta,* and *Coniophora puteana*) compared to commercially accessible polymerized linseed oil [15].

The application of silanes to modify natural oils by means of the reaction of substitution of nucleophilic 3-chloropropyltrimetoxysilane to sodium salts of fatty acids in the presence of potassium iodide as a catalyst appears to be very promising. Rape oil modified with organosilicon compounds obtained in this way was then used to protect pine wood. Due to the formation of chemical bonds between hydroxyl groups located on the surface of wood and the silane (wood–O–Si), a highly durable coating was formed. While the wood was immersed in water, no coating deterioration was observed on pine wood treated with the obtained product. The synthesis of silane derivatives of rape oil determined new prospects for the use of natural products in obtaining silanized natural polymers and their application for manufacturing protective coatings permanently bonded with the wood surface. Such products may provide an alternative for the silanes containing fluorine groups that are more broadly applied as strongly hydrophobic [16–18].

Stable products of the reaction were obtained in soybean oil silylating reaction by means of vinylotrimethoxysilane (VTMOS) in the presence of 5-Bis(tert-butylperoxy)-2.5-dimethylhexane as a catalyst. Srinivasan et al. [19] indicated that the most likely course was the "ene" reaction being a part of the Diels–Alder reaction mechanism. Tambe et al. [20] conducted the silylating reaction for soybean, rape, and abyssinian oils. The researchers concluded that the reactivity of vinylosilane with olefin subjected to the "ene" reaction mechanism did not depend on the double bond position in olefin, and non-terminal unsaturated bonds of fatty acids are also affected by the reaction. The previous tests were supplemented with a detailed description of the VTMOS-modified linseed oil. The products of silylating reactions are characterized by relatively low viscosity and a tendency to form gel without any necessity of introducing catalysts promoting polymer structure netting [21,22].

The course of the linseed oil silylating reaction with VTMOS results in a new product with a high potential to provide resistance to water and, thus, increase the durability of treated wood in response to external impacts. This study aimed to determine the impact of linseed oil parameters on wood utility properties when modified with organosilicon compounds. The following physical properties described the reaction products: non-volatile content/solid content, vapor absorption from humid air, water absorbability, contact angle, testing the change of color in view of aging factors, and exposure to fungi-causing mold growth, both in laboratory and field conditions.

2. Materials and Methods

2.1. Materials

Linseed oil was purchased from Alberdingk boley gmbh (DE). The vinyltrimethoxy silane (VTMOS) [CAS 2768-02-7] with a purity of 99% was purchased from Dow Corning. The catalyst 2.5-Bis(tert-butylperoxy)-2.5-dimethylhexane, (Luperox 101 [CAS 78-63-7]) was purchased from Arkema.

The sapwood of pine (*Pinus sylvestris* L.) and spruce (*Picea abies*), with sizes consistent with specific test requirements, were used for tests.

2.2. Linseed Oil Modification

The catalytic course of the linseed oil silylation reaction with VTMOS was conducted in a high-pressure reactor (PARR Instrument Company, USA, ST Moline, IL). Prior to silylation, the heating reactor was purged with nitrogen. The reaction components were put into the reactor and after 5 min of stirring the mixture at room temperature, the heating started. Depending on the substrates, the proper reaction temperature was achieved after 90–120 min. The reaction was run at the temperature of 280 °C with a rotation speed of 220 rpm ± 2 rpm throughout the entire process. The composition of the amount of the components mixed in a reactor is summarized in Table 1.

Table 1. The chemical composition of the reaction mixture based on linseed oil.

Reaction ID	Luperox 101 [mol]/ 1 mol of Linseed Oil	VTMOS [mol]/ 1 mol of Linseed Oil	Time of VTMOS Addition	Reaction Time in 280 °C [h]
1	0	0	No addition	12
2	0.008	0.6	At the beginning of reaction	3
3	0.04	0.6	3rd h of reaction	6
4	0.04	0.6	3rd h of reaction	9

2.3. Methods

The method and materials used in the paper methodology are graphically presented in Scheme 1.

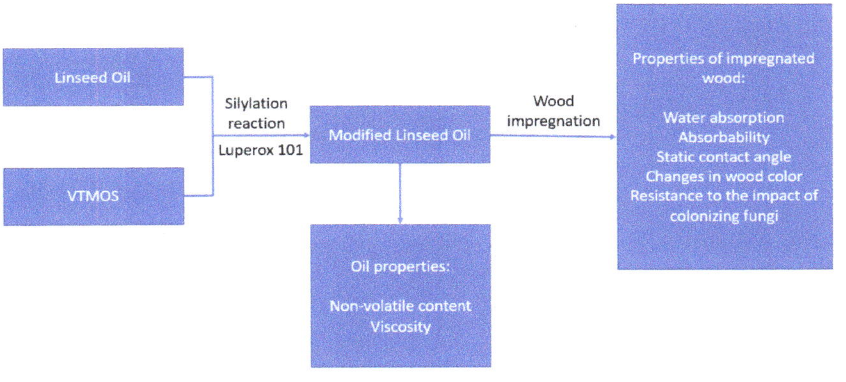

Scheme 1. Graphical picture of the methodology used in the work.

2.3.1. Non-Volatile Content/Solid Content

The solid content of products (2 g) was measured using moisture analyzer equipment (Moisture analyzer MA 40, Satorius Company, Göttingen, Germany) at 105 °C and 125 °C based on the modified PN-EN ISO 3251:2008P method.

2.3.2. Determining the Viscosity of the Obtained Reaction Products

Determining viscosity was performed using the Hoppler apparatus by measuring the times of flow of an adequately selected ball. Then, viscosity was calculated using the following formula:

$$\eta = K * (q1 - q2) * t$$

where:

η—viscosity
K—apparatus calibration constant
q1—density of the ball [g/cubic cm]
q2—density of tested liquid [g/cubic cm]
t—time of falling of the ball between two marks [s]

2.3.3. Water Absorption from Humid Air

Wood samples of the following sizes, 18 × 30 × 30 mm (thickness, width, and length), the latter dimension represents the size along the fibers), were treated by immersion in oil for 10 min (5 samples per product). The average value of the applied preparation was 200 ± 20 g/m^2. The treated samples were dried in the laboratory dryer for 72 h at a temperature of 80 °C. Oversaturated solution of diammonium phosphate was prepared in an airtight glass container (desiccator). The samples were weighed and placed in the desiccator above the solution when dry. The container was tightly closed. Later, the samples were weighed after 1, 2, 4, 24, 48, 72, 96, 120, and 168 h. After each measurement, the samples were put back into the desiccator. Untreated pine wood was used in the reference samples. Water absorption was determined using the following formula:

$$Wa = \frac{[100\%(m_n - m_p)]}{m_p}$$

where:

W_a—water absorption
m_n—humid wood weight
m_p—dry wood weight

2.3.4. Absorbability

The test was performed on the samples of treated pine wood and spruce wood of the following parameters: 20 × 75 × 150 mm (thickness, width, and length, the latter dimension represents the size along the fibers). In order to obtain constant weight, wood samples were conditioned at the temperature of 20 ± 2 °C and a humidity of 65 ± 5%. Then, 200 g (±20 g)/m^2 of the preparations (non-modified and modified oil by silane) were applied to the surface of the wood prepared in this experiment. After conditioning the wood, the side edges and untreated side of samples were blocked with epoxy paint and silicone. The weighed samples prepared in this way were put for 24 h in deionized water by immersing the entire tested surface in water, and then it was conditioning for 3 h at the temperature of 20 ± 2 °C and relative air humidity of 65 ± 5%, and 3 h at the temperature of 50° ± 2 °C, then 18 h at the temperature of 20 ± 2 °C and relative air humidity of 65 ± 5%. Then, the samples were restored at the temperature of 20 ± 2 °C and relative air humidity of 65 ± 5% until a constant weight was achieved. The absorbability test was performed by immersing the treated surface of samples in water so as to ensure that the entire tested surface was wet. After 72 h, the samples were taken out of the water, then water drops were delicately removed with a tissue, and the samples were weighed.

2.3.5. Static Contact Angle

In order to test the hydrophobic properties of the products, the static contact angle was determined on pine wood. The size of the tested samples was 150 ± 2; 75 ± 3, and the thickness of 20 ± 2 mm. Preparations were applied with a brush in the amount of 200 ± 20 g/m^2. The samples prepared in this way were conditioned for 7 days in the environmental chamber (at the temperature of 20 ± 2 °C; RH 65 ± 5%). The measurements were taken by means of the OCA14 optical goniometer (Data Physics). Deionized water was used as a measurement liquid. Water drops with a volume of 5 µL were placed automatically using a pipette directly on the sample located on the measurement table, keeping the same minimum height of the needle above the tested surface. Five seconds after the placement of a drop, at least 10 measurements were taken for each sample. Images of a drop were registered by a camera. The contact angle was measured using the control program after prior determination of a baseline and the contour of a drop.

2.3.6. Determination of Changes in Wood Color

The test was performed in order to check the change in color (yellowing) of oil during the polymerization process and later exposure to wood. The aging procedure was performed in the QUV chamber using water spraying. Pine wood samples were of the following size: 10 × 75 × 150 mm (the latter dimension represents the size along the fibers). Preparations were applied with a brush in the amount of 200 g ± 20 g/m^2. The samples prepared in this way were conditioned for 7 days at a temperature of 20 ± 2 °C and humidity of 65 ± 5%. The color test was performed prior to the exposure to the deteriorating impacts and after 1 and 6 weeks of exposure in the chamber. Measurements were taken in the same areas where the color on the unaged wood was tested. The color was determined on pine wood using the Spectraflash 300 double-beam spectrophotometer featuring a pulsed xenon lamp and the d/8° measurement geometry in the wavelength range of 400–700 nm (accuracy–10 nm). The measurement surface was 18 mm^2. The aging procedure in the QUV chamber was performed in line with the following parameters: EN 927-6, step 1—condensation (45 ± 3 °C, 24 h); step 2—144 h consisting of 48 cycles of 3 h consisting of steps 3 and 4; step 3—UV (60 ± 3 °C, 2.5 h, irradiance set point 0.89 W/(m^2 nm) at 340 nm); step 4—spray (0.5 h), 6 l/min–7 l/min, UV off). Test duration: 12 weeks (2016 h).

2.3.7. Resistance of Treated Wood to the Impact of Colonizing Fungi

Resistance to Colonization by the Fungi Causing Molding

Biological testing was performed on the basis of methodological assumptions of ASTM D 5590. This method involved a visual evaluation of the level of colonization by molds. The mixture of the following testing fungi cultures was applied: *Aspergillus niger* van Tieghem (*A.n.*), *Trichoderma viride* Pers. Ex S.F. Gray aggr. (*T.v.*), and *Paecilomyces variotii* Bainer (*P.v*). Fungi were cultivated on an agar–malt medium (20 g of agar, 30 g of malt extract, 1 L of water) with added Czapek-Dox mineral (4 g/L). The wood was exposed in the cultivation chamber at constant temperature (28 ± 1 °C) and humidity (80 ± 5%) for a period of 28 days. The colonization index (Table 2) of the wood surface by fungi was determined after 7, 14, 21, and 28 days of the test. Part of the samples was aged according to PN-EN 84.

Table 2. The scale of colonization of the wood surface by molds according to ASTM D 5590.

Index	Description
0z	No colonization of the wood surface by fungi, visible zone of inhibition
0	No colonization of the wood surface by fungi
1	1%–10% of the wood surface colonized by fungi
2	11%–30% of the wood surface colonized by fungi
3	31%–60% of the wood surface colonized by fungi
4	More than 60% of wood surface colonized by fungi

Resistance to Decay Caused by *C. puteana*

Impregnated samples in the following size: 7 × 25 × 50 mm (the latter dimension represents the size along the fibers) were exposed to *Coniophora puteana* (Schumacher ex Fries) Karst., BAM Ebw 15 species causing brown rot of wood. The fungus was cultivated on the agar–malt medium (20 g of agar, 50 g of malt extract, 1 L of water). Methodological assumptions of EN 113 and EN 839 were used in the testing procedure. The wood mass loss occurring as a result of exposure to samples of *Coniophora puteana* was determined after 8 weeks of testing. The infected wood was exposed in the cultivation chamber to constant humidity conditions (70 ± 5%) and temperature (21 ± 1 °C). Part of samples were aged according to PN-EN 84.

2.3.8. ANOVA Analysis

GNU PSPP software (version 1.6.2) was used to conduct the statistical and variance analysis. One-way ANOVA study and the Tuckey HSD test with a confidence level of 95% ($p < 0.05$) were used to compare the significance of the difference between the obtained results.

3. Results and Discussion

3.1. Non-Volatile Content

The highest non-volatile content at 105 °C was observed in non-polymerized linseed oil heated for 12 h (Product reaction 1). The addition of VTMOS and catalyst of silylation to reaction slightly reduced the non-volatile content. This change arises from the addition of relatively low molecular weight silane to the reactive mixture. Then, silane is subject to the thermal reaction featuring the emission of volatile low molecular weight compounds [12]. This conclusion is supported by the trend visible in the chart below—the longer the reaction time, the higher the volume of volatile low molecular weight compounds.

Exposure of oil samples at the temperature of 125 °C in the cases of the reaction with VTMOS made it possible to observe the presence of volatile substances, which may suggest the formation of a by-product or decomposition of the resulting product. A likely reason for a growth in the volume of volatile substances at the temperature of 125 °C was the formation of by-products in the form of organosilicon compounds containing ether and ester bonds [12]. Detailed results of the analysis of volatile substance volumes are presented in Figure 1.

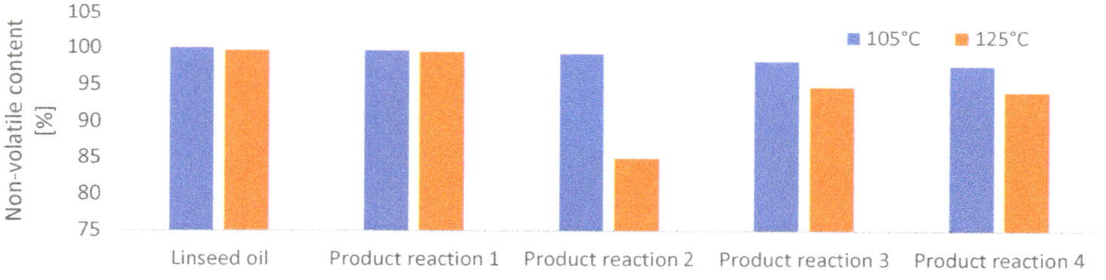

Figure 1. Non-volatile content in modified and unmodified linseed oil.

3.2. Dynamic Viscosity

The viscosity of silylation reaction products was several times lower than the viscosity of the product of the 12 h linseed oil radical polymerization reaction (reaction product 1, Figure 2). Low viscosity, just slightly higher than the viscosity of non-polymerized linseed oil, was the desired effect as it increases the depth of penetration of the product into the wood. The viscosity of VTMOS-modified linseed oil slightly increased during its storage which is most likely related to further curing of non-reacted organosilicon compounds during the thermal reaction. These not-significant viscosity changes allow classifying products as stable during storage. The above observations with the influence of "ene" reaction for oil were described by Tambe et al. [23]. The viscosity of the resulting soybean oil was, opposite to the linseed oil described in this paper, unaffected by the silylation reaction and remained low (32 cPs). Observed differences were carefully described in an article by Depczyńska et al. [22], where the differences were explained through the different compositions of fatty acids in oils.

3.3. Absorption of Vapor from Humid Air

The analysis of vapor absorption from humid air into the pine wood indicates that every protection positively impacted the reduction of this parameter (Figure 3). In the first 4 h of wood exposure against humid air, distinctly higher (approximately 70%) water absorption by unprotected wood compared to the wood protected with reaction products no. 3 and 4 were seen. After 24 h of testing, no significant differences between specific variations of the reactions no. 2–4 were observed. In a longer time perspective of storing wood in humid air conditions, a tendency to absorb less water by the wood impregnated with modified oil became conspicuous. Wood treated with a product of reaction no. 3 or 4 had a reduced water absorption rate. The wood treated with products of reactions no. 1 and 2, after 96 h of testing, had not protected the wood against water. Its absorption was at a similar level to the absorption of water from humid air in untreated pine wood. The results confirmed that the new products that were obtained in reactions no. 3 and 4 increase the hydrophobic nature of wood in long-term exposure to humid air environments.

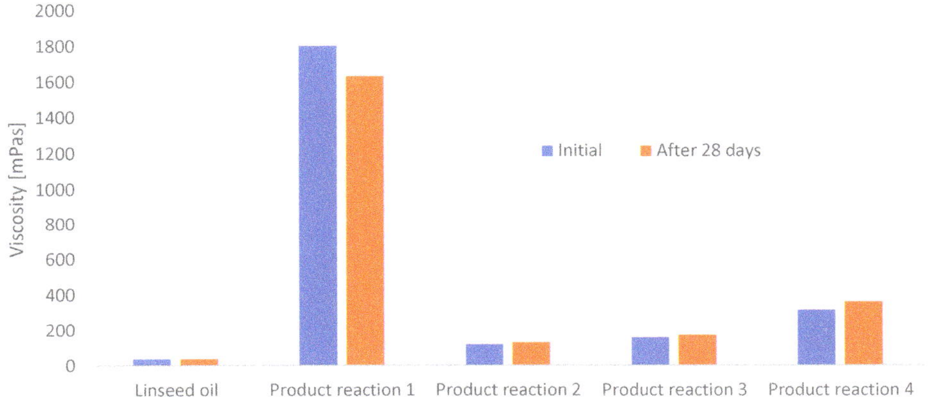

Figure 2. Viscosity of linseed oil and silylation reaction products.

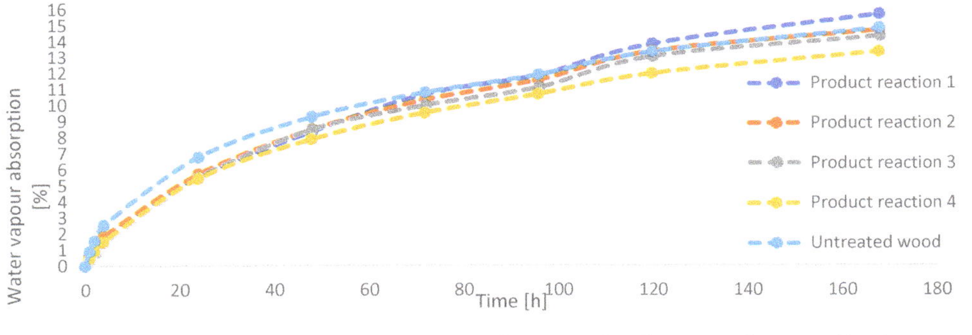

Figure 3. Absorption of water vapor by untreated pine wood and pine wood protected by the tested products.

3.4. Absorbability

The absorbability test confirmed that all tested products of the reactions reduce water absorption in pine wood (Figure 4). An approximately 50% reduction of absorbability relative to untreated wood could be observed for every tested product. The best result was obtained again for the wood protected with the products of reactions no. 3 and 4. The tests performed showed significant differences for pine wood between the following data: Untreated Wood—Product Reaction 1; Untreated Wood—Product Reaction 2; Untreated Wood—Product Reaction 3; Untreated Wood—Product Reaction 4; Product Reaction 1—Product Reaction 3; Product Reaction 2—Product Reaction 3

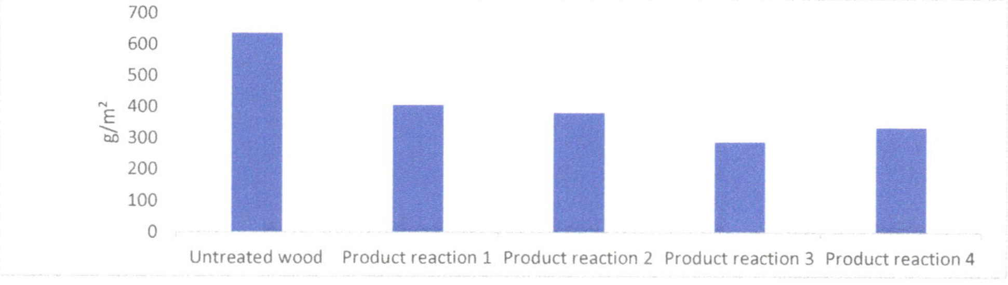

Figure 4. Absorbability of pine wood untreated and treated with the tested products.

Similar to the treatment of pine wood, the treatment of spruce wood reduced the absorption of water compared to untreated wood (Figure 5). Differences were observed in wood impregnation with products of reactions no. 1–4. Polymerized linseed oil reduced water absorption by over 50% compared to unprotected wood. However, the products modified with silane reduced water absorption by approximately 20%. The differences in the tests performed for pine wood and spruce wood are mainly found in comparing two key parameters, i.e., the anatomy of specific species and the viscosity of oils. Spruce wood is a species that is difficult to saturate, which causes the applied preparations to concentrate on its surface, but white parts of pine wood that are easy to saturate will quickly absorb oils with low viscosity. The tests performed showed significant differences between the following data: Untreated Wood—Product Reaction 1; Untreated Wood—Product Reaction 2; Product Reaction 1—Product Reaction 2; Product Reaction 1—Product Reaction 3; Product Reaction 1—Product Reaction 4. Temiz et al. [24] and Humar and Lesar [6] also reported that linseed oil reduced water absorption in wood. The first research team used the oil as a by-product of the ThermoWood thermal modification process. Linseed oil and hydrogen peroxide prepared the treatment agent [24]. The results indicated that the samples treated with bio-oil had lower water absorption than the control group. A second research team confirmed that linseed oil impregnation of spruce and beech wood reduces short and long-term water uptake of liquid water [6].

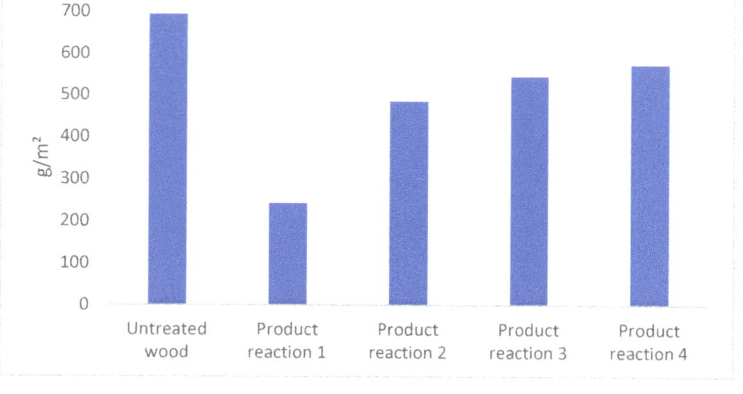

Figure 5. Absorbability of unprotected spruce and the spruce protected by the tested products.

3.5. Static Contact Angle

The static analysis of the measurements of static contact angles (Figure 6) has confirmed that the obtained products after application on wood and drying are characterized by high hydrophobicity. The highest result for the contact angle was observed for pine wood protected by the product of the reaction no. 1. The probable reason of the higher hydrophobic properties was the lower viscosity of the new preparations (reactions no. 2–4) and, thus, better wood penetration. Despite the fact that natural oils do not form a coating, the significantly higher viscosity of the product in reaction no. 1 impeded its penetration into the wood and formed a much more uniform coating on the wood surface.

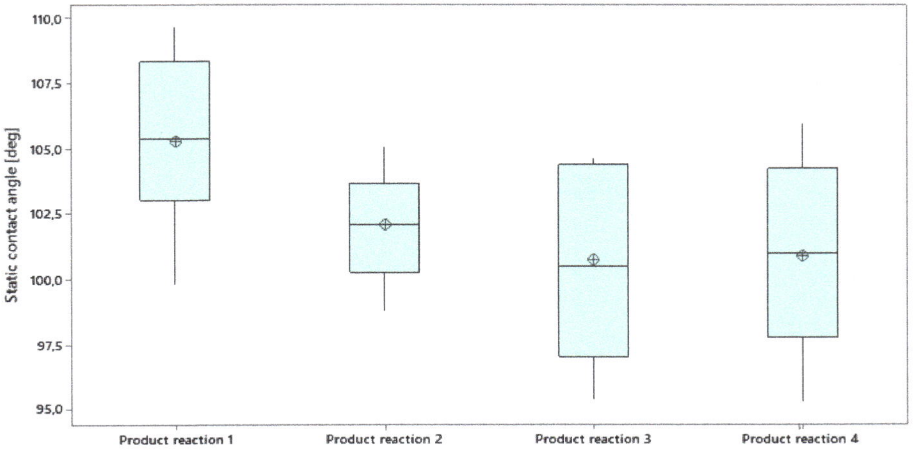

Figure 6. Results of contact angle tests for the protected pine wood.

3.6. Change of Color of the Protected Wood

The impact of wood protection with the products of the reactions no. 1–4 on its change in color has been presented in Figure 7. The oil application caused a color change ($\Delta E > 5$) visible to the naked eye in each case. The product of the silylation reaction (no. 2, 3, 4) affected a change of the wood color to a lower extent than the product of reaction no. 1 (polymerized linseed oil). The observation of the b monochromatic component that demonstrates yellowing of the sample seems to be particularly significant, and it is often perceived as a defect in the industry. The highest values were observed for polymerized linseed oil in this area.

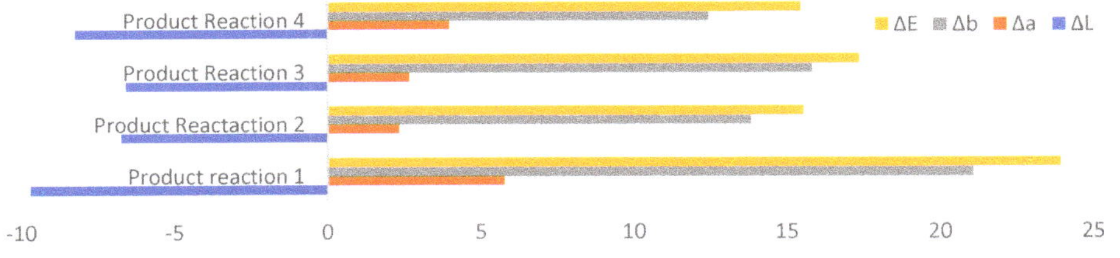

Figure 7. Results of the change of color in the CIELAB system: ΔL, Δa, Δb, ΔE relative to non-impregnated wood.

Figure 8 presents the results of spectrophotometric measurements for pine wood impregnated with oils no. 1–4 and exposed to aging factors for one week and six weeks. Comparison of a change in color of the samples after one week and after six weeks shows that the polymerized linseed oil during its curing process in the wood (the first week) has changed the color of wood ($\Delta E > 20$) distinctly, and then, during the subsequent five weeks, the color did not change significantly. Similarly to the described linseed oil, the products of the reaction of linseed oil with VTMOS changed the color of wood significantly after approximately one week, and then the color of wood changed to the one that was tested on the samples just after their impregnation.

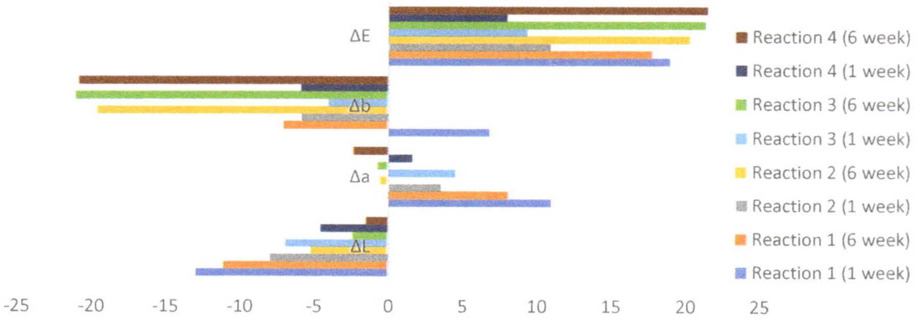

Figure 8. The change in the color of the wood protected with preparations exposed to the impact of aging factors in the periods of 1 week and 6 weeks; ΔL, Δa, Δb monochromatic components and ΔE parameter.

3.7. Resistance of Wood to Biological Impacts

The wood treated with polymerized linseed oil (product of reaction no. 1) was resistant to tested fungi for approximately 7 days. After 14 days, almost the entire surface of the samples was infected. Leaching significantly shortened the period of resistance of wood to mold because just after a week, more than 30% of the wood surface was infected. The wood impregnated with the preparation obtained in reaction no. 4 had the highest resistance. After 7 and 14 days, less than 30% of the surface was infested by tested fungi. Specific attention should be paid to the fact that no significant differences were observed when comparing wood samples that were leached and unleached. It is confirmed by the observations mentioned previously that preparations protected the wood against the impact of water, and, as a result, they were also washed out to a lesser extent. The product of reaction no. 4 limited the development of fungi-causing mold until the end of the test, i.e., to the 28th day of exposure to degrading factors. Detailed results of the test are presented in Figure 9. The tests performed showed significant differences between the following data: Untreated Wood—Product Reaction 4; Untreated Wood—Product Reaction 4 (EN 84); Product Reaction 1—Product Reaction 4 (EN 84); Product Reaction 1 (EN 84)—Product Reaction 4; Product Reaction 1 (EN 84)—Product Reaction 4 (EN 84).

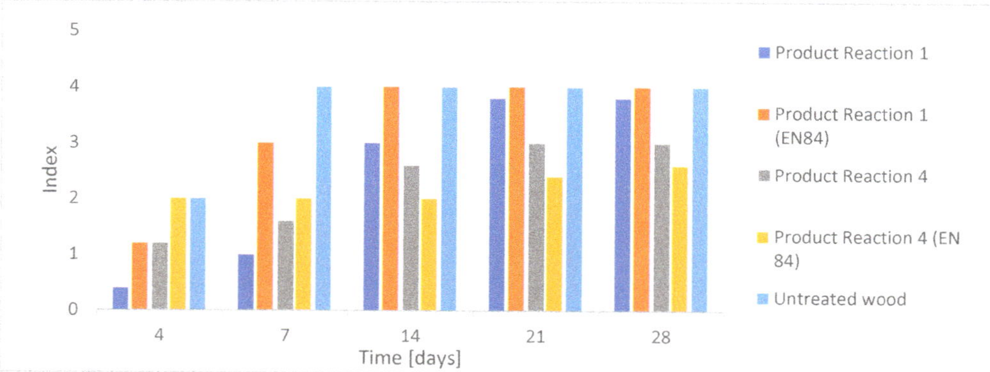

Figure 9. Resistance of the wood protected against the impact of fungi-causing mold.

The mass loss of wood that had not been subjected to impregnation was more than 20%, proving how active *C. puteana* is. In the same conditions of wood exposure, the samples protected with polymerized linseed oil had little resistance to the fungus, and the average mass loss of the washed and unwashed wood was 12.4 and 11.5%, respectively. Impregnation of wood with the product of reaction no. 4 increased wood protection against the decay fungus, and the mass loss was 3.6 and 2%, respectively. Detailed test results are summarized in Table 3. The tests performed showed significant differences between the following data: Product Reaction 1—Product Reaction 4; Product Reaction 1—Product Reaction 4 (EN 84); Product Reaction 1 (EN 84)—Product Reaction 4; Product Reaction 1 (EN 84)—Product Reaction 4 (EN 84). Compared with the literature discussion, product reaction 4 more effectively protected pine wood against C. puteana fungi than non-modified linseed oil. It gives new possibilities for utility linseed oil protection of wood in more hazardous applications than the literature presented [6,24–26].

Table 3. Resistance of the wood protected against *C. puteana*.

Sample ID	Wood Moisture Content [%]	Mass Loss
Product Reaction 1	36.9 ± 6.3	12.4 ± 1.8
Product Reaction 1 (EN84)	30.1 ± 4.2	11.5 ± 9.4
Product Reaction 4	19.9 ± 1.7	3.6 ± 0.7
Product Reaction 4 (EN84)	19.5 ± 1.5	2.0 ± 0.8
Untreated wood/fungi activity	67.9 ± 5.4	21.8 ± 1.8

4. Conclusions

Linseed oil silylation with an organosilicon compound containing a vinyl group (Vinyltrimethoxysilane—VTMOS) that may react with unsaturated bonds contained in fatty acids made it possible to synthesize products showing high stability in storage. Long-term storage of the silylation process products has not changed their viscosity. Impregnation of wood with the resulting products contributed to the increased resistance of protected wood to water exposure compared to unprotected wood, which could translate to the increase of wood resistance to exposure to colonizing and decay fungi. As far as the utility aspect is concerned, it was observed that the process of running the silylating reaction is a crucial parameter limiting the level of wood protection relative to the factors causing biodegradation and biodeterioration. The wood protected with the reaction product in which the linseed oil was prepolymerized for 3 h and silylated for 6 h had the highest durability of all completed syntheses. The resulting synthesized products may show potential as semiproducts used to develop preparations used in wood protection, especially as a stable and hydrofobic substrate for resin and emulsion production. This topic should become a continuation of the efforts described in this study.

Author Contributions: Conceptualization, W.P., E.D., M.F. and B.M.; methodology, W.P., E.D., M.F., K.T., M.K. and B.M.; validation, W.P., E.D., M.F., K.T. and B.M.; formal analysis, W.P., M.F. and B.M.; investigation, W.P., E.D., M.F., K.T. and M.K.; resources, W.P., E.D., M.F., K.T., M.K. and B.M.; writing—original draft preparation W.P., E.D., M.F., K.T., M.K. and B.M.; writing—review and editing, W.P., M.F. and B.M.; visualization, W.P., M.F. and M.K.; supervision, B.M.; project administration, M.F.; funding acquisition, M.F. All authors have read and agreed to the published version of the manuscript.

Funding: The National Centre for Research and Development: POIR.01.02.00-00-0017/16 "Biocide free products and ecological to protection the wood basing silylated plant oil". The project is accomplished by Tikkurila Polska S.A. as part of the Innochem project; Smart Growth Operational Programme, 2014-2020. The project was cofinanced by the European Regional Development Fund.

Data Availability Statement: Data is contained within the article.

Conflicts of Interest: The authors declare no conflict of interest.

References

1. Depczynska, E.; Mazela, B. Modern Bio-Based Wood Coatings-Potential for the Future. *Surf. Coat. Int.* **2019**, *102*, 78–83.
2. Assanvo, E.F.; Gogoi, P.; Dolui, S.K.; Baruah, S.D. Synthesis, Characterization, and Performance Characteristics of Alkyd Resins Based on Ricinodendron Heudelotii Oil and Their Blending with Epoxy Resins. *Ind. Crops Prod.* **2015**, *65*, 293–302. [CrossRef]
3. Athawale, V.D.; Nimbalkar, R.V. Waterborne Coatings Based on Renewable Oil Resources: An Overview. *J. Am. Oil Chem. Soc.* **2011**, *88*, 159–185. [CrossRef]
4. Chang, C.-W.; Chang, J.-P.; Lu, K.-T. Synthesis of Linseed Oil-Based Waterborne Urethane Oil Wood Coatings. *Polymers* **2018**, *10*, 1235. [CrossRef]
5. Hu, W.; Wan, H. Comparative Study on Weathering Durability Property of Phenol Formaldehyde Resin Modified Sweetgum and Southern Pine Specimens. *Maderas. Cienc. Y Tecnol.* **2022**, *24*. [CrossRef]
6. Humar, M.; Lesar, B. Efficacy of Linseed-and Tung-Oil-Treated Wood against Wood-Decay Fungi and Water Uptake. *Int. Biodeterior. Biodegrad.* **2013**, *85*, 223–227. [CrossRef]
7. Hill, C.A.; Farahani, M.M.; Hale, M.D. *The Use of Organo Alkoxysilane Coupling Agents for Wood Preservation*; De Gruyter: Berlin, Germany, 2004. [CrossRef]
8. Lu, K.-T.; Chang, J.-P. Synthesis and Antimicrobial Activity of Metal-Containing Linseed Oil-Based Waterborne Urethane Oil Wood Coatings. *Polymers* **2020**, *12*, 663. [CrossRef]
9. Farahani, M.R.M. *Decay Resistance of Modified Wood*; Bangor University: Bangor, UK, 2003; ISBN 1-68799-660-1.
10. Liu, C.; Li, J.; Lei, W.; Zhou, Y. Development of Biobased Unsaturated Polyester Resin Containing Highly Functionalized Castor Oil. *Ind. Crops Prod.* **2014**, *52*, 329–337. [CrossRef]
11. Veigel, S.; Lems, E.-M.; Grüll, G.; Hansmann, C.; Rosenau, T.; Zimmermann, T.; Gindl-Altmutter, W. Simple Green Route to Performance Improvement of Fully Bio-Based Linseed Oil Coating Using Nanofibrillated Cellulose. *Polymers* **2017**, *9*, 425. [CrossRef]
12. Datta, J.; Głowińska, E. Effect of Hydroxylated Soybean Oil and Bio-Based Propanediol on the Structure and Thermal Properties of Synthesized Bio-Polyurethanes. *Ind. Crops Prod.* **2014**, *61*, 84–91. [CrossRef]
13. Donath, S.; Militz, H.; Mai, C. Weathering of Silane Treated Wood. *Holz Roh Werkst.* **2007**, *65*, 35. [CrossRef]
14. Islam, M.R.; Beg, M.D.H.; Jamari, S.S. Alkyd Based Resin from Non-Drying Oil. *Procedia Eng.* **2014**, *90*, 78–88. [CrossRef]
15. Jebrane, M.; Cai, S.; Sandström, C.; Terziev, N. The Reactivity of Linseed and Soybean Oil with Different Epoxidation Degree towards Vinyl Acetate and Impact of the Resulting Copolymer on the Wood Durability. *Express Polym. Lett.* **2017**, *11*, 383. [CrossRef]
16. Szubert, K. Synthesis of Organofunctional Silane from Rapeseed Oil and Its Application as a Coating Material. *Cellulose* **2018**, *25*, 6269–6278. [CrossRef]
17. Panov, D.; Terziev, N. Study on Some Alkoxysilanes Used for Hydrophobation and Protection of Wood against Decay. *Int. Biodeterior. Biodegrad.* **2009**, *63*, 456–461. [CrossRef]
18. Ratajczak, I.; Szentner, K.; Rissmann, I.; Mazela, B.; Hochmanska, P. Treatment Formulation Based on Organosilanes and Plant Oil Blend—Reactivity to Wood and Cellulose. *Wood Res.* **2012**, *57*, 265–270.
19. Srinivasan, M. *Synthesis, Properties and Applications of Bio-Based Materials*; Materials Science and Engineering. Ph.D. Thesis, Michigan State University, East Lansing, MI, USA, 2010.
20. Tambe, C.; Dewasthale, S.; Shi, X.; Graiver, D.; Narayan, R. Silylation of Non-Terminal Double Bonds of Natural Oils. *Silicon* **2016**, *8*, 87–98. [CrossRef]
21. Depczynska, E.; Furgal, M.; Mazela, B.; Perdoch, W. Silylated Linseed Oil-Invisible Wood Protection? *Surf. Coat. Int.* **2019**, *102*, 22–26.
22. Depczyńska, E.; Perdoch, W.; Mazela, B. Influence of Reaction Parameters on the Gelation of Silanised Linseed Oil. *Materials* **2020**, *13*, 5376. [CrossRef]
23. Tambe, C.; Graiver, D.; Narayan, R. Moisture resistance coating of packaging paper from biobased silylated soybean oil. *Prog. Org. Coat.* **2016**, *101*, 270–278. [CrossRef]
24. Temiz, A.; Kose, G.; Panov, D.; Terziev, N.; Alma, M.H.; Palanti, S.; Akbas, S. Effect of bio-oil and epoxidized linseed oil on physical, mechanical, and biological properties of treated wood. *J. Appl. Polym. Sci.* **2013**, *130*, 1562–1569. [CrossRef]
25. Łucejko, J.J.; La Nasa, J.; McQueen, C.M.; Braovac, S.; Colombini, M.P.; Modugno, F. Protective effect of linseed oil varnish on archaeological wood treated with alum. *Microchem. J.* **2018**, *139*, 50–61. [CrossRef]
26. Terziev, N.; Panov, D. Plant oils as "green" substances for wood protection. In *Minimizing the Environmental Impact of the Forest Products Industries-Ecowood*; Fernando Pessoa University: Porto, Portugal, 2010; pp. 143–149.

forests

MDPI

Article

Influence of Upcycled Post-Treatment Bark Biomass Addition to the Binder on Produced Plywood Properties

Aleksandra Jeżo [1], Anita Wronka [2,*], Aleksander Dębiński [3], Lubos Kristak [4], Roman Reh [4], Janis Rizhikovs [5] and Grzegorz Kowaluk [2]

1 Faculty of Wood Technology, Warsaw University of Life Sciences—SGGW, Nowoursynowska St. 159, 02-787 Warsaw, Poland
2 Institute of Wood Sciences and Furniture, Warsaw University of Life Sciences—SGGW, Nowoursynowska St. 159, 02-776 Warsaw, Poland
3 Institute of Forest Sciences, Warsaw University of Life Sciences—SGGW, Nowoursynowska St. 159, 02-787 Warsaw, Poland
4 Faculty of Wood Sciences and Technology, Technical University in Zvolen, 960 01 Zvolen, Slovakia
5 Latvian State Institute of Wood Chemistry, Dzerbenes 27, LV-1006 Riga, Latvia
* Correspondence: anita_wronka@sggw.edu.pl

Abstract: The valorization of tree bark through chemical treatment into valuable products, such as bark acid, leads to the formation of process residues with a high solids content. Since they are of natural origin and are able to be suspended in water and acid, research was carried out on the recycling of suberic acid residues (SAR) as a bi-functional component of binder mixtures in the production of plywood. The 5%–20% (5%–30% for curing time) mass content of SAR has been investigated with urea-formaldehyde (UF) resin of about 66% of dry content. The results show that the curing time of the bonding mixture can be reduced to about 38% and 10%, respectively, for hot and cold curing, of the initial curing time for the lowest SAR content. The decreasing curing time of the tested binder mixtures with the increase in SAR content was caused by the increasing amount of acidic filler, since amine resins as UF require acidification hardening, and the curing dynamics are strongly dependent, among others, on the content of the acid medium (curing agent). In the case of hot curing, a SAR content of about 20% allowed us to achieve the curing time of bonding mass with an industrial hardener. Investigations into the mechanical properties of examined panels showed a significant modulus of elasticity (MOE) increase with filler content increase. Similar conclusions can be drawn when analyzing the results of the modulus of rupture (MOR) investigations; however, these were only significant regarding hot-pressed samples. The shear strength of the plywood samples increased with the SAR rise for both cold- and hot-pressed panels. The in-wood damage of samples with SAR filler, hot-pressed, rose up to about 30% for the highest SAR filler content. For cold-pressed samples, no in-wood damage was found. The positive effect of veneer impregnation limiter by resin was identified for SAR acting as a filler. Moreover, a higher density of SAR-containing bonding lines was reached for hot-pressed panels. Therefore, the results confirmed the ability to use the SAR as an upcycled component of the bonding mixture for plywood production.

Keywords: plywood; binder; filler; hardener; bark; suberinic acid; residue; upcycling

1. Introduction

Increasingly, attention is being turned to wood waste generated during the harvesting process—branches and bark. Very often these by-products are used as fuel in the incinerators situated next to the factories, or some of these raw materials remain in the forest. Shredded branches were used as an admixture for particleboard [1] or as a face layer of a wood-based composite formed from small slices of branches [2]. When looking for examples of uses for bark, the common use of bark in gardens as mulch is an obvious example [3] but it is also used in the wood-based composites industry, e.g., as an ingredient in insulation panels [4,5] or as an admixture in the production of particleboard [6,7] due to lower strength parameters [8]. For this reason, research has been conducted for several years to extract selected compounds from the bark–tannins [9]. The aforementioned compounds are currently used in the production of natural wood adhesives [10]. An accessible and ecological binder produced from birch outer bark is hydrophobic suberin, which is usually isolated as suberinic acids or partly depolymerized suberin [11].

To approach the topic comprehensively, attention must be paid to the waste associated with tannin production, which is why, in parallel with the development of an ideal tannin-based wood binder, research is being conducted into the use of SAR for, among other things, bonding particleboard [12,13] and plywood [14]. It has been proven that suberinic-acid-based adhesives can cure without any additives or modifiers, remaining a fully ecological product [15], and at the same time resulting in low thickness swelling and high bending strength [16].

The use of fully biodegradable adhesives, e.g., based on gelatin, milk, and casein, is characterized by lower strength parameters and lower moisture resistance and adhesion to wood [17]; for these reasons, high expectations are placed on adhesives based on biopolymers, e.g., polylactide (PLA) [18]; polycaprolactone (PCL); polypropylene (PP) [19]; and, as previously mentioned, the bark [10] and tannins [20]. The ecological aspect is also very important because the emissions of NOx are increased when wood composites with nitrogen-based adhesives, such as urea-formaldehyde (UF), melamine-urea-formaldehyde (MUF), and emulsion polymer isocyanate (EPI), are combusted [21]. As proven, the durability of already mentioned PP, as well as modified polypropylene glues, is equal to that of UF and MUF resins [22].

One of the ingredients in the glue used to make plywood is filler. It has a crucial role as it prevents the glue from soaking into the veneers too much, and thus reduces its viscosity and reduces material costs but can also be used to modify selected properties or improve electrical/thermal conductivity [23]. We distinguish between organic and synthetic fillers [23], which are divided into active and passive fillers. Organic flour is commonly used as a filler. In the research carried out, it was examined whether the type of flour made a difference in the quality of bonding, for which purpose different types of flour were compared: rye flour, hemp flour, coconut flour, rice flour, and pumpkin flour. The tests confirmed that all the flours used could be successfully used as a filler. Against the background of the others, pumpkin flour proved to be distinctive, whereby the adhesive bond strength was the highest. Hemp flour, on the other hand, had a reducing effect on formaldehyde emissions [24], and pine bark had similar properties [25], as well as beech bark [26]. However, it should be pointed out that the main shortcoming of mentioned flour-based fillers is their food origin. Other examples of fillers in plywood and other lignocellulosic composites production are soy flour [27], coffee bean post-extraction residues [28], palm kernel meal [29], oak bark powder [30], thymus plant [31], chestnut shell and coffee waste [32], walnut shell [33], nanocrystalline cellulose [34], and beech bark [35]. In a study using maple bark meal as a filler in the production of three-layer plywood, it was noted that the reduced pH of the bark can lead to the pre-curing of the adhesive. Waste biomass and biomass residues that are abandoned in nature are promising fillers that can be used to replace flour [36]. This knowledge gives hope for the dual use of bark—as a filler and as a curing agent—which can mean considerable savings in the production process [37]. It is not only the chemical properties of the filler that matter but also physical properties, such as the degree of grinding—research confirms that the more ground the filler is, the better the bond quality [38]. Chicken egg shells were used as a bio filler in the other study as well. Due to their non-combustible properties, they can compete with synthetic fillers, and their advantages also include their low price and eco-friendly nature [39]. The examples shown are for commercially used adhesives, but trials can already be found in the literature where fillers are combined with polymers—in this case, wood plastic composites [40]. Starch can also be considered a natural filler used in plywood technology. When combined in the right proportion with citric acid, it produces a biodegradable adhesive bond [41]. It is also used as a binder in wet-molded fiberboard technology [42]. Wood bark has also found its way into plywood technology so far, including the production of plywood without the use of glue as a binder. The role of a binder was played by oxidized bark (*Acacia mangium*). The powdered bark was oxidized using hydrogen peroxide in four variants: 5, 10, 15, and 20%, as well as a catalyst. The best results were obtained for the 20% variant, and the topic was considered forward-looking and developmental [43]. In conclusion, tannin-based adhesives have great potential on the wood market [44], as evidenced by numerous examples of applications, but the by-products must not be forgotten, since they must also be utilized.

The variety of fillers used confirms the potential for substitutes for wheat or rye flour. The bark post-treatment raw material, prepared as flour/powder, can be successfully used as a filler, so this research will test the use of SAR in plywood-production technology.

This research aimed to evaluate the influence of the application of SAR as a bi-functional component, hardener, and filler in the bonding mixture for plywood production. Such a double function of SAR (filler and hardener), as a natural origin, is a new approach in the research of plywood bonding mass fillers. In the scope of research, the curing time of the bonding mixture with various contents of SAR was measured in a room and at elevated temperatures.

2. Materials and Methods

2.1. Materials

The rotary cut birch (*Betula* L.) veneer of an average thickness of 1.5 mm, 5% ± 1% moisture content (MC), and dimensions of 360 × 360 mm^2 were used to produce plywood. As a binder, an industrial UF resin S-120 (Silekol Sp. z o.o., Kędzierzyn—Koźle, Poland) of about 66% of dry content [45] was used with ammonium nitrate water solution (industrial hardener mentioned in Table 1; 20 wt%) as a hardener to reach the curing time of REF 0 (Table 1) gluing mass at 100 °C in about 86 s. The rye flour was used as a reference filler. The mentioned UF resin was also the base of the SAR bonding mixture.

Table 1. Compositions of bonding mixtures and plywood-selected pressing parameters.

Variant Label	Filler	Filler Content	Industrial Hardener Content	Pressing Temp.	Pressing Time
		[pbw [1] per 100 pbw of Solid Resin]		[°C]	[Min]
REF 0	Rye flour	0	2	120	4 min
REF 5	Rye flour	5	2	120	4 min
REF 10	Rye flour	10	2	120	4 min
REF 20	Rye flour	20	2	120	4 min
SAR 5	SAR	5	0	120	12 min
SAR 10	SAR	10	0	120	8 min
SAR 20	SAR	20	0	120	4 min
SAR 5 C	SAR	5	0	Room	24 h
SAR 10 C	SAR	10	0	Room	24 h
SAR 20 C	SAR	20	0	Room	24 h

[1] pbw—parts by weight.

The SAR were used in this research as an alternative filler and hardener. SAR was kindly provided by the Latvian State Institute of Wood Chemistry, Riga, Latvia, made of the residues resulting from the isolation of suberinic acid in the ethanol, in the process described by [12], suspended in water and acidified to pH 2, filtered and rinsed with deionized water, which is described by [13]. The SAR, delivered as a brown paste (Figure 1a), about 25% of dry matter, pH 3.3, was dried at 70 °C to constant mass. The chemical properties are described in [12]—acid number 95.8 mg/KOH, epoxy groups 0.61 mmol/g, cellulose 9.0 wt%, aromatic suberin, lignin 21.4 wt%, ω-hydroxy acids 17.5%, and α, ω-diacids 11.9%. Then, the totally dry matter was crushed in an SM100 cutting mill (RETSCH GmbH,

Haan, Germany), and the fine fraction <0.1 mm (extracted by sieving) was taken for further research (Figure 1b). The pH of the water solution of the achieved SAR powder was 3.7.

Figure 1. The SAR in the wet state (as delivered) (**a**) and dry state (**b**).

2.2. Methods

2.2.1. Bonding Mixtures Curing Time

To measure the curing time of the bonding mixtures of various contents of different fillers, the mixtures of binders were prepared, as given in Table 1. Additionally, the mixture of UF resin and SAR filler in the ratio of 100:30 pbw was prepared. This variant of bonding mixture was not used for further plywood preparation due to extremely high viscosity. The samples of mixtures of a weight of about 8 g each were moved to laboratory glass tubes, and the time to cure was measured during storage of samples in glass tubes in: (1) boiling water (hereinafter called "Hot") or (2) room temperature (about 20 °C; hereinafter called "Cold"; SAR samples only) accompanied by manual slow stirring. As many as 5 repetitions of curing time measurement were performed for every bonding mixture and every temperature mentioned.

2.2.2. Plywood Preparation and Characterization

The three-layer plywood sheets were prepared with use of the bonding mixtures listed in Table 1. The following parameters were applied during plywood preparation: bonding mixture spread 170 g m^{-2} (manually spread) and maximum press pressure 1.4 MPa for hydraulic press (AKE, Mariannelund, Sweden). The pressing time is given in Table 1. All the produced plywood sheets were subjected to conditioning at 20 °C/65% ± 1% R.H. to constant weight before further testing.

The following tests were completed for the produced plywood: bonding quality in a dry state according to [46] (8 repetitions), modulus of elasticity (MOE), and modulus of rupture (MOR) in a parallel direction to the grains of the face veneer layer (8 repetitions), according to [47]. All the mechanical tests were performed on a computer-controlled universal testing machine (Research and Development Centre for Wood-Based Panels Sp. z o.o. Czarna Woda, Poland). The density profile of the tested plywood was measured (3 repetitions) on a GreCon DAX 5000 device (Fagus-GreCon Greten GmbH and Co. KG, Alfeld/Hannover, Germany) with sampling step of 0.02 mm. The density profile measurement results were the representative plots selected after analyses of 3 individual plots for every tested panel.

The images of the cross-cuts of the investigated plywood samples were taken with a Nikon SMZ 1500 (Kabushiki-gaisha Nikon, Minato, Tokyo, Japan) optical microscope. The images of the break zone of the samples after the shear strength test (bonding quality test) were taken with the use of a Canon 550D digital camera (Canon Inc., Tokyo, Japan). The in-wood damage (fiber failure) was estimated, according to [46].

2.3. Statistical Analysis

Analysis of variance (ANOVA) and t-tests calculations were used to a test ($\alpha = 0.05$) for significant differences between factors and levels, where appropriate, using IBM SPSS statistic base (IBM, SPSS 20, Armonk, NY, USA). A comparison of the means was performed when the ANOVA indicated a significant difference by employing the Duncan test. The statistically significant differences for the achieved results are given in the Results and Discussion paragraphs when the data are evaluated. The letters on the plots indicate the homogeneous groups; however, the homogeneity has been tested among the three groups (REF, SAR hot, and SAR cold) in different filler content and curing temperature.

3. Results and Discussion

3.1. Curing Time

The results of the curing time measurement of bonding mixtures of various content and different types of filler are presented in Figure 2. As is shown in the case of hot curing, the curing time rapidly decreases from 225 s for SAR content 5 pbw to 86 s for SAR content 30 pbw. A similar tendency has been found in the case of cold curing, where the curing time was decreasing from 1260 min for SAR content 5 pbw to 120 min for SAR content 30 pbw. All the analyzed average values have been statistically significantly different from one to another. The curing time of the reference bonding mixture (REF 0) was 86 s. The decreasing curing time of the tested bonding mixtures with increasing SAR content was caused by the rising amount of acidic-nature filler. According to [48], the amine resins, including UF, need to be acidified to be cured, and the curing dynamic is strongly dependent, i.a., on the content of the acidic medium (hardener). However, it should be pointed out that the decreasing time has an asymptotic character, which means the intensity of the curing time decrease is lower for higher filler content than for lower filler content. This is also influenced by the phenomena of buffering capacity [49]. The further addition of acidic filler will not affect curing time shortening, as it was found for lower filler content.

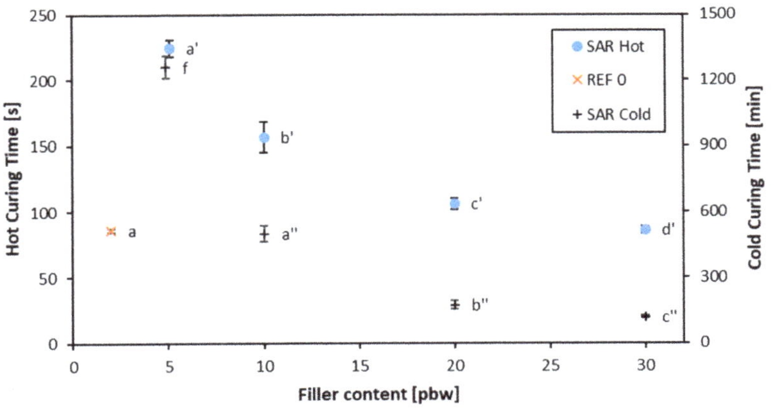

Figure 2. The dependence of the curing time vs. filler content (the letters on the plots indicate the homogeneous groups).

3.2. Bonding Quality

The following features of the produced plywood sheets were investigated to estimate the quality of bonding: shear strength and images of the break zone of the samples after the shear strength test. The results of the shear strength measurement are presented in Figure 3, and the images of the break zone are provided in the pictures collected in Figure 4. As is presented in Figure 3, for all the tested bonding mixtures used, the shear strength increased linearly with the filler content increase. The most intensive increase in shear strength has been found for the reference bonding mixture with rye flour as a filler, where the shear strength increased from 1.06 N mm^{-2} for 0 filler content to 1.71 N mm^{-2} for 20 pbw filler content. The increase in SAR filler content from 5 pbw to 20 pbw caused shear strength to increase from 1.43 N mm^{-2} to 1.57 N mm^{-2}. That means, the SAR content increase did not provide as much dynamic shear strength improvement as it did for the reference bonding mixture. However, it is worth mentioning that for a similar filler content, 5 pbw, better shear strength has been found for SAR filler (SAR hot samples). The SAR filler content increase for the bonding mixture cured at room temperature led to a slight shear strength increase from 0.25 N mm^{-2} to 0.42 N mm^{-2} for the SAR 5 C and SAR 20 C bonding mixtures, respectively. Regarding the statistical significance of differences in average values of shear strength, in the case of the reference binder, the REF 0 was significantly different from the remaining REF panels, as well as REF 20 being significantly different from the remaining samples. The only statistically significant difference in average shear strength for both SAR hot and SAR cold variants was found for SAR 20 compared to the remaining samples.

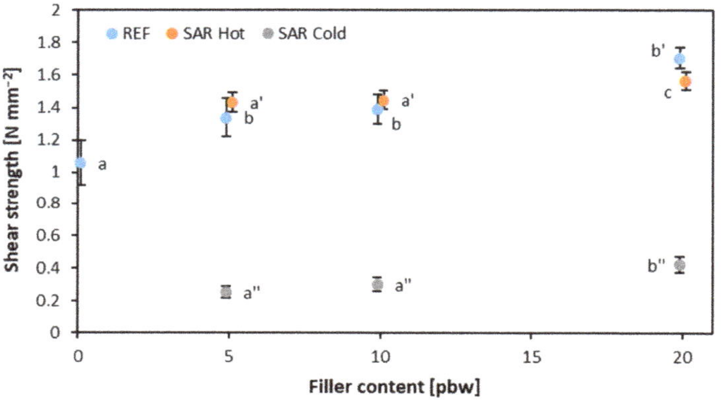

Figure 3. The shear strength of the plywood samples of different filler content (the letters on the plots indicate the homogeneous groups).

Interesting observations can be found when analyzing the images of the break zones of the samples after the shear strength test (Figure 4). For reference panels (REF 0, REF 5, REF 10, and REF 20), the increasing filler content led to the rise of the in-wood destruction area. For SAR hot samples (SAR 5, SAR 10, and SAR 20), the same tendency can be observed. For cold-pressed panels (SAR 5 C, SAR 10 C, and SAR 20 C), only the rising darkness intensity of the bonding line has been found due to increasing SAR filler content without the change in the in-wood damage. As confirmed by [17,50], the in-wood damage area was strongly connected to the shear strength of the bonded wood samples. The significant influence of the bark filler content on the bonding shear strength has also been confirmed by [37,51,52].

The differences in the break zones of the plywood samples bonded with different contents and types of filler pressed in the hot or cold press process after the shear strength test are collected in the pictures that are presented in Figure 4. According to [53,54], the analyses of layered materials break zone help with the evaluation of the properties of materials and allow for the description of the relations between the production parameters and material features. In the case of reference samples (REF 0, REF 5, REF 10, and REF 20), with the increasing filler content, the in-wood damage of the connection significantly rises, especially in the REF 20 sample. There was no in-wood damage for REF 0 and less than 10% for REF 5 and REF 10, while there was about 30% for REF 20. The in-wood damage of samples with SAR filler, hot-pressed, rose from 0 for SAR 5, through 10% for SAR 10 to about 30% for SAR 20 (lowest to highest SAR filler content). For cold-pressed samples (SAR 5 C, SAR 10 C, and SAR 20 C), except for the progressively darker surface, caused by an increasing content of the brown filler, no in-wood damage has been found.

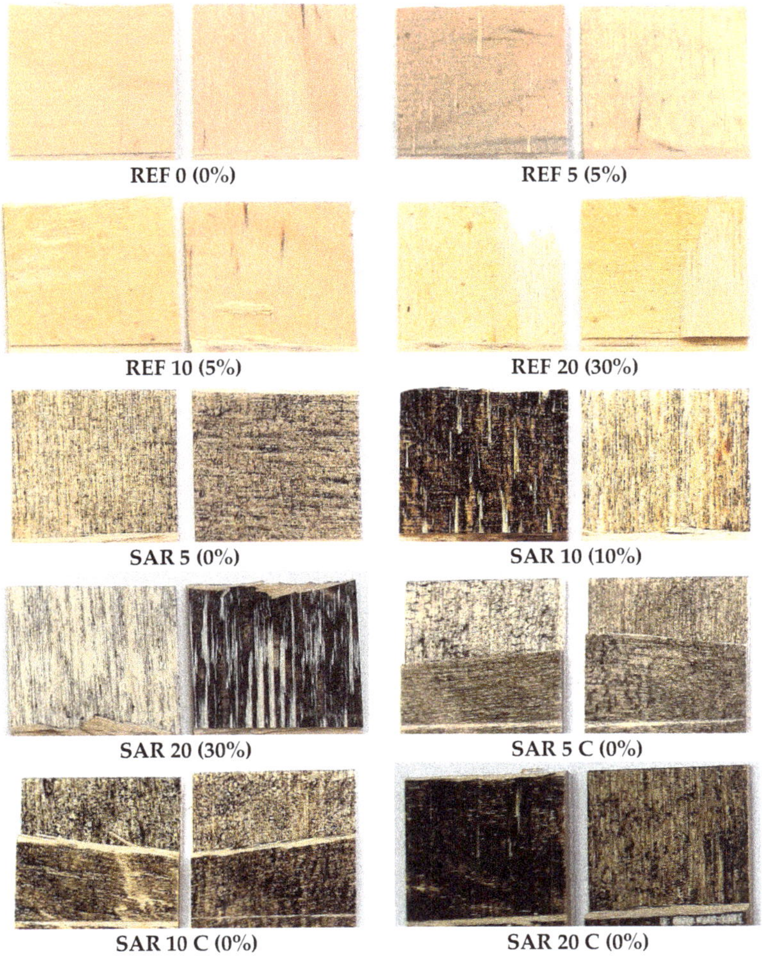

Figure 4. The images of the break zones of the samples after the shear strength test (dimensions: 25 × 25 mm^2; numbers in brackets are the estimation of in-wood damage).

3.3. Modulus of Elasticity and Modulus of Rupture

The modulus of elasticity results are presented in Figure 5. A significant MOE increase has been found with filler content increase; however, the maximum MOE has been noted for about 10% of filler content. Further rise of filler content reduces MOE. The difference between MOE for 10% filler content compared to the remaining samples of reference binder were statistically significant. According to the state of the art [35], the filler addition of more than 15% led to a radical binder viscosity increase. As was stated for bark-based filler for plywood bonding mass [55], the increase in filler content of about 20% significantly increases the viscosity of the binder. Thus, the possible problems with the even spreading of the binder over the veneer surface can negatively influence the bending strength of the plywood sheets. When analyzing the results for SAR hot-pressed panels, it can be concluded that within the tested filler content of 5%–20%, the filler content increase caused the rise in MOE. What is more, all the achieved results are statistically significantly different from one to another, and high repeatability of the MOE results has been found (low standard deviation values indicated on the plots as the error bars). The increasing tendency of MOE for SAR cold-pressed panels has also been found, and it can also be seen that all the achieved results are statistically significantly different from one to another.

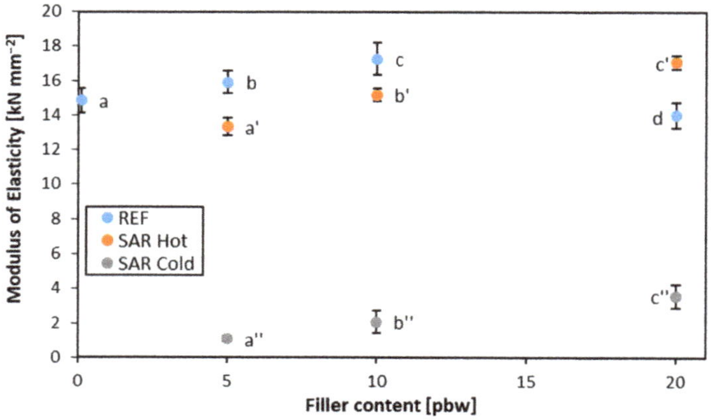

Figure 5. The modulus of elasticity of the plywood samples of different filler content (the letters on the plots indicate the homogeneous groups).

Similar conclusions can be drawn when analyzing the results of MOR investigations (Figure 6). However, in the case of cold-pressed panels, no significant influence of SAR filler content has been found. Again, for industrial filler, the optimal content, leading to maximum MOR, was about 10%. This is in line with the findings of [29], where the filler content in the range of 10%–15% was indicated as optimal for the mechanical properties of the layered wood composite.

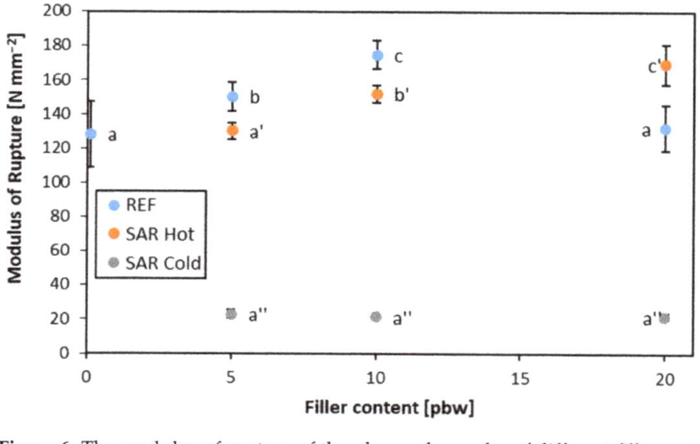

Figure 6. The modulus of rupture of the plywood samples of different filler content (the letters on the plots indicate the homogeneous groups).

The lower content of the filler decreases MOE and MOR due to the insufficient bonding line creation, caused by too high of an impregnation of the veneer by the bonding mixture. Thus, the break of the bent sample occurs in the bonding line (between veneers), as is shown in Figure 7. In the same figure, the destruction of the bent sample by the break of the bottom layer has been shown, where the highest values of MOE and MOR were reached.

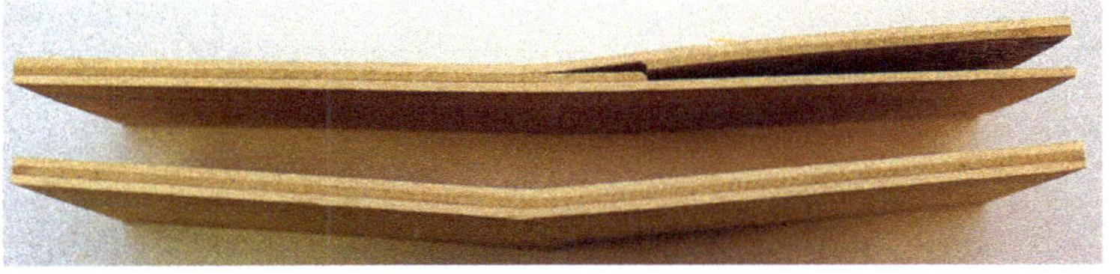

Figure 7. The image of samples after bending test: REF 0 delamination in bonding line (**top**) and REF 10 bottom layer break (**bottom**).

3.4. Density Profile

Figure 8 shows the course of density profiles for control samples, without SAR. The shape of the density profiles is generally symmetrical. The density of the veneers differed between the variants, ranging from 600 kg m^{-3} to 800 kg m^{-3}, without a special relation to production parameters. The average thickness of the samples was about 4.4 mm. The bonding lines showed a significant increase in density across the sample, reaching 1000 kg m^{-3} for REF 0 (without filler addition) and about 1100 kg m^{-3} for the other variants. One can see here that the relationship between the density of the binder and the mass fraction of the filler—the higher its share, the greater the density of the bonding line.

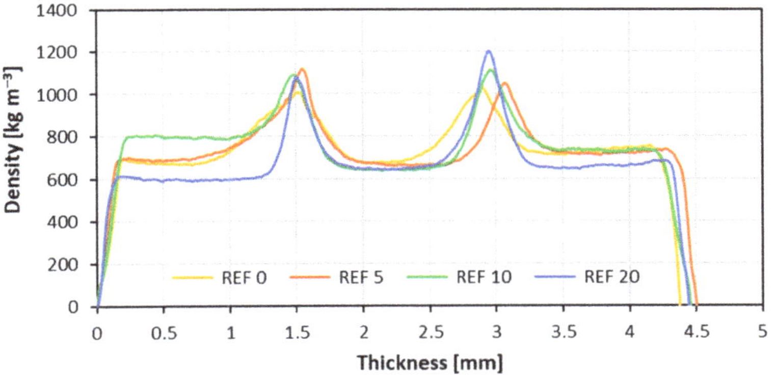

Figure 8. The density profiles of the reference plywood samples of different filler content.

The density profiles of the hot-pressed samples prepared with SAR as the filler (Figure 9) showed a symmetrical course. The density of the veneers was, depending on the sample, 700–800 kg m^{-3}, while the density of the bonding line ranged from about 1100 kg m^{-3} for samples with a mass fraction of SAR of 5 to about 1300 kg m^{-3} for samples with a mass fraction of SAR of 20. That means an increase in bonding line maximum density has been found with a SAR filler content increase. A slight rise in the panels' densification can be found, since the mean thickness of the samples was about 4.3 mm.

In Figure 10, the density profiles of cold-pressed plywood with SAR as the filler have been presented. The density profile of three-layer cold-pressed plywood is symmetrical for all three variants. The veneer layers showed a density of about 600 kg m^{-3}. For samples with the share of SAR as a filler of 20 parts by mass, the adhesive joint showed a higher density of about 900 kg m^{-3}. However, for the remaining variants—5 and 10 mass parts of SAR—the test showed a decrease in density at the point of the adhesive joint, most clearly visible for the smallest share of SAR—about 600 kg m^{-3}, below the density values achieved by the veneer. Between the decrease in density and the veneer layers, there was a brief increase in density—presumably at the point of contact between the wood and the glue, i.e., in the layer of wood where the glue soaked. A smaller decrease in density is also seen for samples with a mass fraction of SAR of 10—here, after an increase in the value to over 820 kg m^{-3}, a decrease to lower values, closer to 800 kg m^{-3}, can be observed. The described observations are confirmed by the research of [56], which proved that in the case of layered materials, the glue line bonds for all the dry-pressed plies (glue-formulated and control) are stronger than those for the cold-pressed samples before and after field exposure.

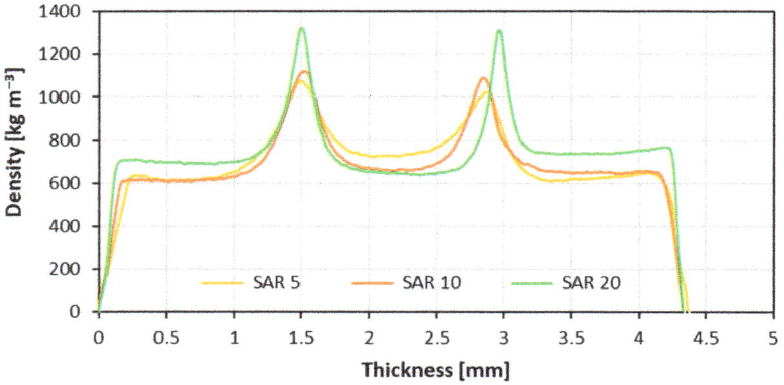

Figure 9. The density profiles of the plywood samples of different SAR filler content, hot-pressed.

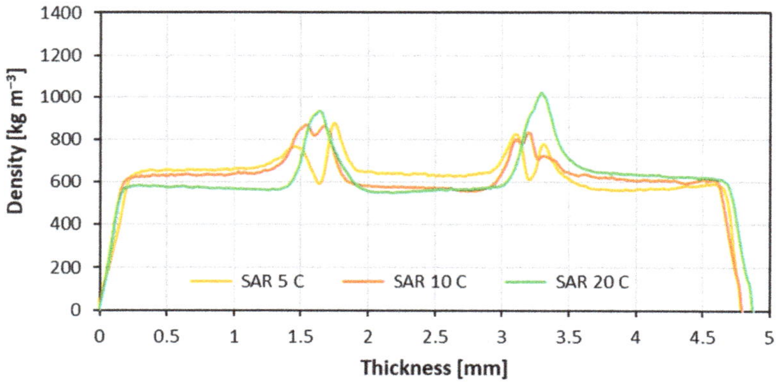

Figure 10. The density profiles of the plywood samples of different SAR filler content, cold-pressed.

What is more, since the cured resin has a density higher than wood, the density profiles allow estimations of the depth of wood penetration by resin. As can be seen in Figure 8, the penetration depth has been the highest for REF 0 samples, which can be explained by the lack of filler. As has been confirmed by [29], the rise in filler content significantly reduces the depth of soaking of resin by veneers. In the case of REF 0 panels, the depth of penetration was about 0.7 mm. With raising filler content, the penetration depth decreases to about 0.3 mm. In the case of the bonding mixture with SAR filler and hot-pressed panels, the depth of penetration of veneers by resin decreased from 0.5 mm for SAR 5 (lowest filler content) to about 0.3 mm (similar to industrial filler) for SAR 20 (highest filler content). The specific relation between the bonding mixture and veneer has been found for the bonding mixture with SAR filler and cold-pressed panels. The potential depth of penetration should be investigated along with the analysis of the cross-cut of samples. Due to the relatively high viscosity of the bonding mixtures with fillers, especially those of bark origin, the viscosity of the bonding mixture is high and rises over time [24]. The hot pressing, due to rising temperature, helps to decrease the viscosity of that mixture [29]. Thanks to this, the bonding mixture of lower viscosity can penetrate deeper into the wood structure. That was the case with hot-pressed samples. This way, better adhesion can be achieved, where the adsorption phenomenon is strongly supported by mechanical anchorage [57]. When cold-pressing is conducted, there is no factor leading to lower viscosity. This is why the penetration of the bonding mixture of cold-pressed panels was extremely low. This can be visible in Figure 11, where the comparison of the bonding lines is shown. It is hard to estimate the thickness of the bonding line for the REF 10 sample when the bonding line thickness for SAR 10 sample (hot-pressed) was almost 0, but the resin penetration is well visible and is about 0.4 mm, which is in line with the findings that came out from the Figure 9 analysis. In the case of the SAR 10 C sample presented in Figure 11, there is no penetrated (dark) zone in the wood structure. In that case, the thickness of the bonding line is about 0.1–0.15 mm. According to [29], the increasing content of the filler in the plywood bonding mixture leads to an increase in bonding line thickness. Here, the only adhesion interface is based on the adsorption phenomenon, with weak mechanical anchorage due to low penetration.

Figure 11. The images of the plywood samples bonding line of the same filler content (vertical bonding line in the middle of pictures; depth of binder penetration highlighted by blue frames).

4. Conclusions

In this research, an attempt has been made to upcycle the bark post-treatment residues that came from suberinic acid production to a bi-functional component of bonding mixture in plywood production. The acid-character SAR acts as a filler and hardener of a bonding mixture based on urea-formaldehyde resin. The 5%–20% (5%–30% for curing time) mass content of SAR has been investigated. The activity of SAR filler has been tested in a room at elevated temperatures.

The results show that in the case of hot-curing, the curing time of the bonding mixture can be reduced to about 38% of the initial curing time for the lowest SAR content. The curing time of samples cured at room temperature has been reduced to less than 10% of the initial curing time. In the case of hot curing, the SAR content of about 20% allows one to achieve the curing time of bonding mass with an industrial hardener.

The shear strength of the plywood samples increases with the SAR rise for both cold- and hot-pressed panels. The positive effect of veneer impregnation limiter by resin has been identified for SAR acting as a filler. Additionally, a higher density of SAR-containing bonding lines has been reached for hot-pressed panels. In the case of bending, strength, and modulus of elasticity, the increase in both parameters has been found when increasing the SAR filler content within the above-mentioned range.

The results confirmed the ability to use the SAR as an upcycled component of the bonding mixture for plywood production. Further, the details concerning binder viscosity tuning by either temperature and/or water addition should be investigated to improve the modified binder features.

Author Contributions: Conceptualization, A.J., A.W., L.K., J.R. and G.K.; Data curation, A.W. and G.K.; Formal analysis, L.K., R.R., J.R. and G.K.; Funding acquisition, J.R. and G.K.; Investigation, A.J., A.W., A.D. and G.K.; Methodology, A.J., L.K., J.R. and G.K.; Project administration, A.W. and G.K.; Resources, A.W., J.R. and G.K.; Supervision, G.K.; Validation, L.K. and R.R.; Visualization, A.J. and A.D.; Writing—original draft, A.J., A.W. and G.K.; Writing—review and editing, A.J., A.W. and G.K. All authors have read and agreed to the published version of the manuscript.

Funding: This research has been funded by the National Science Centre, Poland, under the Forest-Value 2021 Programme, reg. no. 2021/03/Y/NZ9/00038, project acronym BarkBuild, as well as being supported by the Slovak Research and Development Agency under contracts APVV-19-0269 and No. SK-CZ-RD-21-0100.

Data Availability Statement: Not applicable.

Acknowledgments: The mentioned research has been completed with the support of the Student Furniture Scientific Group (Koło Naukowe Meblarstwa), Faculty of Wood Technology, Warsaw University of Life Sciences—SGGW.

Conflicts of Interest: The authors declare no conflict of interest.

References

1. Wronka, A.; Kowaluk, G. Upcycling Different Particle Sizes and Contents of Pine Branches into Particleboard. *Polymers* **2022**, *14*, 4559. [CrossRef] [PubMed]
2. Pinchevska, O.; Šmidriaková, M. Wood Particleboard Covered with Slices Made of Pine Tree Branches. *Acta Fac. Xylologiae* **2016**, *58*, 67–74. [CrossRef]
3. Khan, A.U. Home Garden and Women Participation: A Mini Review. *Curr. Res. Agric. Farming* **2021**, *2*, 46–52. [CrossRef]
4. Barbu, M.C.; Lohninger, Y.; Hofmann, S.; Kain, G.; Petutschnigg, A.; Tudor, E.M. Larch Bark as a Formaldehyde Scavenger in Thermal Insulation Panels. *Polymers* **2020**, *12*, 2632. [CrossRef] [PubMed]
5. Tudor, E.M.; Kristak, L.; Barbu, M.C.; Gergel', T.; Němec, M.; Kain, G.; Réh, R. Acoustic Properties of Larch Bark Panels. *Forests* **2021**, *12*, 887. [CrossRef]
6. Medved, S.; Tudor, E.M.; Barbu, M.C.; Jambreković, V.; Španić, N. Effect of Pine (Pinus Sylvestris) Bark Dust on Particleboard Thickness Swelling and Internal Bond. *Drv. Ind.* **2019**, *70*, 141–147. [CrossRef]
7. Jiang, W.; Adamopoulos, S.; Hosseinpourpia, R.; Žigon, J.; Petrič, M.; Šernek, M.; Medved, S. Utilization of Partially Liquefied Bark for Production of Particleboards. *Appl. Sci.* **2020**, *10*, 5253. [CrossRef]
8. Nemli, G.; Gezer, E.D.; Yildiz, S.; Temiz, A.; Aydin, A. Evaluation of the Mechanical, Physical Properties and Decay Resistance of Particleboard Made from Particles Impregnated with Pinus Brutia Bark Extractives. *Bioresour. Technol.* **2006**, *97*, 2059–2064. [CrossRef]
9. Fedorov, V.S.; Ryazanova, T.V. Bark of Siberian Conifers: Composition, Use, and Processing to Extract Tannin. *Forests* **2021**, *12*, 1043. [CrossRef]
10. Navarrete, P.; Mansouri, H.R.; Pizzi, A.; Tapin-Lingua, S.; Benjelloun-Mlayah, B.; Pasch, H.; Rigolet, S. Wood Panel Adhesives from Low Molecular Mass Lignin and Tannin without Synthetic Resins. *J. Adhes. Sci. Technol.* **2010**, *24*, 1597–1610. [CrossRef]
11. Paze, A.; Rizhikovs, J.; Brazdausks, P.; Puke, M.; Grinins, J.; Tupciauskas, R.; Plavniece, A. Processing Possibilities of Birch Outer Bark into Green Bio-Composites. *Vide. Tehnol. Resur.-Environ. Technol. Resour.* **2017**, *3*, 249–253. [CrossRef]
12. Makars, R.; Rizikovs, J.; Godina, D.; Paze, A.; Merijs-Meri, R. Utilization of Suberinic Acids Containing Residue as an Adhesive for Particle Boards. *Polymers* **2022**, *14*, 2304. [CrossRef] [PubMed]
13. Jeżo, A.; Wronka, A. Post-Extraction Birch Bark Residues as a Potential Binder in Particleboards. *Ann. Wars. Univ. Life Sci.-SGGW For. Wood Technol.* **2022**, *118*, 35–47. [CrossRef]
14. Janceva, S.; Andersone, A.; Spulle, U.; Tupciauskas, R.; Papadopoulou, E.; Bikovens, O.; Andzs, M.; Zaharova, N.; Rieksts, G.; Telysheva, G. Eco-Friendly Adhesives Based on the Oligomeric Condensed Tannins-Rich Extract from Alder Bark for Particleboard and Plywood Production. *Materials* **2022**, *15*, 3894. [CrossRef] [PubMed]
15. Rizhikovs, J.; Brazdausks, P.; Paze, A.; Tupciauskas, R.; Grinins, J.; Puke, M.; Plavniece, A.; Andzs, M.; Godina, D.; Makars, R. Characterization of Suberinic Acids from Birch Outer Bark as Bio-Based Adhesive in Wood Composites. *Int. J. Adhes. Adhes.* **2022**, *112*, 102989. [CrossRef]

16. Rižikovs, J.; Zandersons, J.; Paže, A.; Tardenaka, A.; Spince, B. Isolation of Suberinic Acids from Extracted Outer Birch Bark Depending on the Application Purposes. *Balt. For.* **2014**, *20*, 98–105.
17. Kowaluk, G.; Wronka, A. Bonding of Sawmill Birch Wood with Selected Biopolymer-Based Glues. *Ann. Wars. Univ. Life Sci.-SGGW For. Wood Technol.* **2020**, *109*, 32–36. [CrossRef]
18. Travieso-Rodriguez, J.A.; Zandi, M.D.; Jerez-Mesa, R.; Lluma-Fuentes, J. Fatigue Behavior of PLA-Wood Composite Manufactured by Fused Filament Fabrication. *J. Mater. Res. Technol.* **2020**, *9*, 8507–8516. [CrossRef]
19. Gumowska, A.; Kowaluk, G. Bonding of Birch Solid Wood of Sawmill Surface Roughness with Use of Selected Thermoplastic Biopolymers. *Ann. Wars. Univ. Life Sci.-SGGW For. Wood Technol.* **2019**, *106*, 9–15. [CrossRef]
20. Sepahvand, S.; Doosthosseini, K.; Pirayesh, H.; Maryan, B.K. Supplementation of Natural Tannins as an Alternative to Formaldehyde in Urea and Melamine Formaldehyde Resins Used in Mdf Production. *Drv. Ind.* **2018**, *69*, 215–221. [CrossRef]
21. Risholm-Sundman, M.; Vestin, E. Emissions during Combustion of Particleboard and Glued Veneer. *Eur. J. Wood Wood Prod.* **2005**, *63*, 179–185. [CrossRef]
22. Kajaks, J.; Reihmane, S.; Grinbergs, U.; Kalnins, K. Use of Innovative Environmentally Friendly Adhesives for Wood Veneer Bonding. *Proc. Est. Acad. Sci.* **2012**, *61*, 207–211. [CrossRef]
23. Sanghvi, M.R.; Tambare, O.H.; More, A.P. *Performance of Various Fillers in Adhesives Applications: A Review*; Springer: Berlin/Heidelberg, Germany, 2022; Volume 79, ISBN 0028902104022.
24. Kawalerczyk, J.; Dziurka, D.; Mirski, R.; Trociński, A. Flour Fillers with Urea-Formaldehyde Resin in Plywood. *BioResources* **2019**, *14*, 6727–6735. [CrossRef]
25. Antov, P.; Savov, V.; Neykov, N. Reduction of Formaldehyde Emission from Engineered Wood Panels by Formaldehyde Scavengers—A Review. In Proceedings of the 13th International Scientific Conference Wood EMA 2020 and 31st International Scientific Conference ICWST 2020, Sustainability of Forest-Based Industries in the Global Economy, Vinkovci, Croatia, 28–30 September 2020; pp. 289–294.
26. Bekhta, P.; Sedliačik, J.; Noshchenko, G.; Kačík, F.; Bekhta, N. Characteristics of Beech Bark and Its Effect on Properties of UF Adhesive and on Bonding Strength and Formaldehyde Emission of Plywood Panels. *Eur. J. Wood Wood Prod.* **2021**, *79*, 423–433. [CrossRef]
27. Hand, W.G.; Ashurst, W.R.; Via, B.; Banerjee, S. Curing behavior of soy flour with phenol-formaldehyde and isocyanate resins. *Int. J. Adhes. Adhes.* **2018**, *87*, 105–108. [CrossRef]
28. Daniłowska, A.; Kowaluk, G. The Use of Coffee Bean Post-Extraction Residues as a Filler in Plywood Technology. *Ann. Wars. Univ. Life Sci.-SGGW For. Wood Technol.* **2020**, *109*, 24–31. [CrossRef]
29. Ong, H.R.; Khan, M.M.R.; Prasad, D.M.R.; Yousuf, A.; Chowdhury, M.N.K. Palm kernel meal as a melamine urea formaldehyde adhesive filler for plywood applications. *Int. J. Adhes. Adhes.* **2018**, *85*, 8–14. [CrossRef]
30. Mirski, R.; Kawalerczyk, J.; Dziurka, D.; Siuda, J.; Wieruszewski, M. The Application of Oak Bark Powder as a Filler for Melamine-Urea-Formaldehyde Adhesive in Plywood Manufacturing. *Forests* **2020**, *11*, 1249. [CrossRef]
31. Mohamed Abdoul-Latif, F.; El Montassir, Z.; Ainane, A.; Gharby, S.; Sakar, E.H.; Merito, A.; Mohamed, J.; Ainane, T. Use of Thymus Plants as an Ecological Filler in Urea-Formaldehyde Adhesives Intended for Bonding Plywood. *Processes* **2022**, *10*, 2209. [CrossRef]
32. Oh, Y.-S. Evaluation of Chestnut Shell and Coffee Waste with Phenol-Formaldehyde Resin for Plywood Filler X1—Avaliação de Cascas de Castanha e Resíduos de Café Como Material de Enchimento Do Adesivo Fenol-Formaldeído Para a Produção de Compensados. *Ciência Florest.* **2021**, *31*, 1991–2001. [CrossRef]
33. Khanjanzadeh, H.; Pirayesh, H.; Sepahvand, S. Influence of Walnut Shell as Filler on Mechanical and Physical Properties of MDF Improved by Nano-SiO2. *J. Indian Acad. Wood Sci.* **2014**, *11*, 15–20. [CrossRef]
34. Moradpour, P.; Behnia, M.; Pirayesh, H.; Shirmohammadli, Y. The Effect of Resin Type and Strand Thickness on Applied Properties of Poplar Parallel Strand Lumber Made from Underutilized Species. *Eur. J. Wood Wood Prod.* **2019**, *77*, 811–819. [CrossRef]
35. Réh, R.; Igaz, R.; Krišťák, L.; Ružiak, I.; Gajtanska, M.; Božíková, M.; Kučerka, M. Functionality of Beech Bark in Adhesive Mixtures Used in Plywood and Its Effect on the Stability Associated with Material Systems. *Materials* **2019**, *12*, 1298. [CrossRef]
36. Liu, J.; Li, Y.; Mo, H.; Xie, E.; Fang, J.; Gan, W. Current Utilization of Waste Biomass as Filler for Wood Adhesives: A Review. *J. Ind. Eng. Chem.* **2022**, *115*, 48–61. [CrossRef]
37. Walkiewicz, J.; Kawalerczyk, J.; Mirski, R.; Dziurka, D. The Application of Various Bark Species as a Fillers for UF Resin in Plywood Manufacturing. *Materials* **2022**, *15*, 7201. [CrossRef]
38. Xu, J.Y.; Niu, Q.; Liao, R.; Zhong, Z. Study on the Bonding Characteristics and Mechanism of Plant-Based Powder. In Proceedings of the 2012 International Conference on Biobase Material Science and Engineering, Changsha, China, 21–23 October 2012; pp. 28–31. [CrossRef]
39. Yew, M.C.; Ramli Sulong, N.H.; Yew, M.K.; Amalina, M.A.; Johan, M.R. Eggshells: A Novel Bio-Filler for Intumescent Flame-Retardant Coatings. *Prog. Org. Coat.* **2015**, *81*, 116–124. [CrossRef]
40. Benthien, J.T.; Sommerhuber, P.F.; Heldner, S.; Ohlmeyer, M.; Seppke, B.; Krause, A. Influence of Material Origin on the Size Distribution of Wood Particles for Wood-Plastic Composite (WPC) Manufacture. *Eur. J. Wood Wood Prod.* **2017**, *75*, 477–480. [CrossRef]

41. Kusumah, S.S.; Jayadi; Wibowo, D.T.; Pramasari, D.A.; Widyaningrum, B.A.; Darmawan, T.; Ismadi; Dwianto, W.; Umemura, K. Investigation of Eco-Friendly Plywood Bonded with Citric Acid—Starch Based Adhesive. *IOP Conf. Ser. Earth Environ. Sci.* **2020**, *460*, 012009. [CrossRef]
42. Wronka, A.; Rdest, A.; Kowaluk, G. Influence of Starch Content on Selected Properties of Hardboard. *Ann. Wars. Univ. Life Sci.-SGGW For. Wood Technol.* **2020**, *109*, 48–52. [CrossRef]
43. Suhasman, S. Agussalim Resinless Plywood Production by Using Oxidized Acacia Bark Powder as a Binder. *IOP Conf. Ser. Mater. Sci. Eng.* **2019**, *593*, 012010. [CrossRef]
44. Pizzi, A.; Papadopoulos, A.N.; Policardi, F. Wood Composites and Their Polymer Binders. *Polymers* **2020**, *12*, 1115. [CrossRef] [PubMed]
45. *EN 827*; Adhesives—Determination of Conventional Solids Content and Constant Mass Solids Content. European Committee for Standardization: Brussels, Belgium, 2005.
46. *EN 314-1*; Plywood—Bonding Quality—Part 1: Test Methods. European Committee for Standardization: Brussels, Belgium, 2004.
47. *EN 310*; Wood-Based Panels. Determination of Modulus of Elasticity in Bending and of Bending Strength. European Committee for Standardization: Brussels, Belgium, 1993.
48. Fink, J. Gelling Agents. *Pet. Eng. Guid. Oil F. Chem. Fluids* **2021**, 393–417. [CrossRef]
49. Król, P.; Toczyłowska-Mamińska, R.; Mamiński, M. A Critical Role for the Presence of Lignocellulosic Material in the Determination of Wood Buffering Capacity. *J. Wood Chem. Technol.* **2017**, *37*, 478–484. [CrossRef]
50. Bartoszuk, K.; Kowaluk, G. Moisture Influence on Solid Wood Bonded with Modified Starch. *Ann. Wars. Univ. Life Sci.-SGGW For. Wood Technol.* **2022**, *118*, 67–73. [CrossRef]
51. Aydin, I.; Demirkir, C.; Colak, S.; Colakoglu, G. Utilization of Bark Flours as Additive in Plywood Manufacturing. *Eur. J. Wood Wood Prod.* **2017**, *75*, 63–69. [CrossRef]
52. Mirski, R.; Kawalerczyk, J.; Dziurka, D.; Wieruszewski, M.; Trociński, A. Effects of Using Bark Particles with Various Dimensions as a Filler for Urea-Formaldehyde Resin in Plywood. *BioResources* **2020**, *15*, 1692–1701. [CrossRef]
53. Gumowska, A.; Wronka, A.; Borysiuk, P.; Robles, E.; Sala, C.M.; Kowaluk, G. Production of Layered Wood Composites with a Time- Saving Layer-by-Layer Addition. *BioResources* **2018**, *13*, 8089–8099. [CrossRef]
54. Gumowska, A.; Kowaluk, G. The Quality of the Wood Bonding Depending on the Method of Applying the Selected Thermoplastic Biopolymers. *Ann. WULS For. Wood Technol.* **2021**, *116*, 78–85. [CrossRef]
55. Réh, R.; Krišťák, L.; Sedliačik, J.; Bekhta, P.; Božiková, M.; Kunecová, D.; Vozárová, V.; Tudor, E.M.; Antov, P.; Savov, V. Utilization of Birch Bark as an Eco-Friendly Filler in Urea-Formaldehyde Adhesives for Plywood Manufacturing. *Polymers* **2021**, *13*, 511. [CrossRef]
56. Antwi-Boasiako, C.; Appiah, J.K. Glue-Line Durability of Organo-Chemical/Urea Formaldehyde Resin Joints of Ceiba Pentandra (L.) Gaertn. Plywood. *J. Indian Acad. Wood Sci.* **2017**, *14*, 49–59. [CrossRef]
57. Mamiński, M.Ł.; Sedliacik, J. *Kleje i Procesy Klejenia*; Wydawnictwo SGGW: Warsaw, Poland, 2016; ISBN 978-83-7583-702-5.

forests

MDPI

Article

Impact of Sugar Beet Pulp Share on Selected Physical and Mechanical Properties of Particleboards

Radosław Auriga [1,*], Piotr Borysiuk [1], Maciej Latos [1], Alicja Auriga [2], Łukasz Kwaśny [3] and Joanna Walkiewicz [4]

1 Institute of Wood Sciences and Furniture, Warsaw University of Life Sciences—SGGW, ul. Nowoursynowska 159, 02-776 Warsaw, Poland
2 Faculty of Biotechnology and Animal Husbandry, West Pomeranian University in Szczecin, Janickiego 33 St., 71-270 Szczecin, Poland
3 Institute of Forest Sciences, Warsaw University of Life Sciences—SGGW, ul. Nowoursynowska 159, 02-776 Warsaw, Poland
4 Department of Mechanical Wood Technology, Poznań University of Life Sciences, ul. Wojska Polskiego 28, 60-637 Poznań, Poland
* Correspondence: radoslaw_auriga@sggw.edu.pl; Tel.: +48-22-5938-656

Abstract: The aim of this study was to investigate the applicability of sugar beet pulp (SBP) in particleboard production as a substitute for wood material. Two variants of board density, 650 kg/m^3 and 550 kg/m^3, containing 0%, 25% and 50% of SPB, were evaluated. During the study the following features were measured: modulus of rupture (MOR), modulus of elasticity (MOE), thickness swelling (TS) and water absorption (WA). The results showed that boards containing up to 25% of SBP meet standards for boards for general-purpose used in dry conditions. Further increases in SBP content deteriorated some of the investigated properties of the boards.

Keywords: sugar beet pulp; particleboards; mechanical properties; physical properties

1. Introduction

Global production of wood-based materials has been following a constant upward trend for years, and in 2020 it reached 102 million cubic meters [1]. Such high value is determined by the constantly growing demand for this type of board. Their mechanical and physical properties as well as their relatively low price make them widely applied in such industries as furniture, building, construction and packaging [2].

High production binds with high demand for the feedstock, and the main raw materials in this case are wood particles. Due to the finite possibilities of obtaining wood raw material from forests, more and more attention is being devoted to attempts to manufacture innovative particleboards using alternative feedstock. Recent research has focused on boards made with post-waste paper [3], cement-particle boards [4,5], basalt-particle boards [6] and with the use of post-consumer thermoplastic materials [7]. However, it seems that the most promising direction for the production of particleboards may be obtaining raw material from agricultural lingo-cellulosic waste [8–10].

In recent years a large number of scientific works describing use of alternative lignocellulosic sources in the production of particleboards have been released. The studies included among others, such materials as: vine prunings [11,12], eggplant stalks [13], kenaf stalks [14,15], straw [16], tomato stalks [17], sunflower stalks [18], apple prunings [19], corn cobs [20], oil camellia [21,22] and others. One such agricultural waste is sugar beet pulp (SBP), which is a by-product of sugar production. It is mainly composed of arabinians, hemicelluloses, pectins, and cellulose microfibrils [23–25]. SBP is commonly used as animal feed. The low pectin content, high drying costs and large availability of this material led to attempts to use it as an alternative feedstock for different purposes. Studies conducted so far have been focused on the use of SBP as, e.g., a source of cellulosic microfibrils [26], a source of polyols for the production of urethanes and polyurethanes [27], a source of fiber in biodegradable composites [28–30] or in paper manufacture [31], and in the production of bioethanol [32].

SBP was also tested as a raw material in the process of particleboard production. Borysiuk et al. [33] demonstrated the possibility of producing particleboards with up to 30% SBP in the core layer without significant deterioration of their mechanical properties. However, the authors did not discuss the potential impact of panel density within the research. Present studies concentrate on the prospect of producing low-density particleboards [34,35]; however, the density is an important factor determining the properties of wood-based panels [2]. The aim of the present study was to determine the influence of the proportion of SBP and the density of particleboards on selected mechanical and physical properties.

2. Materials and Methods

2.1. Material

SBP used in the experiment was provided by a Polish sugar producer (Krajowa Grupa Spożywcza S.A. w Toruniu, Oddział Cukrownia Krasnystaw, Poland). The sugar beet pulp consisted of flakes with dimensions of approximately 5–7 mm in diameter and 1–2 mm in thickness. The SBP used in the particleboard manufacturing process was characterized by a moisture content of 4%.

The industrial wood particles were applied in the face and core layers. The pine wood particles were delivered by a Polish factory producing particleboards. Particles for the face layers were characterized by a moisture content of 7%, and for the core layer of 4%.

The adhesive used was based on urea formaldehyde resin (UF) (Silekol 120, Silekol Sp, z o.o., Kędzierzyn-Koźle, Poland) with a dry matter content of 67%, a relative density of 1.30 g/cm^3 and a dynamic viscosity of about 500 mPas. The hardener was a 10% aqueous solution of ammonium sulfate. The unit composition of the adhesive was 50:15.5:1.5 UF resin, water, and hardener, respectively.

2.2. Particleboard Manufacturing

Three-layer particleboards with a thickness of 16 mm were produced in two variants of density: 550 kg/m^3 and 650 kg/m^3. The face layers constituted 35% of the panel. The content of the binder in these layers was 12% and in the core layer 10%. The core layer of board was varied in the beet pulp content as follows: 0% (control variant), 25%, 50%, and the face layers were made with 100% wood particles. Table 1 shows variants of particleboards used in the experiment.

Table 1. Variants of manufactured boards.

Variant	Core Layer SBP Share (%)	Density (kg/m^3)
A	0	650
B	0	550
C	25	650
D	25	550
E	50	650
F	50	550

The boards were produced using a ZUP-NYSA PH-1LP25 hydraulic press at a unit pressing pressure of 2.5 MPa, a temperature of 180 °C and pressing time of 288 s (pressing factor 18 s/mm). Four boards with dimensions of 320 × 320 mm were produced for each variant. The manufactured particleboards were conditioned under normal conditions (20 ± 2 °C, 65 ± 5% air humidity) for at least seven days.

2.3. Experimental

The modulus of rupture (MOR) and modulus of elasticity (MOE) were determined according to EN 310 [36]. Internal bonding was determined through the tensile test perpendicular to the surface of the board, according to EN 319 [37]. All mechanical tests were conducted via a laboratory-testing machine custom-made by the Research & Development Center for Wood-Based Panels Sp. z.o.o. in Czarna Woda, Poland. At least 10 samples were used to determine each property.

The density of particleboards was measured according to EN 323 [38]. Thickness swelling (TS) after 2 h and 24 h of soaking in water were measured according to EN 317 [39]. Briefly, samples were completely immersed in water at room temperature for 2 h and 24 h in such a way that their possible dimensional changes were not limited. Water absorption (*WA*) was measured during TS and calculated as follows:

$$WA = \frac{m_2 - m_1}{m_1} \cdot 100, \quad (1)$$

where *WA* is the water absorption (%), m_1 is the sample weight before soaking (g) and m_2 is the sample weight after soaking (g).

2.4. Statistical Analysis

Statistical analysis of the results was carried out in Statistica (data analysis software system), Version 13 (TIBCO Software Inc., Palo Alto, CA, USA). Analysis of variance (MANOVA) were used to test (significant level $\alpha = 0.05$) for significant differences between factors. A comparison of the means was performed by Tukey test, with a 0.05 significance level.

3. Results and Discussion

The particleboards produced in the study had an average density in the range of 644–663 kg/m^3 for variants A, C, E and 553–562 kg/m^3 for variants B, C, F (Table 2). It should be noted that the difference in the particleboards' density in relation to their assumed value did not exceed 3%. In addition, statistical analysis did not reveal statistically significant differences in density values between board variants with the same assumed density.

Table 2. Average density of manufactured particleboards.

Variant	Particleboards Density (kg/m^3)	
	Average	Std. Dev.
A	663 ^{a}	32
B	553 ^{b}	27
C	648 ^{a}	34
D	566 ^{b}	22
E	644 ^{a}	34
F	562 ^{b}	31

^{a,b}—homogenous group by Tukey test ($\alpha = 0.05$).

Particleboards with 25% of SBP content and the density of 650 kg/m^3 revealed a significant decline in modulus of rupture (about 40% lower) compared to the boards with no SBP (Figure 1). The same parameter was somewhat lower in the case of boards with the density of 550 kg/m^3 and 25% SPB compared to the boards of the same density and no SBP additive.

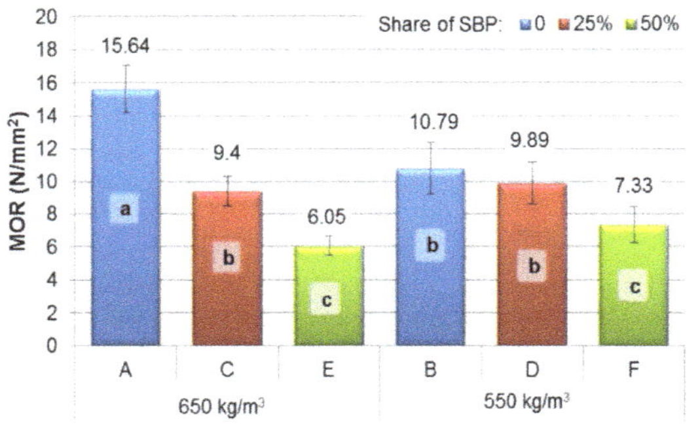

Figure 1. Modulus of rupture of particleboards. a,b,c—homogenous group by Tukey test (α = 0.05).

The technical requirements of particleboards, depending on their intended use, are compiled and described in the EN 312 standard [40]. The minimum value of modulus of rupture for particleboards intended for general purpose and use in dry conditions (P2 type boards) is 10 N/mm^2. When analyzing the results presented in Figure 1, it can be concluded that none of the particleboards made with SBP met these requirements. However, it should be noted that the variants with a 25% SBP share were characterized by a strength of at least 9.4 N/mm^2 regardless of the density. Similar dependencies were noted by Borysiuk et al. [33].

An analysis of the MOE results demonstrates that the manufactured boards with 25% SBP share met the minimum requirements of the EN 312 standard (1600 N/mm^2) [40]. A significant decrease in the MOE value for particleboards with 25% SBP share compared to boards without SBP was observed only for boards with a density of 650 kg/m^3 (Figure 2). Interestingly, the decrease in the parameter occurs much more smoothly along with the increase in the SBP share in the case of boards with the lower density. This may be related to pectins and sugars in the beet pulp, which may behave as additional adhesives due to pressing. In addition, the proportion of SBP may influence how the board is compacted during pressing. The analysis of variance showed a significant influence of the board density and SBP share on the properties of the manufactured particleboards (Table 3). The percentage impact coefficient of SBP share was 67.68% for MOR and 74.13% for MOE, while the density values were 2.51% for MOR and 2.15% for MOE. This proves that it is mainly the SBP share that determines the MOR and MOE of the manufactured particleboards, thus evidencing that the SBP share can be expectedly to affect all the other panel properties. If the SBP share predominantly determines the properties of the MOR and MOE, its properties will largely affect the properties of the manufactured particleboards. When comparing the results of the research on the use of agriculture residues, a general trend of a decrease in the mechanical properties of particleboards with an increase in the share of agriculture residues can be observed [19,33]. The observed relationships are consistent with the literature data.

Table 3. MANOVA results of mechanical properties of particleboards.

	MOR		MOE		IB	
	p	X	*p*	X	*p*	X
Share	0.000	67.68	0.000	74.13	0.000	38.98
Density	0.001	2.51	0.001	2.15	0.212	1.48
Share × Density	0.000	17.66	0.000	13.83	0.009	9.46
Error		12.15		9.89		50.07

p—significant with α = 0.05; X—percentage of contribution.

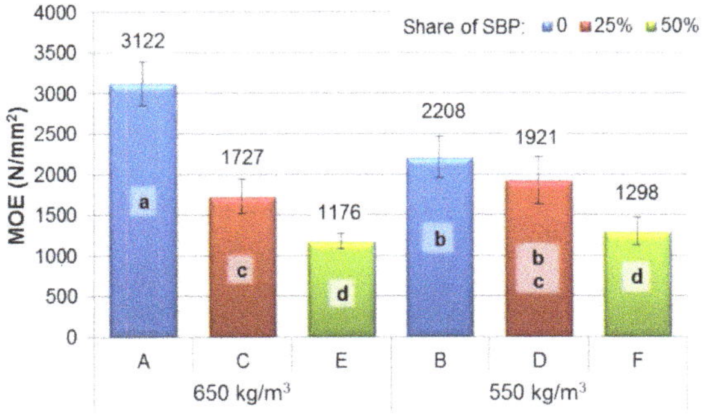

Figure 2. Modulus of elasticity of particleboards. a,b,c,d—homogenous group by Tukey test ($\alpha = 0.05$).

A significant increase in the IB values was observed in the case of boards with 25% SBP compared to reference boards regardless of their density (Figure 3). It should be noted that all the particleboards produced met at least the minimum requirements of the EN 312 standard (0.24 N/mm^2) for P2 type boards. The analysis of variance showed a significant influence of the SBP share (X = 38.98%). However, the error value (X = 50.07%) was higher than the X values for SBP content and the density; this indicates that untested factors had a greater influence on IB.

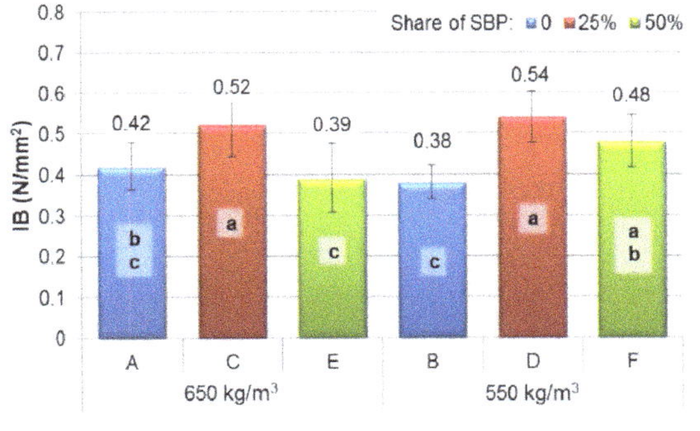

Figure 3. Internal bound of particleboards. a,b,c—homogenous group by Tukey test ($\alpha = 0.05$).

Figure 4 illustrates the results of the physical properties of the manufactured particleboards following soaking. When analyzing the results of thickness swelling and water absorption, it should be noted that no hydrophobic agents were used in the process of manufacturing the particleboards. SBP particleboards generally showed a higher TS value than reference boards. Analyzing Figure 4, it can be seen that the increase in the value of the TS after 24 h of immersion in water is almost proportional to the share of SBP in the particleboard. The analysis of variance showed that both the density of the manufactured particleboards and the share of SBP are statistically significant (Table 4). However, the analysis of the percentage of factors' impact showed that the determining factor in the case of TS was the SBP share. On the other hand, the density played a less significant role in this case, and its value of X was lower than that of the errors.

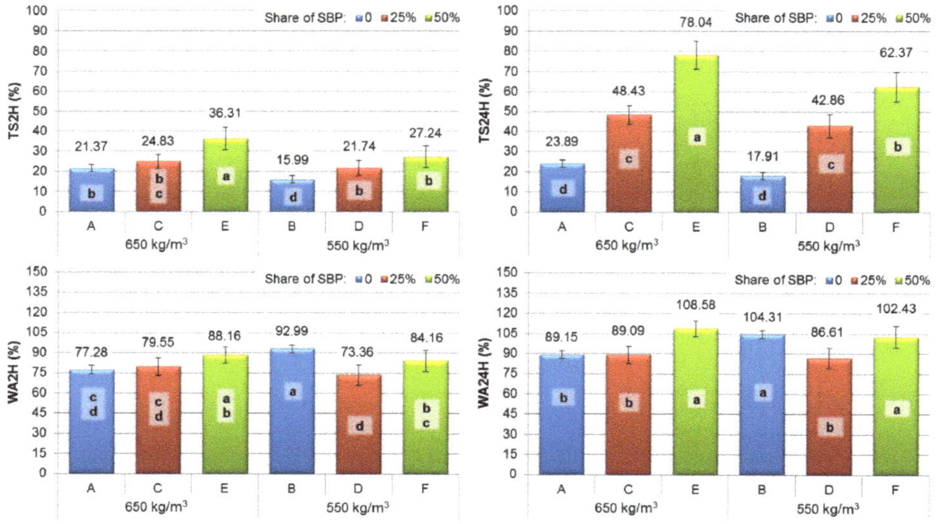

Figure 4. Thickness swelling and water absorption after 2 and 24 h of soaking in water. a,b,c,d—homogenous group by Tukey test (α = 0.05).

Table 4. MANOVA results of physical properties of particleboards.

	TS2H		TS24H		WA2H		WA24H	
	p	X	*p*	X	*p*	X	*p*	X
Share	0.000	55.31	0.000	88.97	0.000	24.90	0.000	44.88
Density	0.000	16.07	0.000	4.52	0.237	1.11	0.216	1.02
Share × Density	0.058	2.86	0.004	1.20	0.000	31.87	0.000	18.71
Error		25.75		5.31		42.11		35.38

p—significant with α = 0.05; X—percentage of contribution.

In the case of water absorption, only the SBP share was statistically significant. However, the X value for the SBP share after 2 h of soaking in water was lower than that recorded for the error, which may indicate that in the first 2 h of immersion in water, the absorption is mainly affected by other factors than the tested ones.

Then again, analysis of variance demonstrated that water absorption after 24 h of soaking in water was significantly influenced by the SBP share (X = 44.88%) and other factors untested in the experiment (error X = 35.38%).

4. Conclusions

Based on the conducted research, it can be concluded that the SBP share significantly influenced the mechanical and physical properties of manufactured boards. MOR and MOE properties significantly decreased as the content of SBP increased. However, addition of SBP improved IB. The tests revealed that the SBP share, when not exceeding 25%, enables the production of boards with strength properties which are close to meeting the requirements of the EN 312 standard for boards for general-purpose use in dry conditions, that may find applicability in the construction industry as non-structural filling elements. SBP addition is one of the main factors determining the thickness swelling and water absorption. At the same time, the density in this study did not significantly affect the water absorption. In order to reduce the swelling and water absorption values, the use of hydrophobic additives during the production of boards should be considered.

Author Contributions: Conceptualization, R.A., P.B. and A.A.; methodology, J.W. and R.A.; validation, R.A., P.B.; formal analysis, R.A. and Ł.K.; investigation, R.A., P.B., M.L.; data curation, R.A.; writing—original draft preparation, R.A., P.B., A.A., J.W., Ł.K. and M.L.; writing—review and editing, A.A.; visualization, R.A. and A.A.; supervision, R.A.; project administration, R.A. All authors have read and agreed to the published version of the manuscript.

Funding: This research received no external funding.

Data Availability Statement: Not applicable.

Conflicts of Interest: The authors declare no conflict of interest.

References

1. FAOSTAT. Available online: https://www.fao.org/faostat/en/#home (accessed on 10 October 2022).
2. Thoemen, H.; Irle, M.; Sernek, M. *Wood-Based Panels An Introduction for Specialists*; Brunel University Press: London, UK, 2010; ISBN 9781902316826.
3. Lertsutthiwong, P.; Khunthon, S.; Siralertmukul, K.; Noomun, K.; Chandrkrachang, S. New Insulating Particleboards Prepared from Mixture of Solid Wastes from Tissue Paper Manufacturing and Corn Peel. *Bioresour. Technol.* **2008**, *99*, 4841–4845. [CrossRef] [PubMed]
4. Wang, L.; Chen, S.S.; Tsang, D.C.W.; Poon, C.-S.; Shih, K. Recycling Contaminated Wood into Eco-Friendly Particleboard Using Green Cement and Carbon Dioxide Curing. *J. Clean. Prod.* **2016**, *137*, 861–870. [CrossRef]
5. Wang, L.; Yu, I.K.M.; Tsang, D.C.W.; Yu, K.; Li, S.; Poon, C.S.; Dai, J.G. Upcycling Wood Waste into Fibre-Reinforced Magnesium Phosphate Cement Particleboards. *Constr. Build. Mater.* **2018**, *159*, 54–63. [CrossRef]
6. Kramár, S.; Mayer, A.K.; Schöpper, C.; Mai, C. Use of Basalt Scrim to Enhance Mechanical Properties of Particleboards. *Constr. Build. Mater.* **2020**, *238*, 117769. [CrossRef]
7. Borysiuk, P.; Wilkowski, J.; Krajewski, K.; Auriga, R.; Skomorucha, A.; Auriga, A. Selected Properties of Flat-Pressed Wood-Polymer Composites for High Humidity Conditions. *BioResources* **2020**, *15*, 5141–5155. [CrossRef]
8. Müller, C.; Schwarz, U.; Thole, V. Zur Nutzung von Agrar-Reststoffen in Der HolzwerkstoffindustrieOn the Utilization of Agricultural Residues in the Wood-Based Panel Industry. *Eur. J. Wood Wood Prod.* **2011**, *70*, 587–594. [CrossRef]
9. Pędzik, M.; Janiszewska, D.; Rogoziński, T. Alternative Lignocellulosic Raw Materials in Particleboard Production: A Review. *Ind. Crops Prod.* **2021**, *174*, 114162. [CrossRef]
10. Lee, S.H.; Lum, W.C.; Boon, J.G.; Kristak, L.; Antov, P.; Pędzik, M.; Rogoziński, T.; Taghiyari, H.R.; Lubis, M.A.R.; Fatriasari, W.; et al. Particleboard from Agricultural Biomass and Recycled Wood Waste: A Review. *J. Mater. Res. Technol.* **2022**, *20*, 4630–4658. [CrossRef]
11. Ntalos, G.A.; Grigoriou, A.H. Characterization and Utilisation of Vine Prunings as a Wood Substitute for Particleboard Production. *Ind. Crops Prod.* **2002**, *16*, 59–68. [CrossRef]
12. Auriga, R.; Auriga, A.; Borysiuk, P.; Wilkowski, J.; Fornalczyk, O.; Ochmian, I. Lignocellulosic Biomass from Grapevines as Raw Material for Particleboard Production. *Polymers* **2022**, *14*, 2483. [CrossRef]
13. Guntekin, E.; Karakus, B. Feasibility of Using Eggplant (Solanum Melongena) Stalks in the Production of Experimental Particleboard. *Ind. Crops Prod.* **2008**, *27*, 354–358. [CrossRef]
14. Kalaycioglu, H.; Nemli, G. Producing Composite Particleboard from Kenaf (*Hibiscus cannabinus* L.) Stalks. *Ind. Crops Prod.* **2006**, *24*, 177–180. [CrossRef]
15. Paridah, M.T.; Juliana, A.H.; El-Shekeil, Y.A.; Jawaid, M.; Alothman, O.Y. Measurement of Mechanical and Physical Properties of Particleboard by Hybridization of Kenaf with Rubberwood Particles. *Meas. J. Int. Meas. Confed.* **2014**, *56*, 70–80. [CrossRef]
16. Tabarsa, T.; Jahanshahi, S.; Ashori, A. Mechanical and Physical Properties of Wheat Straw Boards Bonded with a Tannin Modified Phenol-Formaldehyde Adhesive. *Compos. Part B Eng.* **2011**, *42*, 176–180. [CrossRef]
17. Taha, I.; Elkafafy, M.S.; El Mously, H. Potential of Utilizing Tomato Stalk as Raw Material for Particleboards. *Ain Shams Eng. J.* **2018**, *9*, 1457–1464. [CrossRef]
18. Zeleniuc, O.; Brenci, L.M.; Cosereanu, C.; Fotin, A. Influence of Adhesive Type and Content on the Properties of Particleboard Made from Sunflower Husks. *BioResources* **2019**, *14*, 7316–7331. [CrossRef]
19. Auriga, R.; Borysiuk, P.; Gumowska, A.; Smulski, P. Influence of Apple Wood Waste from the Annual Care Cut on the Mechanical Properties of Particleboards. *Ann. WULS For. Wood Technol.* **2019**, *105*, 47–53. [CrossRef]
20. Akinyemi, A.B.; Afolayan, J.O.; Ogunji Oluwatobi, E. Some Properties of Composite Corn Cob and Sawdust Particle Boards. *Constr. Build. Mater.* **2016**, *127*, 436–441. [CrossRef]
21. Chaydarreh, K.C.; Lin, X.; Dandan, L.; Zhang, W.; Guan, L.; Hu, C. Developing 3-Layer Tea Oil Camellia (*Camellia oleifera* Abel.) Shells-Based Particleboard with Systematic Study on Particle Geometry and Distribution. *Ind. Crops Prod.* **2022**, *179*, 114682. [CrossRef]
22. Choupani Chaydarreh, K.; Lin, X.; Guan, L.; Hu, C. Interaction between Particle Size and Mixing Ratio on Porosity and Properties of Tea Oil Camellia (*Camellia oleifera* Abel.) Shells-Based Particleboard. *J. Wood Sci.* **2022**, *68*, 43. [CrossRef]

23. Dinand, E.; Chanzy, H.; Vignon, M.R. Parenchymal Cell Cellulose from Sugar Beet Pulp: Preparation and Properties. *Cellulose* **1996**, *3*, 183–188. [CrossRef]
24. Sun, R.; Hughes, S. Fractional Extraction and Physico-Chemical Characterization of Hemicelluloses and Cellulose from Sugar Beet Pulp. *Carbohydr. Polym.* **1998**, *36*, 293–299. [CrossRef]
25. Oosterveld, A.; Beldman, G.; Schols, H.A.; Voragen, A.G.J. Characterization of Arabinose and Ferulic Acid Rich Pectic Polysaccharides and Hemicelluloses from Sugar Beet Pulp. *Carbohydr. Res.* **2000**, *328*, 185–197. [CrossRef] [PubMed]
26. Dinand, E.; Chanzy, H.; Vignon, R.M. Suspensions of Cellulose Microfibrils from Sugar Beet Pulp. *Food Hydrocoll.* **1999**, *13*, 275–283. [CrossRef]
27. Pavier, C.; Gandini, A. Oxypropylation of Sugar Beet Pulp. 1. Optimisation of the Reaction. *Ind. Crops Prod.* **2000**, *12*, 1–8. [CrossRef]
28. Leitner, J.; Hinterstoisser, B.; Wastyn, M.; Keckes, J.; Gindl, W. Sugar Beet Cellulose Nanofibril-Reinforced Composites. *Cellulose* **2007**, *14*, 419–425. [CrossRef]
29. Liu, L.S.; Finkenstadt, V.L.; Liu, C.-K.; Coffin, D.R.; Willett, J.L.; Fishman, M.L.; Hicks, K.B. Green Composites from Sugar Beet Pulp and Poly(Lactic Acid): Structural and Mechanical Characterization. *J. Biobased Mater. Bioenergy* **2008**, *1*, 323–330. [CrossRef]
30. Mohamed, A.A.; Finkenstadt, V.L.; Palmquist, D.E. Thermal Properties of Extruded/Injection-Molded Poly(Lactic Acid) and Biobased Composites. *J. Appl. Polym. Sci.* **2008**, *107*, 898–908. [CrossRef]
31. Fišerová, M.; Gigac, J.; Butaš, R. Influence of Sugar Beet Pulp on Bond Strength and Structure of Paper. *Wood Res.* **2007**, *52*, 59–74.
32. Vučurović, V.M.; Razmovski, R.N. Sugar Beet Pulp as Support for Saccharomyces Cerivisiae Immobilization in Bioethanol Production. *Ind. Crops Prod.* **2012**, *39*, 128–134. [CrossRef]
33. Borysiuk, P.; Jenczyk-Tolloczko, I.; Auriga, R.; Kordzikowski, M. Sugar Beet Pulp as Raw Material for Particleboard Production. *Ind. Crops Prod.* **2019**, *141*, 111829. [CrossRef]
34. Boruszewski, P.; Borysiuk, P.; Jankowska, A.; Pazik, J. Low-Density Particleboards Modified with Blowing Agents—Characteristic and Properties. *Materials* **2022**, *15*, 4528. [CrossRef] [PubMed]
35. Boruszewski, P.; Borysiuk, P.; Jankowska, A.; Pazik, J. Low-Density Particleboards Modified with Expanded and Unexpanded Fillers—Characteristics and Properties. *Materials* **2022**, *15*, 4430. [CrossRef] [PubMed]
36. *EN 310*; Wood-Based Panels-Determination of Modulus of Elasticity in Bending and of Bending Strength. European Committee for Standardisation: Brussels, Belgium, 1994.
37. *EN 319*; Particleboards and Fibreboards-Determination of Tensile Strength Perpendicular to the Plane of the Board. European Committee for Standardisation: Brussels, Belgium, 1999.
38. *EN 323*; Wood-Based Panels-Determination of Density. European Committee for Standardisation: Brussels, Belgium, 1999.
39. *EN 317*; Particleboards and Fibreboards-Determination of Swelling in Thickness after Immersion in Water. European Committee for Standardisation: Brussels, Belgium, 1999.
40. *EN 312*; Particleboards-Specyfications. European Committee for Standardisation: Brussels, Belgium, 2011.

forests MDPI

Article

The Effect of Ethanol Extracts and Essential Oils Obtained from Different Varieties of Mint on Wood Molding

Izabela Betlej [1,*], Bogusław Andres [1], Krzysztof Krajewski [1], Anna Kiełtyka-Dadasiewicz [2,3], Piotr Boruszewski [1], Dominika Szadkowska [1], Janusz Zawadzki [1], Andrzej Radomski [1] and Piotr Borysiuk [1,*]

1 Institute of Wood Sciences and Furniture, Warsaw University of Life Sciences—SGGW, 159 Nowoursynowska St., 02-776 Warsaw, Poland; boguslaw_andres@sggw.edu.pl (B.A.); krzysztof_krajewski@sggw.edu.pl (K.K.); piotr_boruszewski@sggw.edu.pl (P.B.); dominika_szadkowska@sggw.edu.pl (D.S.); janusz_zawadzki@sggw.edu.pl (J.Z.); andrzej_radomski@sggw.edu.pl (A.R.)

2 Department of Plant Production Technology and Commodity Science, University of Life Sciences in Lublin, Akademicka 15 St., 20-950 Lublin, Poland; anna.kieltyka-dadasiewicz@up.lublin.pl

3 Garden of Cosmetic Plants and Raw Materials, Research and Science Innovation Center, 20-819 Lublin, Poland

* Correspondence: izabela_betlej@sggw.edu.pl (I.B.); piotr_borysiuk@sggw.edu.pl (P.B.)

Abstract: This paper presents the results of research on the effect of essential oils and ethanol extracts on the growth of mold fungi on Scots pine (*Pinus sylvestris* L.) wood. The analysis of fungal growth on the microbial medium showed that the degree of inhibition of microorganism growth depends on the amount of the extract added to the medium. At the same time, it was found that the highest dose of the extract, amounting to 5.0 cm^3, almost completely inhibited the growth of fungi. In addition, it was found that mint ethanol extracts, the application of which in the wood is at least 40 g/m^2, have a fungistatic effect at the initial stage of fungal development. Solutions of essential oils turned out to be more active against fungi, although also in this case the desired biocidal effect was not achieved. Essential oils significantly slowed down the growth of the fungus *Ch. globosum*, with the strongest fungistatic effect found for 'Morocco' spearmint oil (*Mentha spicata* L.). Despite the fact that in tests on agar-maltose medium, the strongest biocidal activity against *Ch. globosum* was found for spearmint 'Crispa' (*Mentha spicata* L.) oil, the effect of growth inhibition was not so clearly visible in studies on wood. Essential oils applied to the surface of the wood slowed down the growth of *T. viride* fungus, but not to the extent that it was found in the case of *Ch. globosum*. The qualitative and quantitative composition of substances belonging to the group of terpenes and their derivatives was characterized using the GCMS technique. It was shown that the ethanol extracts of mints were dominated by substances belonging to the oxygen-containing monoterpenoid and monoterpene groups. In terms of quality, the composition of essential oils turned out to be richer.

Keywords: *Mentha* sp.; essential oil; ethanol extracts; wood; mold growth

1. Introduction

Treating wood with preservatives is intended to increase the durability of the material. Extending the durability of wood in various working environments is possible only as a result of treating it with biocides containing active substances, permitted for use in European Union countries, in accordance with the guidelines of the Regulation of the European Parliament and of the Council No. 528/2012 on the provision and use of products biocides [1]. The regulation in question greatly limited the market of active substances permitted for use in biocidal products, which was related to the imposition of restrictions on EU Member States to ensure the maximum safety of biocides. Among the biocidal substances used to protect wood against biodegradation, arsenic and chromium compounds, fluorine compounds, phenols, naphthalenes, and numerous petroleum distillation products have been withdrawn from the market. There are only those substances for which it has been shown in appropriate toxicological and ecotoxicological studies, as well as risk and exposure assessment, that they are safe for humans, animals, and the environment while treating or using treated wood [2]. Other active substances to be used in wood preservation are subject to a continuous process of review and testing, and when new information on possible harmfulness is obtained, confirmed by the latest state of knowledge, these substances are withdrawn from wood preservatives. Taking into account the safety of users of biocides and environmental protection considerations, more and more research is undertaken to assess the biocidal effectiveness of substances or chemical compounds of natural origin [3–5]. Compounds such as essential oils [6–9], alkaloids [10], and natural resins [11], as well as substances of animal origin or from microbial cultures [12,13], and even the microorganisms themselves [14] are evaluated. It should also be mentioned that Regulation No. 528/2012 does not limit the possibility of extending the list of active substances approved for use. Article 56 of this regulation guarantees the possibility of development and research into new substances and biocidal products [1]. Taking into account the legal provisions of Regulation No. 528/2012, it is worth looking for new, safe, and effective biocides naturally occurring in nature, which can be an alternative to a number of synthetic substances currently used in wood protection. Due to the increasing burden on the environment, decisive steps should be taken towards ecological wood treatment.

Plants are the source of many substances with fungicide activity; thus, they are of growing interest in the field of wood protection and preservation. The literature data indicate that various plant components have been tested in wood preservation against the harmful effects of fungi [15,16]. Essential oils in particular can constitute an important group of diverse substances that could be included in the formulation of traditional biocides, reducing their harmfulness to a minimum level. In the literature on the subject, the most frequently evaluated oils of plant origin in protecting wood against brown decay and mold were thyme [17,18], oregano [19–21], lavender [22], and clove [23–25]. The high fungicide potential of Lavandula angustifolia Mill oils and Lavandula latifolia Medik, as well as good wood preservation against termites, was shown by Šimůnková et al. [26]. The authors of the study showed the biocidal activity of lavender oils similar to the level of effectiveness of commercial biocides based on trivalent boron and quaternary ammonium salt. Xie et al. [27] found that Origanum vulgare L. oil is highly toxic to white and brown rot fungi. An interesting study of the literature on the effectiveness of essential oils as active ingredients in wood preservation was presented by Woźniak [28]. The author of the publication has compiled a set of results of the evaluation of the fungicide properties of various plant oils (lemongrass, neem, peppermint, rosemary, thyme), indicating very good results in protecting wood against fungi.

The biocidal potential of plant oils largely depends on the method and place of their extraction. Ludwiczuk et al. [29] report that the availability and quality of essential oils depend on the quality of the soil, the climate in which the plant grew, the amount of annual atmospheric precipitation, the time of harvesting, the method of drying, and the morphological part. The same authors indicate that the exact identification of the plant is of key importance in its further use. Studies using gas chromatography have shown that mint leaves may contain over 70 active substances belonging to the group of terpenes [30], while oregano leaves may contain up to 40 potential active substances [31]. The composition, and thus the biocidal potential of essential oils, is also determined by the genetic factors of plants, which is important in the case of successively emerging new cultivars, created by man in order to obtain the appropriate plant characteristics. An example of such diversity is the separation of different mint chemotypes even within the species [32].

The development of innovative, natural biocides as an alternative to synthetic fungicides should be one of the methods of modern development of wood preservatives. The fungicide potential of plant components could be used for research activities in the development of new, pro-ecological biocides. The interaction between natural ingredients and synthetic ingredients of wood preservative preparations is little studied. Obtaining the effect of enhancing the biocidal effect by introducing substances of natural origin into the formulation of wood protection products, while reducing the content of active chemical substances, often harmful to the environment, would be a great prospect for the development of a new group of impregnations, safer and more environmentally friendly. Wood preservatives containing biocides of natural origin are undoubtedly important for the development of innovation and a knowledge-based economy. In addition, they clearly fit into the concepts of sustainable development, based on technological progress, but maintaining natural balance and the durability of basic natural processes. By creating new environmentally friendly biocides, we minimize problems related to the biodegradation of preparations and the residue of harmful waste in the environment.

While in the literature we can find a large group of studies describing the antifungal properties of various mint extracts, these studies mainly concern experiments conducted on agar media. The assessment of fungal growth on a porous material such as wood has not been the subject of deeper analyses so far. In addition, the possibility of evaluating the antifungal properties of different varieties of mint and identifying a variety with high fungicidal potential allows for a better characterization of the fungicide variability of different varieties of this plant and the indication of a variety that has significant indications for use in the production of new environmentally friendly biocides.

In this study, an attempt was made to determine the effect of ethanol extracts and essential oils obtained from various varieties of mint on the growth of mold fungi. An analysis of the content of ingredients that may have a biocidal effect was carried out. The intention of the authors was to determine the potential possibilities of using extracts from plant raw materials as ingredients in the development of environmentally friendly wood impregnation agents.

2. Materials and Methods

2.1. The Plant Material

Four varieties of mint were used in the research: peppermint 'Almira' (*Mentha piperita* L.) (No. 6), spearmint 'Crispa' (*Mentha spicata* L.) (No. 3) spearmint 'Morocco' (*Mentha spicata* L.) (No. 5), pineapple mint 'Variegata' (*Mentha suaveolens* Ehrh.) (No. 8) (Figure 1).

Figure 1. *Mentha* sp. used in the experiment: No. 3. *Mentha spicata* 'Morocco', No. 5. *M. spicata* 'Crispa', No. 6. *M. piperita* 'Almira', No. 8. *M. suaveolens* 'Variegata' (photographed by Anna and Marcin Dadasiewicz).

All mint varieties were cultivated on the experimental fields of the Research and Science Innovation Center in Wola Zadybska near Lublin (Poland) (51°44′49″ N 21°50′38″ E). The plants were cultivated on lessive soil, of light mineral agronomic category, with the granulometric composition of loamy sand. The soil was slightly acidic (pHKCl 6.1–6.2), with medium humus content, very high content of assimilable forms of phosphorus and potassium, high content of magnesium, average content of manganese, zinc and iron, and low content of copper and boron. The plantations were located in the temperate climate zone, where the average daily air temperature in the growing season (May-September) was, respectively: 14.2; 17.2; 19.2; 18.1 and 12.9 °C, and the average monthly amount of precipitation ranged from 62.4 to 91.8 mm (determined on the basis of data from the Institute of Meteorology and Water Management—National Research Institute (IMiGW-PIB) from the measuring station located in Jarczew (51°48′52″ N 21°58′21″ E). Plants were harvested for testing just prior to flowering at BBCH 29 (Biologische Bundesanstalt, Bundessortenamt und Chemische Industrie) growth stage. Immediately after cutting, the mint herb was dried in a laboratory drier with forced air circulation at a temperature of 32 °C. After drying the raw material, the leaves were separated from the stems. Only leaves were used for further analysis.

2.1.1. Preparation of Ethanol Extracts

A total of 10 g of mint leaves were flooded with 200 cm^3 of 60% ethanol (Polmos, Poland) and shaken for 72 h on a laboratory shaker (IKA KS 3000 icontrol, IKA-Werke GmbH & Co.KG, Staufen, Germany). The extracts were cleaned of leaves and then sterilized. Syringe filters were used for sterilization. The pore diameter of the filters was 0.22 μm.

2.1.2. The Method of Obtaining Essential Oils

The oils used in this study were obtained by distilling the dried leaves in a Deryng apparatus. After weighing 30 g (with an accuracy of 0.01 g) of the dried leaves of the tested mint cultivars, they were placed in turn in a 1000 cm^3 round-bottomed flask, covered with 300 cm^3 of distilled water and connected to the Deryng apparatus. The flask, heater, and condenser were connected and distilled for 3 h from the start of condensation. The operation was repeated for each variety of mint until obtaining samples with a volume of at least 2 cm^3.

2.2. Methods

2.2.1. Evaluation of Fungicide Properties on Agar Medium

The effect of essential oils and ethanol extracts on the growth of mold fungi was carried out on an agar medium. Two species of fungi, significant in wood preservatives testing, were used in the study: *Trichoderma viride* Pers., strain A-102, and *Chaetomium globosum* Kunze, strain A-141 (ATCC 6205), from the collection of pure cultures of the Institute of Wood Sciences and Furniture, at the Warsaw University of Life Sciences. Ethanol extracts of mint, in the amounts of 0.1, 0.5, 1.0, 2.5, and 5 cm^3, were added to a sterile Petri dish and then covered with 10 cm^3 of maltose-agar medium. Essential oils were diluted in ethyl alcohol (60 μL of oil per 10 cm^3 of 60% ethanol). Oil solutions were dosed onto the plate in the amount of 0.05, 0.1, 0.25, 0.5, 1.0, 1.5, 2.0 cm^3. The inoculum of mold fungi, 5–6 mm in size, was inoculated centrally into Petri dishes. Cultures were performed in a Thermolyne Type 42,000 model thermal incubator (ThermoFisher Scientific, Waltham, MA, USA). The temperature of the culture was 26 ± 2 °C, and the relative air humidity was 63 ± 2%. At 48-h intervals, the growth diameter of mold fungi was measured. The diameter of mycelial growth was measured in two perpendicular directions. The tests were considered completed when the Petri dish was completely covered in the control samples. The analysis of variance using the Snedecor statistics was used to verify the statistical analysis. Statistical inference was carried out for the significance level $\alpha = 0.05$. In the case of rejection of the null hypothesis, Tukey's test was performed. The statistical hypothesis was as follows: H0: Ø 0.5 = Ø 5 = Ø 10 = Ø 15 = Ø 20 = Ø 40 = Ø 60 = ØK, H1: There are at least two means that differ significantly.

2.2.2. Wood Treatment and Assessment of Fungi Overgrowth

Samples of sapwood from pine wood (*Pinus sylvestris* L.) were selected for surface treatment. The dimensions of the wood samples were 40 × 40 × 4 mm. The density of the wood at 12% humidity was 400 kg/m^3. Wood samples were sterilized in a steam autoclave

(SMS, Warsaw, Poland). The impregnation of wood samples was carried out under sterile conditions in a laminar chamber. Using a sterile pipette, ethanol extracts of mints and ethanol solutions of essential oils were applied to one of the larger surfaces of the wood and then spread over this surface. Ethanol was applied to the control samples. Treatment was carried out by applying an appropriate amount of ethanol extract or essential oil to the largest surface of the wood samples. In the case of the ethanol extract, the spreads were 24.0, 32.0, and 40.0 g/m^2, in the case of the essential oil solution, the spreads were 140 i 180 g/m^2. The impregnated samples were left for 24 h in empty sterile vessels, and after that time they were transferred to Petri dishes with overgrown mycelium.

Treated wood samples were placed on a maltose-agar medium overgrown with mycelium, on glass spacers, avoiding the direct contact of wood with the substrate. The degree of overgrowth of the wood sample by fungi was determined on the basis of high-resolution photographic images taken periodically for each tested sample. The growth of fungi on the sample was determined as the percentage of the mycelium overgrown area of the sample to the total area of the sample tested. The percentage of overgrowth of the samples was determined with an accuracy of 5% with the help of ImageJ2 (Fiji v.1.52i) image analysis software [33]. The attack of mold fungi on wood is superficial, especially in the first period of this attack. The assumed duration of the test (14 days) corresponded to this first stage—the attack of mold on the wood surface [5].

2.2.3. GC-MS Analysis

The analysis of components in both types of samples (essential oils and ethanol extracts) was carried out on the GC-2010 gas chromatograph (Shimadzu, Kyoto, Japan) coupled with the GCMS-QP2010 mass spectrometer. An AQUATIC-2 capillary column (Phenomenex, Torrance, CA, USA) with a length of 60 m, diameter of 0.25 mm, and a film thickness of 0.25 μm was used. Samples were introduced to the column with the AOC-20i autosampler. GCMS software version 2.72 (Shimadzu, Kyoto, Japan) was used for sample analysis and the peaks were identified with the NIST11 and NIST11b mass spectrum libraries.

GC-MS separation conditions were as follows:

- Starting temperature: 75 °C, hold time 6 min, ramp rate 5 °C/min to 160 °C, hold time 4 min, ramp rate 5 °C/min to 180 °C, hold time 4 min; ramp rate 5 °C/min to 200 °C, hold time 4 min; ramp rate 5 °C/min to 220 °C, hold time 4 min; ramp rate 5 °C/min to 240 °C, hold time 4 min;
- Carrier gas: helium 5.0 (PGNiG, Warsaw, Poland);
- Carrier gas flow: 1.55 cm^3/min;
- Injection mode: split—1.8;
- Temperature of the injection: 250 °C;
- Detector voltage: 0.7 kV;
- Ion source temperature: 200 °C;
- Interface temperature: 250 °C.

3. Results and Discussion

3.1. Antifungal Properties

The analysis of fungal growth on the maltose-agar medium showed that the degree of growth inhibition depends on the amount of extract added to the medium (Tables 1 and 2). Ethanol extracts showed weak fungicidal activity. At the same time, it was found that the highest dose of the extract, amounting to 5.0 cm^3, either completely inhibited the growth of the fungi or significantly reduced the growth. Due to the fact that the solvent used also has a fungicidal effect [34], an analysis of the effect of ethanol on the growth of fungi was carried out. In order to confirm or exclude the biocidal effect of 60% ethanol on the growth of fungi, control tests were performed. It was not found that the ethanol used in amounts from 0.5 to 5 cm^3 had an inhibitory effect on the growth of the test fungi.

Table 1. Diameter of mycelium *T. viride* on a microbial medium with ethanol mint extracts.

Plant Materials	Concentration of Mint Extracts in Growth Medium ($cm^3/100\ cm^3$)	Day of Observation 2	4	6	p-Value	α
		Growth (mm)			Tukey's Test	
		statistics F			2.02×10^{-24}	0.05
(No. 3) *M. spicata* 'Morocco'	0 (control)	65.4	90.0	-	a	
	0.5	31.8	86.7	-	a	
	1.0	29.0	74.8	-	b	
	0.1	32.0	72.0	-	b	
	2.5	17.8	55.8	-	c	
	5.0	0.0	0.0	-	d	
		statistics F			2.88×10^{-96}	0.05
(No. 5) *M. spicata* 'Crispa'	0 (control)	32.7	65.0	90.0	a	
	0.1	28.3	65.3	90.0	a	
	0.5	29.5	73.7	90.0	a	
	1.0	25.8	64.2	90.0	a	
	2.5	19.5	55.7	90.0	a	
	5.0	0.0	0.0	0.0	b	
		statistics F			2.88×10^{-96}	0.05
(No. 6) *M. piperita* 'Almira'	0 (control)	32.7	65.0	90.0	a	
	0.1	28.0	67.2	90.0	a	
	0.5	27.5	71.2	90.0	a	
	1.0	25.3	63.2	90.0	a	
	2.5	18.0	51.2	90.0	a	
	5.0	0.0	0.0	0.0	b	
		statistics F			2.34×10^{-31}	0.05
(No. 8) *M. suaveolens* 'Variegata'	0 (control)	32.7	65.0	90.0	a	
	0.1	25.7	60.0	90.0	a	
	0.5	28.3	69.2	90.0	a	
	1.0	28.2	74.2	90.0	a	
	2.5	18.3	56.5	90.0	a	
	5.0	0.0	0.0	7.2	b	

abcd is homogeneous groups by the Tukey test, *p*-value—significance of the F statistic, α—statistical significance level.

Table 2. Diameter of mycelium *Ch. globosum* on a microbial medium with ethanol mint extracts.

Plant Materials	Concentration of Mint Extracts in Growth Medium ($cm^3/100\ cm^3$)	Day of Observation 2	4	6	8	10	p-Value	α
		Growth (mm)					Tukey's Test	
		statistics F					2.50×10^{-35}	0.05
(No. 3) *M. spicata* 'Morocco'	0 (control)	19.3	34.5	58.0	80.3	90.0	a	
	0.1	28.0	64.2	90.0	90.0	90.0	a	
	0.5	28.2	63.7	90.0	90.0	90.0	a	
	1.0	23.2	49.0	76.3	90.0	90.0	a	
	2.5	11.5	17.7	22.5	33.7	48.8	b	
	5.0	0.0	0.0	0.0	0.0	0.0	c	
		statistics F					1.49×10^{-26}	0.05
(No. 5) *M. spicata* 'Crispa'	0 (control)	20.8	51.3	90.0	-	-	a	
	0.1	24.2	54.3	88.5	-	-	ab	
	0.5	23.2	50.7	85.0	-	-	abc	

Table 2. *Cont.*

Plant Materials	Concentration of Mint Extracts in Growth Medium ($cm^3/100\ cm^3$)	Day of Observation 2	4	6	8	10	p-Value	α
		Growth (mm)					Tukey's Test	
(No. 5) *M. spicata* 'Crispa'	1.0	22.2	48.8	79.7	-	-	abc	
	2.5	17.2	21.0	26.5	-	-	d	
	5.0	0.0	0.0	0.0	-	-	e	
	statistics F						1.44×10^{-27}	0.05
(No. 6) *M. piperita* 'Almira'	0 (control)	20.8	51.5	90.0	-	-	a	
	0.1	20.2	47.5	76.8	-	-	b	
	0.5	20.0	44.0	72.7	-	-	b	
	1.0	19.3	39.2	64.2	-	-	c	
	2.5	15.7	20.2	24.0	-	-	d	
	5.0	0.0	0.0	11.7	-	-	e	
	statistics F						6.40×10^{-31}	0.05
(No. 8) *M. suaveolens* 'Variegata'	0 (control)	20.8	51.3	90.0	-	-	a	
	0.5	21.5	52.2	90.0	-	-	a	
	0.1	22.7	54.5	82.7	-	-	b	
	1.0	19.8	44.5	73.2	-	-	c	
	2.5	16.2	20.0	24.8	-	-	d	
	5.0	0.0	0.0	12.5	-	-	e	

abcde is homogeneous groups by the Tukey test, *p*-value—significance of the F statistic, α—statistical significance level.

On the basis of the obtained results, different sensitivity of the fungi to the applied extracts was found. The growth rate of *Ch. globosum* on media with ethanol extracts was definitely lower than the growth rate of *T. viride* at the same time of the study (Tables 1 and 2). The dominant biocidal effect of a particular variety of mint on the growth of fungi was not demonstrated.

The *T. viride* fungus was insensitive to the addition of essential oil solutions. The doses used in the tests did not show fungicide activity against one of the most dangerous fungi—causing gray decay of wood. On the fourth day of culture, complete overgrowth of the microbial media was observed (Table 3). Different biocidal activity of the solutions of the tested essential oils was demonstrated against the fungus *Ch. globosum*. Growth of the fungus was completely inhibited in medium containing 2.0 cm^3 and 1 cm^3, respectively, of an essential oil solution obtained from *M. piperita* 'Almira' and *M. spicata* 'Moroc-co' (Table 4).

Table 3. Diameter of mycelium *T. viride* on a microbial medium with essential oils solution.

Plant Materials	Concentration of Mint Extracts in Growth Medium ($cm^3/100\ cm^3$)	Day of Observation 2	4	6	p-Value	α
		Growth (mm)			Tukey's Test	
	statistics F				5.23×10^{-9}	0.05
(No. 3) *M. spicata* 'Morocco'	0 (control)	54.7	90.0	-	a	
	0.05	54.5	90.0	-	a	
	0.1	46.8	90.0	-	a	
	0.25	49.8	90.0	-	a	
	0.5	50.0	90.0	-	a	
	1.0	42.8	90.0	-	a	
	1.5	26.2	90.0	-	a	
	2.0	22.0	90.0	-	a	

Table 3. *Cont.*

Plant Materials	Concentration of Mint Extracts in Growth Medium ($cm^3/100\ cm^3$)	Day of Observation 2	4	6	p-Value	α
		Growth (mm)			Tukey's Test	
(No. 5) *M. spicata* 'Crispa'		statistics F			1.00	0.05
	0 (control)	54.7	90.0	-	a	
	0.05	46.5	90.0	-	a	
	0.1	44.5	90.0	-	a	
	0.25	49.7	90.0	-	a	
	0.5	47.8	90.0	-	a	
	1.0	53.7	90.0	-	a	
	1.5	44.8	90.0	-	a	
	2.0	48.5	90.0	-	a	
(No. 6) *M. piperita* 'Almira'		statistics F			3.82×10^{-58}	0.05
	0 (control)	54.7	90.0	-	a	
	0.05	48.3	90.0	-	a	
	0.1	51.5	90.0	-	a	
	1.0	51.2	90.0	-	a	
	0.25	49.3	90.0	-	a	
	1.0	51.2	90.0	-	a	
	1.5	45.7	90.0	-	a	
	2.0	10.7	90.0	-	a	
(No. 8) *M. suaveolens* 'Variegata'		statistics F			0.45	0.05
	0 (control)	54.7	90.0	-	a	
	0.05	52.7	90.0	-	a	
	0.1	52.3	90.0	-	a	
	0.25	48.7	90.0	-	a	
	0.5	48.5	90.0	-	a	
	1.0	40.5	90.0	-	a	
	1.5	33.2	90.0	-	a	
	2.0	47.7	88.3	-	a	

a is homogeneous groups by the Tukey test, p-value—significance of the F statistic, α—statistical significance level.

Table 4. Diameter of mycelium *Ch. globosum* on a microbial medium with essential oils solution.

Plant Materials	Concentration of Mint Extracts in Growth Medium ($cm^3/100\ cm^3$)	Day of Observation 2	4	6	p-Value	α
		Growth (mm)			Tukey's Test	
(No. 3) *M. spicata* 'Morocco'		statistics F			3.85×10^{-70}	0.05
	0 (control)	25.2	61.7	90.0	a	
	0.05	27.7	67.7	90.0	ab	
	0.1	25.7	64.3	90.0	ab	
	0.25	11.0	12.0	13.5	c	
	0.5	6.5	7.8	9.2	d	
	1.0	0.0	0.0	0.0	e	
	1.5	0.0	0.0	0.0	e	
	2.0	0.0	0.0	0.0	e	
(No. 5) *M. spicata* 'Crispa'		statistics F			2.88×10^{-96}	0.05
	0 (control)	25.2	61.7	90.0	a	
	0.05	30.0	69.8	90.0	a	
	0.1	27.2	68.3	90.0	a	

Table 4. *Cont.*

Plant Materials	Concentration of Mint Extracts in Growth Medium (cm^3/100 cm^3)	Day of Observation 2	4	6	p-Value	α
		Growth (mm)			Tukey's Test	
(No. 5) *M. spicata* 'Crispa'	0.25	21.3	50.3	77.6	ab	
	0.5	14.3	32.5	61.3	bc	
	1.0	15.5	27.8	50.2	c	
	1.5	9.5	12.8	17.2	d	
	2.0	0.0	2.5	3.7	d	
		statistics F			2.88×10^{-96}	0.05
(No. 6) *M. piperita* 'Almira'	0 (control)	25.2	61.7	90.0	a	
	0.05	25.7	62.5	90.0	a	
	0.1	25.5	56.5	90.0	a	
	0.25	27.0	65.7	88.3	ab	
	0.5	23.7	49.3	78.3	ab	
	1.0	14.8	19.3	57.0	c	
	1.5	14.2	30.3	25.2	d	
	2.0	0.0	0.0	0.0	e	
		statistics F			2.34×10^{-31}	0.05
(No. 8) *M. suaveolens* 'Variegata'	0 (control)	25.2	61.7	90.0	a	
	0.05	28.3	67.3	90.0	a	
	0.1	24.7	62.5	90.0	a	
	0.25	26.8	65.5	90.0	a	
	1.0	21.7	48.2	78.7	ab	
	0.5	21.2	44.7	71.5	b	
	1.5	13.2	28.8	48.7	c	
	2.0	13.0	20.3	32.2	d	

abcde is homogeneous groups by the Tukey test, *p*-value—significance of the F statistic, α—statistical significance level.

Figures 2–4 show the growth activity of mold fungi on the surface of wood samples treated with mint ethanol extracts. None of the extracts used was found to be effective in protecting wood against biodegradation. Despite the lack of fungicidal activity, it was noticed that some extracts clearly slow down the growth of the fungus *Ch. globosum*. This effect was particularly visible when the extract was applied to the wood surface in the amount of 40 g/m^2. With such application, the growth of the fungus *Ch. globosum* relative to the control was delayed by about 72 h (Figure 4b). It must therefore be concluded that mint ethanol extracts, the application of which in wood is at least 40 g/m^2, have a fungistatic effect in the initial stage of fungal development. Ethanol extracts in the doses used not only did not inhibit the growth of the *T. viride* fungus, but also stimulated the growth of the fungus, especially at lower doses (Figures 2a and 3a).

Solutions of essential oils turned out to be more active against fungi; although, in this case, the desired biocidal effect was not achieved. Essential oils significantly slowed down the growth of the fungus *Ch. globosum*, the strongest fungistatic effect was found for 'Marocco' spearmint oil (*Mentha spicata* L.) (No. 5) (Figures 5b and 6b). Although in tests on agar-maltose medium the strongest biocidal activity against *Ch. globosum* was found for oil from spearmint 'Crispa' (*Mentha spicata* L.) (No. 3), the growth inhibition effect was not so clearly visible in the studies on wood. Essential oils applied to the wood surface slowed down the growth of *T. viride* fungus, but not as markedly as was found in the case of *Ch. globosum*. In addition, it was noticed that a solution of pineapple mint 'Variegata' oil clearly stimulates the growth of the fungus (Figures 5a and 6a).

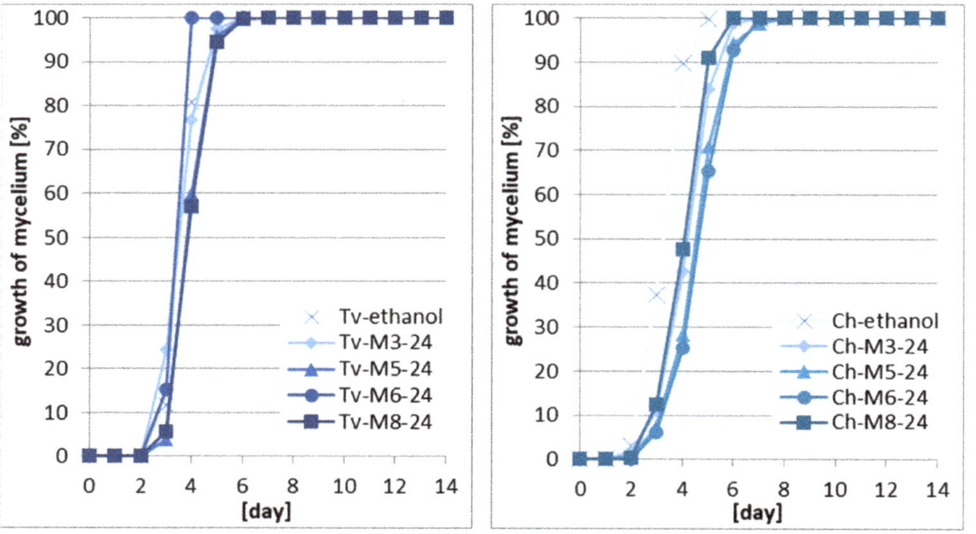

(**a**) (**b**)

Figure 2. Mold growth on surface of wood: (**a**) *T. viride;* (**b**) *Ch. globosum* (the legend of symbols: Tv-ethanol/Ch-ethanol—control, 24—sample code indicating the application of 24 g of extract per m^2, M3—*M. spicata* 'Morocco', M5—*M. spicata* 'Crispa', M6—*M. piperita* 'Almira', M8—*M. suaveolens* 'Variegata').

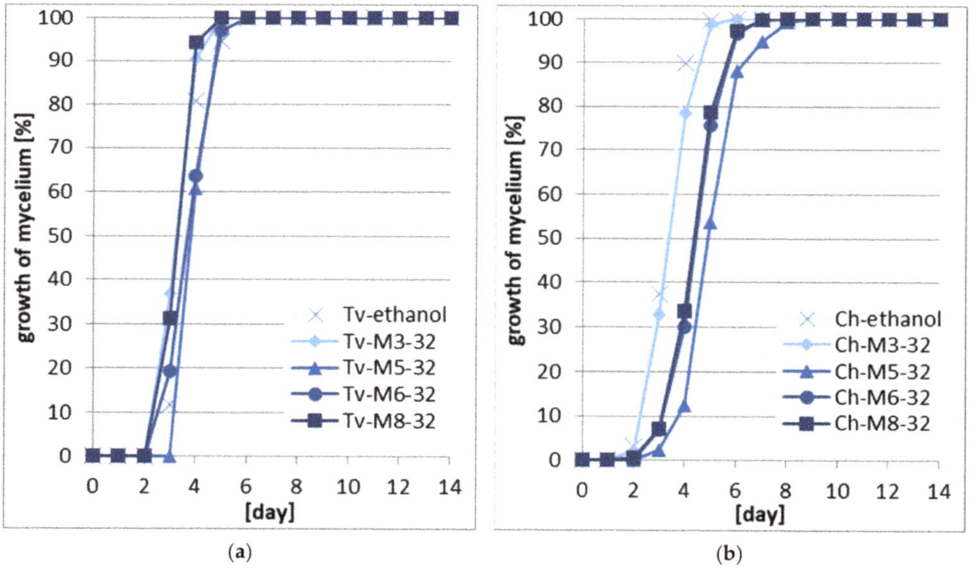

(**a**) (**b**)

Figure 3. Mold growth on surface of wood: (**a**) *T. viride;* (**b**) *Ch. globosum* the legend of symbols: Tv-ethanol/Ch-ethanol—control, 32—sample code indicating the application of 32 g of extract per m^2, M3—*M. spicata* 'Morocco', M5—*M. spicata* 'Crispa', M6—*M. piperita* 'Almira', M8—*M. suaveolens* 'Variegata').

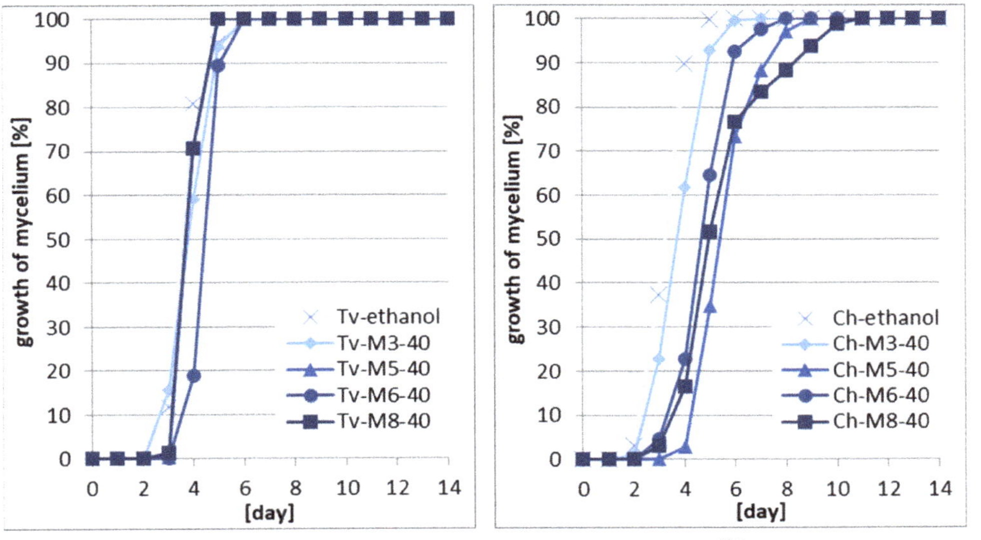

(a) (b)

Figure 4. Mold growth on surface of wood: (**a**) *T. viride*; (**b**) *Ch. globosum* (the legend of symbols: Tv-ethanol/Ch-ethanol—control, 40—sample code indicating the application of 40 g of extract per m^2, M3—*M. spicata* 'Morocco', M5—*M. spicata* 'Crispa', M6—*M. piperita* 'Almira', M8—*M. suaveolens* 'Variegata').

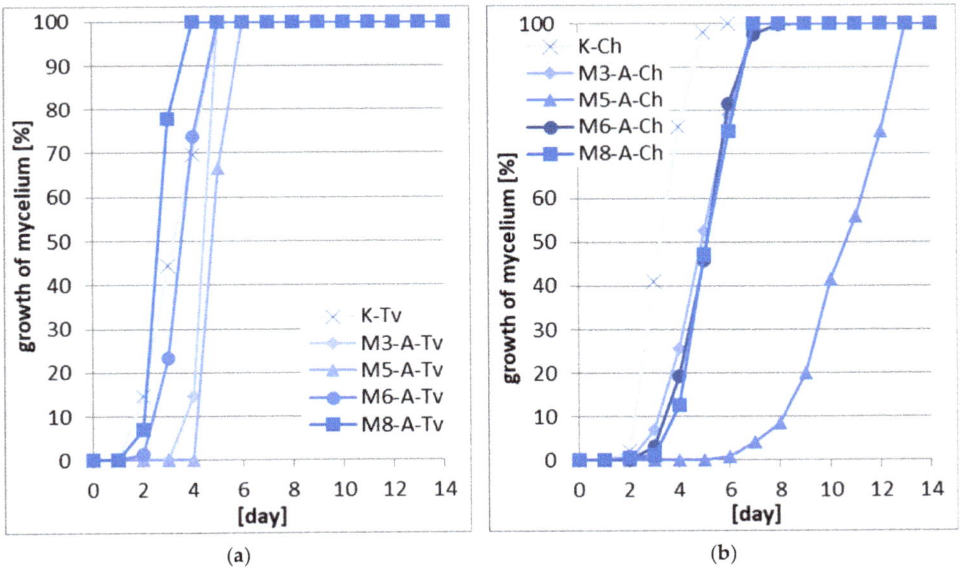

(a) (b)

Figure 5. Mold growth on surface of wood: (**a**) *T. viride*; (**b**) *Ch. globosum* (the legend of symbols: K-Tv/K-Ch—control, A—sample code indicating the application of 140 g of essential oils solution per m^2, M3—*M. spicata* 'Morocco', M5—*M. spicata* 'Crispa', M6—*M. piperita* 'Almira', M8—*M. suaveolens* 'Variegata').

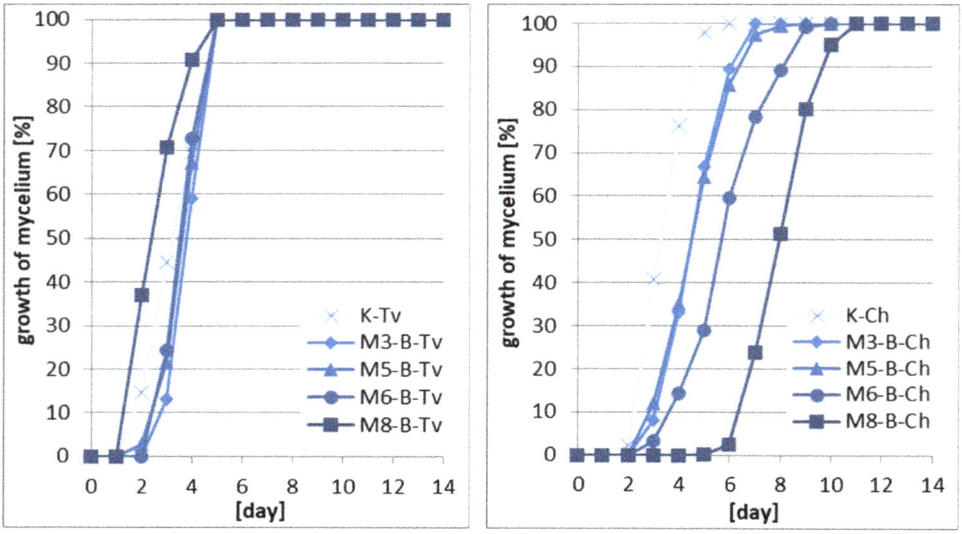

Figure 6. Mold growth on surface of wood: (**a**) *T. viride*; (**b**) *Ch. globosum* (the legend of symbols: K-Tv/K-Ch—-ontrol, B—sample code indicating the application of 180 g of essential oils solution per m^2, M3—*M. spicata* 'Morocco', M5—*M. spicata* 'Crispa', M6—*M. piperita* 'Almira', M8—*M. suaveolens* 'Variegata').

The fungicide effectiveness of mint extracts largely depends on the dose of the substances used and the method of obtaining the active ingredients. Despite the failure to achieve high biocidal activity in the assumed experiment, there are examples in the scientific literature of research indicating that mint extracts can be used as products that completely inhibit the growth of mold fungi. Ali et al. [35] showed that oils obtained from *M. longifolia* cause 100% growth inhibition of *Aspergillus flavus* and *A. fumigatus*. The authors of the study, however, used much higher doses of extracts of 250–500 µL/mL. The fungicide properties of mint extracts are also confirmed by the studies of Singh et al. [36] and Desam et al. [37]. Džamić et al. [38] proved the fungicidal properties of mint essential oils against the fungi *Aspergillus flavus, A. fumigatus, Alternaria alternaria*, and *Fusarium oxyporum*. However, in this case, the authors of the study used much higher effective doses than in the presented studies. The research by Gulluce et al. [39] indicated that the method of obtaining active ingredients has a decisive influence on the effectiveness of plant extracts against mold fungi. The authors of the study found that the essential oil extracted from *M. longifolia* showed strong antimicrobial activity against the 30 microorganisms tested, while the methanol extract remained almost inactive. If mint extracts are to be a component of wood preservatives in the future, it should also be considered whether the application of such a product in doses at which full fungicidal effect was achieved would be acceptable to the producers of wood preservatives, taking into account the costs of obtaining, processing the raw material and proper wood preservation.

3.2. Chemical Composition

The qualitative and quantitative composition of substances belonging to the group of terpenes and their derivatives was characterized using the GCMS technique. These substances are known for their antimicrobial properties [40,41]. Studies have shown that mint ethanol extracts are dominated by substances belonging to the oxygen-containing monoterpenoid and monoterpene. Menthol dominated in extracts from spearmint 'Crispa'

(*Mentha spicata* L.) (No. 3) and spearmint 'Morocco' (*Mentha spicata* L.) (No. 5). On the other hand, in the ethanol extracts of peppermint 'Almira' (*Mentha piperita* L.) (No. 6) the most isomenthols were identified. Substances such as borneol, gamma-terpinene, or camphene were present only in *M. piperita* extracts. gamma-terpineol, 4-terpinenol, homocatechol, isophorone, and 3(10)-Caren-4-ol were identified only in *M. suaveolens* extracts (Table 5).

Table 5. Chemical composition of the mint ethanol extracts.

Compound	Class *	Concentration (%) Calculated Relative to the Area of Peaks Identified			
		(No. 3) *M. spicata* 'Morocco'	(No. 5) *M. spicata* 'Crispa'	(No. 6) *M. piperita* 'Almira'	(No. 8) *M. suaveolens* 'Variegata'
3-Carene	MH	0.29	-	0.51	-
Camphene	MH	-	-	2.16	-
alpha-Pinene	MH	0.92	-	1.54	5.95
beta-Pinene	MH	1.26	3.45	-	-
p-Cymene	NH	0.11	0.56	-	5.75
D-Limonene	MH	7.69	-	1.14	-
beta-Myrcene	MH	0.68	0.57	-	-
beta-Phellandrene	MH	0.67	-	1.92	2.41
Eucalyptol	OM	5.77	7.16	3.24	6.16
gamma-Terpinene	MH	-	-	1.87	-
Isopulegol	OM	0.17	-	-	-
Linalol	OM	0.6	1.6	3.34	-
Borneol	OM	-	-	0.62	-
Caproaldehyde	ON	-	1.97	-	-
Menthol	OM	69.39	58.41	-	16.97
Isomenthol	OM	-	1.97	8.66	-
alpha-Terpineol	OM	-	0.87	1.6	25.50
beta-Terpineol	OM	2.52	1.79	-	8.7
gamma-Terpineol	OM	-	-	-	7.01
4-Terpinenol	OM	-	-	-	2.55
Carveol	OM	4.93	15.67	2.15	-
Carvone	OM	1.05	28.31	-	5.46
Dihydrocarvone	OM	10.4	0.55	-	-
Pulegone	OM	0.13	3.01	1.93	5.85
Geraniol	OM	-	1.76	-	-
beta-Citral	OM	-	-	-	3.53
Piperitone	OM	0.94	0.98	-	-
Homocatechol	ON	-	-	-	15.34
Isophorone	ON	-	-	-	3.52
3(10)-Caren-4-ol	OM	-	-	-	5.46

* OM—oxygen-containing monoterpenoids, MH—monoterpenes, SH—sesquiterpenes, NH—non-terpene hydrocarbons, OS—oxygen-containing sesquiterpenoids, ON—oxygen-containing non-terpenoids.

In terms of quality, the composition of essential oils turned out to be richer, although oxygen-containing monoterpenoids also dominated (Table 6). Six substances belonging to sesquiterpenes were identified in essential oils, which were not found in ethanol extracts. In oils of spearmint 'Crispa' (*Mentha spicata* L.) (No. 3) and spearmint 'Morocco' (*Mentha spicata* L.) (No. 5). Large amounts of carvone have been identified, which is consistent with studies by other authors [42]. Chrysanthenone was identified only in 'Almira' peppermint oils (Table 6). The fungicidal activity of these substances was described in studies conducted by Reddy et al. [43].

Table 6. Chemical composition of the essential oils.

Compound	Class *	Concentration (%) Calculated Relative to the Area of Peaks Identified			
		(No. 3) *M. spicata* 'Morocco'	(No. 5) *M. spicata* 'Crispa'	(No. 6) *M. piperita* 'Almira'	(No. 8) *M. suaveolens* 'Variegata'
3-Carene	MH	0.78	0.57	-	-
alpha-Pinene	MH	-	-	1.54	5.95
beta-Pinene	MH	-	-	0.51	-
p-Cymene	NH	-	-	1.92	2.41
D-Limonene	MH	5.29	5.94	3.24	6.16
beta-Myrcene	MH	0.67	3.45	-	-
beta-Phellandrene	MH	-	-	-	-
Eucalyptol	OM	2.94	15.71	-	-
Linaol	OM	0.55	1.6	2.16	-
Borneol	OM	-	-	3.29	-
Caproaldehyde	ON	-	-	1.18	-
Ocimene	MH	-	2.22	-	-
Menthol	OM	-	2.11	-	-
Menthone	OM	-	13.56	-	-
Isomenthol	OM	-	1.97	-	-
Menthyl acetate	OM	-	1.76	-	-
alpha-Terpineol	OM	-	0.55	-	-
beta-Terpineol	OM	3.6	-	-	-
Carveol	OM	-	0.98	1.04	8.79
Carvone	OM	64.84	40.49	8.66	-
Dihydrocarvone	OM	7.24	-	-	-
p-Menth-1-en-9-al	OM	-	-	-	-
Valeraldehyde	ON	-	-	-	7.01
D-Verbenone	OM	-	-	-	8.7
Chrysanthenone	OM	-	-	67.73	-
Sabinene	MH	1.35	-	-	-
Piperitone	OM	-	1.38	-	-
Catechol	ON	0.46	-	-	2.55
Levulinic acid	ON	-	-	1.87	5.75
Coumarin	ON	1.65	0.87	1.6	16.71
p-Vinylguaiacol	ON	1.42	2.21	1.66	-
Hydroquinone	ON	-	-	1.05	5.46
Eugenol	ON	-	-	0.62	-
cis-Jasmone	ON	-	1.63	1.93	5.85
beta-Bourbonene	SH	3.36	1.79		-
Caryophyllene	SH	3.6	-	-	3.53
Humulene	SH	0.66	-	-	-
beta-Copaene	SH	1.59	-	-	15.34
beta-Elemene	SH	-	-	-	3.52
Farnesene	SH	-	-	-	5.46

* OM—oxygen-containing monoterpenoids, MH—monoterpenes, SH—sesquiterpenes, NH—non-terpene hydrocarbons, OS—oxygen-containing sesquiterpenoids.

4. Conclusions

The results obtained in the study indicate that with the applied doses of extracts and the method of wood impregnation, the full fungicidal effect was not achieved, which would indicate that mint ethanol extracts and essential oils completely protect wood against the development of mold fungi. Despite the lack of a fungicide effect on wood, several very important conclusions can be drawn:

1. The degree of fungi growth inhibition depends on the amount of extract added to the medium.
2. The highest dose of the extract, 5.0 cm^3, either completely inhibited or severely limited the growth of the fungi.

3. Ethanol extracts from the tested mints, the application of which in the wood is at least 40 g/m^2, have a fungistatic effect in the initial stage of fungal development.
4. Solutions of essential oils turned out to be more active against fungi; although, in this case, the desired biocidal effect was not achieved.
5. Essential oils significantly slow down the growth of the fungus *Ch. globosum*, with the strongest fungistatic effect found in 'Morocco' spearmint oil (*Mentha spicata* L.).
6. It was shown that ethanol extracts from mints were dominated by substances belonging to the oxygen-containing monoterpenoid and monoterpene groups. In terms of quality, the composition of essential oils is richer, although oxygen-containing monoterpenoids also dominated.
7. Six substances belonging to sesquiterpenes were identified in essential oils, which were not found in ethanol extracts.

At this stage of the findings, it is impossible to talk about the practical aspect of the obtained results. The extracts used in the proposed concentrations and the proposed application did not protect the wood against growth by mold fungi; however, the obtained results allow us to observe some differences in the biocidal activity that occurred for individual varieties of mint and individual types of extracts. This gives a starting point for further research and the search for ways to enhance the fungicide effect. If mint extracts were to be used in wood protection, further research should be undertaken to search for synergistic effects with other types of plant extracts or with synthetic fungicides. The latter research concept seems to be more promising and prospective.

Author Contributions: Conceptualization, I.B. and B.A.; methodology, I.B., B.A., K.K., A.K.-D. and D.S.; software, I.B. and B.A.; validation, B.A. and I.B.; formal analysis, I.B., B.A., P.B. (Piotr Boruszewski), P.B. (Piotr Borysiuk) and A.R.; investigation, I.B.; resources, I.B.; data curation, I.B.; writing—original draft preparation, I.B.; writing—review and editing, P.B. (Piotr Boruszewski), J.Z., A.K.-D. and A.R.; visualization, I.B. and B.A.; supervision, I.B.; project administration, P.B. (Piotr Borysiuk); funding acquisition, P.B. (Piotr Borysiuk). All authors have read and agreed to the published version of the manuscript.

Funding: This research received no external funding.

Data Availability Statement: No new data were created in these studies.

Acknowledgments: The research was carried out as part of a scientific internship carried out by Izabela Betlej in the Department of Plant Production Technology and Commodity Science, University of Life Sciences in Lublin.

Conflicts of Interest: The authors declare no conflict of interest.

References

1. Regulation (EU) No 528/2012 of the European Parliament and of the Council of 22 May 2012 Concerning the Making Available on the Market and Use of Biocidal Products. Available online: https://echa.europa.eu/pl/information-on-chemicals/biocidal-products (accessed on 20 July 2023).
2. ESD for PT 8: Revised Emission Scenario Document for Wood Preservatives (OECD Series No. 2, 2013). Available online: https://echa.europa.eu/pl/guidance-documents/guidance-on-biocides-legislation/emission-scenario-documents (accessed on 26 June 2023).
3. Pop, D.M.; Timar, M.C.; Varodi, A.M. Comparative assessment of the biocidal potential of 3 essential oils. *Bull. Transilv. Univ. Braşov* **2019**, *12*, 83–96. [CrossRef]
4. Bahmani, M.; Schmidt, O. Plant Essential Oils for Environment-Friendly Protection of Wood Objects against Fungi. *Maderas Cienc. Tecnol.* **2018**, *20*, 325–332. [CrossRef]
5. Betlej, I.; Andres, B.; Krajewski, K.; Kiełtyka-Dadasiewicz, A.; Szadkowska, D.; Zawadzki, J. Effect of Various *Mentha* sp. Extracts on the Growth of *Trichoderma viride* and *Chaetomium globosum* on Agar Medium and Pine Wood. *Diversity* **2023**, *15*, 152. [CrossRef]
6. Pánek, M.; Reinprecht, L.; Hulla, M. Ten essential oils for beech wood protection—Efficacy against wood-destroying fungi and moulds, and effect on wood discoloration. *BioResources* **2014**, *9*, 5588–5603. [CrossRef]
7. Souza, M.A.A.; Lemos, M.J.; Brito, D.M.C.; Fernandes, M.S.; Castro, R.N.; Souza, S.R. Production and quality of menthol mint essential oil and antifungal and antigerminative activity. *Am. J. Plant Sci.* **2014**, *5*, 3311–3318. [CrossRef]

8. Galgano, M.; Mrenoshki, D.; Pellegrini, F.; Capozzi, L.; Cordisco, M.; Del Sambro, L.; Trotta, A.; Camero, M.; Tempesta, M.; Buonavoglia, D.; et al. Antibacterial and Biofilm Production Inhibition Activity of *Thymus vulgaris* L. Essential Oil against *Salmonella* spp. Isolates from Reptiles. *Pathogens* **2023**, *12*, 804. [CrossRef]
9. Russo, R.; Palla, F. Plant Essential Oils as Biocides in Sustainable Strategies for the Conservation of Cultural Heritage. *Sustainability* **2023**, *15*, 8522. [CrossRef]
10. Adedeji, G.A.; Ogunsanwo, O.Y.; Elufioye, T.O. Quantifications of phytochemicals and biocide actions of *Lawsonia inermis* linn. Extracts against wood termites and fungi. *Int. Biodeterior. Biodegrad.* **2017**, *116*, 155–162. [CrossRef]
11. González-Laredo, R.F.; Rosales-Castro, M.; Rocha-Guzmán, N.E.; Gallegos-Infante, J.A.; Moreno-Jiménez, M.R.; Karchesy, J.J. Wood preservation using natural products. *Madera Y Bosques* **2015**, *21*, 63–76. [CrossRef]
12. Betlej, I.; Andres, B.; Szadkowska, D.; Krajewski, K.; Ościłowska, A. Fungicidal properties of the medium from SCOBY microorganism cultivation in saturated wood against *Coniophora puteana* fungus. *BioResources* **2021**, *16*, 1287–1295. [CrossRef]
13. Betlej, I.; Andres, B.; Krajewski, K. Evaluation of fungicidal effects of post-culture medium of selected mold fungi and bacteria in relation to Basidiomycetes fungi, causing wood destruction. *BioResources* **2020**, *15*, 2471–2482. [CrossRef]
14. Lee, J.; Huh, N.; Hong, H.J.; Kim, B.S.; Kim, G.H.; Kim, J.J. The antagonistic properties of Trichoderma spp. inhabiting woods for potential biological control of wood-damaging fungi. *Holzforschung* **2012**, *66*, 883–887. [CrossRef]
15. Zhang, Z.; Yang, T.; Mi, N.; Wang, Y.; Li, G.; Wang, L.; Xie, Y. Antifungal Activity of Monoterpenes against Wood White-Rot Fungi. *Int. Biodeterior. Biodegrad.* **2016**, *106*, 157–160. [CrossRef]
16. Singh, T.; Singh, A.P. A Review on Natural Products as Wood Protectant. *Wood Sci. Technol.* **2012**, *46*, 851–870. [CrossRef]
17. Tari, S.M.M.; Tarmian, A.; Azadfallah, M. Improving fungal decay resistance of solvent and waterborne polyurethane-coated wood by free and microencapsulated thyme essential oil. *J. Coat. Technol. Res.* **2022**, *19*, 959–966. [CrossRef]
18. Sparacello, S.; Gallo, G.; Faddetta, T.; Megna, B.; Nicotra, G.; Bruno, B.; Giambra, B.; Palla, F. Thymus vulgaris Essential Oil and Hydro-Alcoholic Solutions to Counteract Wooden Artwork Microbial Colonization. *Appl. Sci.* **2021**, *11*, 8704. [CrossRef]
19. Reinprecht, L.; Vidholdová, Z. Growth Inhibition of Moulds on Wood Surfaces in Presence of Nano-Zinc Oxide and Its Combinations with Polyacrylate and Essential Oils. *Wood Res.* **2017**, *62*, 37–44.
20. Reinprecht, L.; Pop, D.A.; Vidholdová, Z.; Timar, M.C. Anti-decay potential of five essential oils against the wood-decaying fungi *Serpula lacrymans* and *Trametes versicolor*. *Acta Fac. Xylologiae* **2019**, *61*, 63–72. [CrossRef]
21. Shakama, H.M.; Mohamed, A.A.; Salem, M.Z.M. Down-regulatory effect of essential oils on fungal growth and Tri4 gene expression for some Fusarium oxysporum strains: GC-MS analysis of essential oils. *Arch. Phytopathol.* **2022**, *55*, 951–972. [CrossRef]
22. Sen, S.; Yalcin, M. Activity of commercial still waters from volatile oils production against wood decay fungi. *Maderas Cienc. Tecnol.* **2010**, *12*, 127–133. [CrossRef]
23. Pop, D.M.; Timar, M.C.; Beldean, E.C.; Varodi, A.M. Combined Testing Approach to Evaluate the Antifungal Efficiency of clove (*Eugenia caryophyllata*) Essential Oil for Potential Application in Wood Conservation. *BioResources* **2020**, *15*, 9474–9489. [CrossRef]
24. Pop, D.M.; Timar, M.C.; Varodi, A.M.; Beldean, E.C. An evaluation of clove (*Eugenia caryophyllata*) essential oil as a potential alternative antifungal wood protection system for cultural heritage conservation. *Maderas Cienc. Tecnol.* **2022**, *24*, 1–6. [CrossRef]
25. Yingprasert, W.; Matan, N.; Matan, N. Effects of surface treatment with cinnamon oil and clove oil on mold resistance and physical properties of rubberwood particleboards. *Eur. J. Wood Wood Prod.* **2015**, *73*, 103–109. [CrossRef]
26. Šimůnková, K.; Hýsek, Š.; Reinprecht, L.; Šobotník, J.; Lišková, T.; Pánek, M. Lavender oil as eco-friendly alternative to protect wood against termites without negative effect on wood properties. *Sci. Rep.* **2022**, *12*, 1909. [CrossRef] [PubMed]
27. Xie, Y.; Wang, Z.; Huang, Q.; Zhang, D. Antifungal Activity of Several Essential Oils and Major Components against Wood-Rot Fungi. *Ind. Crops Prod.* **2017**, *108*, 278–285. [CrossRef]
28. Woźniak, M. Antifungal Agents in Wood Protection—A Review. *Molecules* **2022**, *27*, 6392. [CrossRef]
29. Ludwiczuk, A.; Asakawa, Y. Chemotaxonomic value of essential oil components in liverwort species. A review. *Flavour Fragr. J.* **2015**, *30*, 189–196. [CrossRef]
30. Łyczko, J.; Kiełtyka-Dadasiewicz, A.; Skrzyński, M.; Klisiewicz, K.; Szumny, A. Chemistry behind quality—The usability of herbs and spices essential oils analysis in light of sensory studies. *Food Chem.* **2023**, *411*, 135537. [CrossRef]
31. Teixeira, B.; Marques, A.; Ramos, C.; Serrano, C.; Matos, O.; Neng, N.R.; Nogueira, J.M.F.; Saraiva, J.A.; Nunes, M.L. Chemical composition and bioactivity of different oregano (*Origanum vulgare*) extracts and essential oil. *J. Sci. Food Agric.* **2013**, *93*, 2702–2714. [CrossRef]
32. Ludwiczuk, A.; Kiełtyka-Dadasiewicz, A.; Sawicki, R.; Golus, J.; Ginalska, G. Essential oils of some Mentha species and cultivars, their chemistry and bacteriostatic activity. *Nat. Prod. Commun.* **2016**, *11*, 1015–1018. [CrossRef]
33. Borysiuk, P.; Krajewski, K.; Auriga, A.; Auriga, R.; Betlej, I.; Rybak, K.; Nowacka, M.; Boruszewski, P. PLA Biocomposites: Evaluation of Resistance to Mold. *Polymers* **2022**, *14*, 157. [CrossRef]
34. Dantigny, P.; Guilmart, A.; Radoi, F.; Bensoussan, M.; Zwietering, M. Modelling the effect of ethanol on growth rate of food spoilage moulds. *Int. J. Food Microbiol.* **2005**, *98*, 261–269. [CrossRef]
35. Ali, H.M.; Elgat, W.A.A.A.; EL-Hefny, M.; Salem, M.Z.M.; Taha, A.S.; Al Farraj, D.A.; Elshikh, M.S.; Hatamleh, A.A.; Abdel-Salam, E.M. New Approach for Using of *Mentha longifolia* L. and Citrus reticulata L. Essential Oils asWood-Biofungicides: GC-MS, SEM, and MNDO Quantum Chemical Studies. *Materials* **2021**, *14*, 1361. [CrossRef]

36. Singh, P.; Kumar, R.; Prakash, O.; Kumar, M.; Pant, A.K.; Isidorov, V.A.; Szczepaniak, L. Reinvestigation of Chemical Composition, Pharmacological, Antibacterial and Fungicidal Activity of Essential oil from *Mentha longifolia* (L.) *Huds. Res. J. Phytochem.* **2017**, *11*, 129–141.
37. Desam, N.R.; Al-Rajab, A.J.; Sharma, M.; Mylabathula, M.M.; Gowkanapalli, R.R.; Mohammed, A. Chemical composition, antibacterial and antifungal activities of Saudi Arabian Mentha longifolia L. essential oil. *J. Coast. Life Med.* **2017**, *5*, 441–446. [CrossRef]
38. Džamić, A.M.; Soković, M.D.; Ristić, M.S.; Novaković, M.; Grujić-JovanoviC, S.; Tešević, V.; Marin, P.D. Antifungal and antioxidant activity of *Mentha longifolia* (L.) Hudson (*Lamiaceae*) essential oil. *Bot. Serb.* **2010**, *34*, 57–61.
39. Gulluce, M.; Sahin, F.; Sokmen, M.; Ozer, H.; Daferera, D.; Sokmen, A.; Polissiou, M.; Adiguzel, A.; Ozkan, H. Antimicrobial and antioxidant properties of the essential oils and methanol extract from *Mentha longifolia* L. ssp. Longifolia. *Food Chem.* **2007**, *103*, 1449–1456. [CrossRef]
40. Bomfim de Barros, D.; de Oliveira e Lima, L.; Alves da Silva, L.; Cavalcante Fonseca, M.; Ferreira, R.C.; Diniz Neto, H.; da Nóbrega Alves, D.; da Silva Rocha, W.P.; Scotti, L.; de Oliveira Lima, E.; et al. α-Pinene: Docking Study, Cytotoxicity, Mechanism of Action, and Anti-Biofilm Effect against Candida albicans. *Antibiotics* **2023**, *12*, 480. [CrossRef]
41. Pirabalouti, A.G.; Firoznezhad, M.; Craker, L.; Akbarzadeh, M. Essential oil compositions, antibacterial and antioxidant activities of various populations of *Artemisia chamaemelifolia* at two phenological stages. *Rev. Bras. Farmacogn.* **2013**, *23*, 861–869. [CrossRef]
42. Hawrył, M.A.; Skalicka-Woźniak, K.; Świeboda, R.; Niemiec, M.; Stępak, K.; Waksmundzka-Hajnos, M.; Hawrył, A.; Szymczak, G. GC-MS fingerprints of mint essential oils. *Open Chem.* **2015**, *13*, 1326–1332. [CrossRef]
43. Reddy, D.N.; Al-Rajab, A.J.; Sharma, M.; Moses, M.M.; Reddy, G.R.; Albratty, M. Chemical constituents, in vitro antibacterial and antifungal activity of *Mentha* × *piperita* L. (peppermint) essential oils. *J. King Saud Univ. Sci.* **2019**, *31*, 528–533. [CrossRef]

forests

MDPI

Article

Influence of Phenol–Formaldehyde Resin Oligomer Molecular Weight on the Strength Properties of Beech Wood

Qian Lang [1,*], Vladimirs Biziks [2] and Holger Militz [2]

[1] Faculty of Engineering, Northeast Agricultural University, Haerbin 150030, China
[2] Wood Biology and Wood Product, Burckhardt-Institute, Georg-August University of Göttingen, Büsgenweg 4, 37077 Göttingen, Germany
* Correspondence: langqian4521@126.com; Tel.: +86-451-5519-1606

Abstract: The objective of this study was to determine the effects of four phenol–formaldehyde (PF) resin treatments with different molecular weights at four different concentrations (5, 10, 15, and 20%) in treated beech wood. The mechanical properties of untreated and treated beech wood were evaluated. After impregnation with PF resin, all modified beech wood at all PF resin concentrations exhibited an increase in weight percent gain compared with that in untreated beech samples. PF resins with lower molecular weights more easily penetrate the wood cell wall, leading to increased bulking of the wood structure, which in turn improves the dimensional stability of the wood. The PF resin treatment with a molecular weight of 305 g/mol showed better impregnation ability than that of the other PF resins. The impact bending strength of PF-treated wood was considerably reduced because PF-cured resins formed inside the wood and are rigid and brittle. Additionally, PF resin treatments at all concentrations decreased the modulus of elasticity of the wood. Scanning electron microscopy and light microscopy revealed that the PF resins were comparatively well fixed in the wood samples. The results indicate that the large molecular weight PF resins are more uniformly distributed in the fiber lumens.

Keywords: phenol–formaldehyde resin; average molecular weight; wood modification; mechanical properties

1. Introduction

Wood is a hygroscopic and polymeric material due to its abundance of hydroxyl groups associated with cell wall polymers. Wood also has two types of internal voids: relatively large voids, such as cell lumina and pit openings, and cell wall microvoids [1,2]. The chemical modification of wood involves the use of chemical and/or chemi-physical techniques to improve its dimensional stability, which occurs mainly through the covalent bonding of chemicals with the OH groups of cell wall polymers or the deposition of sterically fixed compounds in the cell wall [3]. With chemical modification, hygroscopic hydroxyl groups of the cell walls become partly blocked, and the nanopores (nanocapillaries) of the cell walls are filled with the chemicals. The water sorption sites decrease when the hydroxyl groups are blocked. Additionally, the deposition of chemicals in the cell walls reduces the space available for moisture from the environment, thereby making them more dimensionally stable [4,5].

To enhance the dimensional stability of solid wood, the modification of wood with water soluble and low molecular weight (M_w) phenol–formaldehyde (PF) resins has been practiced since the 1930s. The mechanical properties of commercial plywood composites, i.e., impreg and compreg plywood, are also enhanced by PF modification [6]. PF resins are known to penetrate and bulk the cell walls of solid wood. Additionally, PF resin modification has been shown to enhance the resistance of wood to white rot and brown rot fungi.

The use of water soluble and low M_W PF resin groups to enhance the dimensional stability and physical mechanical performance of wood has been intensively studied as a means of chemical modification [7–9]. PF resin is an aqueous mixture of oligomers that differ in terms of their M_W and shape. These PF resins can increase the dimensional stability of wood because they penetrate and swell the cell wall, thereby reducing hygroscopicity and forming a rigid crosslinked network upon curing. Furthermore, PF resins have a softening effect, similar to that of steam or heat, which plasticizes the cell walls of wood.

The reticular structure of cellulose fibrils, hemicelluloses, and lignin that comprises the cell walls of wood fibers form a three-dimensional structure in which the spacing between the various components defines the porous structure. Moreover, the porous structure determines the outside accessible surface area and upper size limit of molecules at which the cell wall can exchange with an external medium [10]. Accordingly, the particle size of the chemicals used for modification must be smaller than the minute openings of the small pores in the polymeric network structure of the cell walls.

The aim of the present study was to evaluate the physical properties of beech wood samples treated with four PF resins with different M_W values at different concentrations. The properties studied in the wood included the modulus of elasticity (MOE) and impact bending strength (IBS), which were chosen to characterize the application of wood in outdoor structures.

2. Materials and Methods

2.1. Wood Samples

To evaluate the effectiveness of impregnation of PF resins in the inner structure of the wood, two sample types with different dimensions, i.e., 25 × 25 × 10 mm^3 and 10 × 10 × 150 mm^3 (R × T × L), were prepared from beech (*Fagus sylvatica*) wood and used to investigate the bulking coefficient and mechanical properties. In total, 629 beech wood samples were selected for use in the study. The variance in density between the samples was <10%. As shown in Figure 1, the density was distributed mainly from 620 to 660 kg/m^3. In total, 17 groups of beech wood samples (16 treated wood sample groups and 1 untreated wood sample group) were selected. The oven dry weights of samples were recorded after drying at 103 °C ± 2 °C for 28 h. The dried wood specimens were then impregnated using a vacuum on the same day. Ten blocks per group and treatment were used.

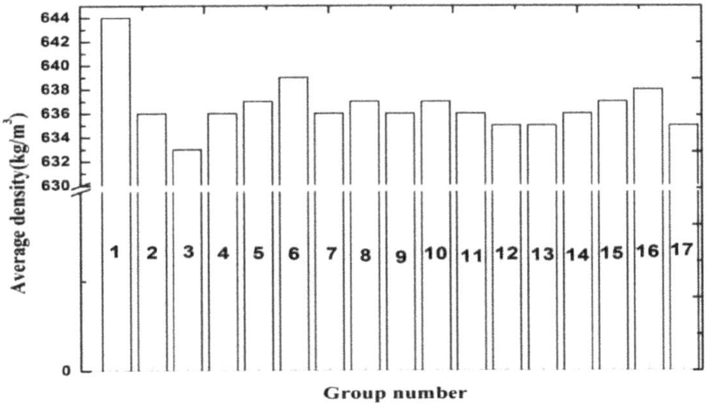

Figure 1. Density distribution of each group.

2.2. PF Resins

Four different types of PF (Hexion Oy, Puhos, Finland) resin were used to prepare aqueous solutions and impregnate the beech wood samples. For each of the four types of resin assessed, 10 samples were impregnated with an aqueous solution containing 5, 10, 15, and 20% (*w/w*) PF resin solid content. The physical–chemical properties of the four types of PF resin are listed in Table 1. The purity grades of the four PF resins and catalyst were 99%.

Table 1. Properties of the PF resins used for beech wood modification.

Resin	Molecular Weight Mn (g/mol)	Solid Content (%)	Catalyst	Amount of Formaldehyde (%)	Free Phenol (%)
A	237	44.1	NaOH	<1	<4
B	305	45.8	NaOH	<1	<4
C	403	47.02	NaOH	<1	<4
D	520	46.6	NaOH	1.88	0.24

2.3. Wood Treatment

Oven-dried wood samples with the aforementioned dimensions of 25 × 25 × 10 mm^3 and 10 × 10 × 150 mm^3 (R × T × L) were impregnated in the above-listed aqueous solutions in an autoclave under 0.008 kPa gage pressure in a vacuum for 45 min using a two-step vacuum impregnation method. During impregnation, the samples were subjected to room temperature (20 °C) followed by 101.32 kPa gage pressure for 60 min. They were removed from the modified solution after vacuum-pressure impregnation. Control samples were not modified with PF and were VTC (viscoelastic thermal compressed)-processed directly from their conditioned state.

2.4. Weight Percent Gain

Weight percent gain (WPG) was characterized using 10 replicates of treated beech wood with a dimension of 25 × 25 × 10 mm^3 (R × T × L). The average WPG of untreated and treated beech wood samples was estimated using the dry mass according to the following equation:

$$WPG = ((m_t - m_u)/m_u) \times 100 \quad (1)$$

where $m_{treated}$ is the oven-dried mass of the treated wood samples and $m_{untreated}$ is the oven-dried mass of the untreated wood samples.

2.5. Bulking Coefficient

The volume difference (bulking coefficient (BC)) due to impregnation was determined by comparing the dry specimen volumes before and after modification. Wood samples with dimensions of 25 × 25 × 10 mm^3 (R × T × L) were used to determine cell wall bulking. The BC was calculated using the following equation:

$$BC = ((V_t - V_u)/V_u) \times 100 \quad (2)$$

where V_t is the oven-dried volume of the treated wood samples and V_u is the oven-dried volume of the untreated wood samples.

2.6. MOE

The MOE was determined following EN 408 (2011) in a conditioned atmosphere (50% relative humidity and 20 °C). Ten specimens from each formula were analyzed using a universal testing machine (Zwick GmbH & Co., KG, Ulm, Germany) equipped with a 10 kN load head and testing software Xpert II. The loading direction was perpendicular to the grain with a loading speed of 10 mm/min. The distance between the supporters was 100 mm with 0.5% accuracy, and the rate of loading was adjusted to ensure the maximum load was reached within 60 ± 30 s. Specimens were kept in sealed plastic bags before the experiment.

2.7. Bending and Impact Bending Tests

IBS was assessed according to DIN 52,189 (1981) with modified specimen dimensions of 10 × 10 × 150 mm^3 (Resil Impactor, Instron, Norwood, MA, USA). The capacity of the machine used was 30 kgw/cm and bearing width was 120 mm. The machine was equipped with a 25 J hammer and an integrated force-measuring device. To calculate energy (A_W in $kJ.m^{-2}$), the following equation was used:

$$A_W = 1000 \cdot Q / b \cdot h \quad (3)$$

where Q is the energy (J) required to fracture the test piece, and b and h are the dimensions of the test samples in the radial and tangential directions (mm) [11].

2.8. Morphology Analysis

Scanning electron microscopy (SEM; Supra 45, Leo Elektronenmikroskopie GmbH, Oberkochen, Germany) was used to observe the untreated and treated beech wood samples, which were first sputter-coated with carbon. The observations were conducted with an accelerating voltage of 15 kV and a 70 Pa vacuum.

Light microscopy was performed using an Ellipse E600 microscope (Nikon, Tokyo, Japan) equipped with a digital camera (Nikon DS-Fi1c). The transverse section of the untreated and treated beech wood samples was cut using Leica microtome blades (Leica DB80 HS (Leica, Wetzlar, Germany). The samples were then stained with 0.09% safranin solution for 6 min, after which they were dehydrated using two ethanol solutions (50 and 99%). All specimens were dried at 42 °C for 25 min before they were embedded in glass.

3. Results

3.1. WPG

Figure 2 demonstrates the average WPG of PF resin-treated beech wood samples (treated with PF resin modifiers with four different M_W values at four different concentrations, i.e., 5, 10, 15, and 20%). As expected, the average WPG increased linearly as the concentration of the PF resin modifier increased, confirming that the PF resin was able to impregnate the wood samples at all concentrations. The same PF resins have the same distances between the points, which also leads to a linear increase in WPG. However, significant differences in WPG were not detected between the untreated and PF resin-treated beech wood samples. The samples treated with oligomers with a M_W of 237 g/mol exhibited higher WPGs compared with those of samples treated with oligomers with M_W values of 305, 403, and 520 g/mol. For the PF resin treatment with a M_W of 237 g/mol, the WPG of the wood was 6.7, 12.3, 18.3, and 24.7% with the 5, 10, 15, and 20% resin concentration (*w/w*) treatments. For the PF resin treatment with the highest M_W (520 g/mol), the WPGs of wood were lower, indicating that larger oligomers could not easily penetrate the cell wall of beech wood because less PF resin was found in the cell wall. Lower WPG might also be due to the macroporosity of the beech wood (influenced by the early and late wood within the year ring, number of year rings, and quantity of rays, vessels, and fibers) [12].

3.2. Bulking Effect

The bulking effect also tended to increase with the concentration of the PF resin treatment. Figure 3 shows average BC results of beech wood samples.

Comparing the four resin types at similar concentration levels, the bulking effect of the PF resin with a M_W of 305 g/mol was higher than that of the PF resins with M_W values of 237, 403, and 520 g/mol. For example, at the 20% concentration level, PF resins with M_W values of 237, 305, 403, and 520 g/mol led to bulking percentages of 12.2, 12.8, 11.0, and 7.9%, respectively, in treated wood. PF resins with low M_W values can more easily penetrate the cell walls of wood and have a higher bulking effect on the inner structure of the wood, which improves the wood's dimensional stability [13]. The increased volume of the wood specimens after PF resin treatment implies the presence of cell wall bulking due to the incorporation of PF resin polymers into the cell wall [2]. The step-wise increase in concentration led to an unequal increase in bulking, and an increase in PF resin M_w markedly reduced the BC [14,15]. Bulking is known to reduce the ability of the cell wall to shrink and swell; thus, bulking enhances the dimensional stability of beech wood [11]. The present BC results also demonstrate that the M_w of a resin is important for its ability to impregnate resin polymers into the cell walls of beech wood.

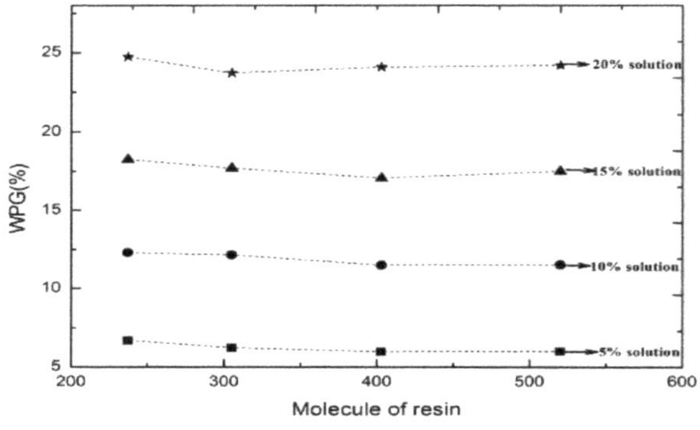

Figure 2. The weight percent gain (WPG) of wood samples treated with PF resins.

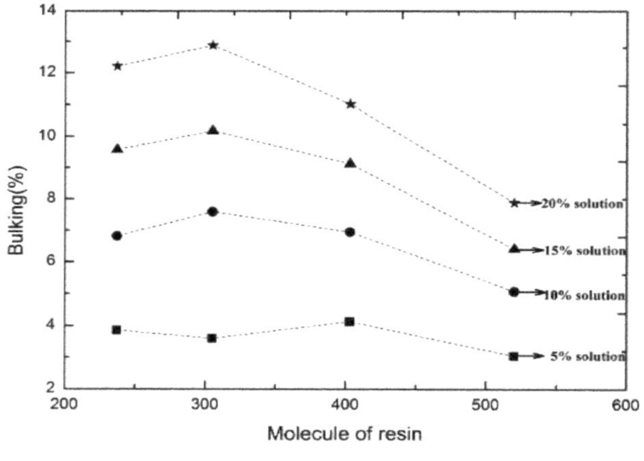

Figure 3. Bulking effect in wood samples treated with PF resins.

3.3. MOE

Figure 4 shows the MOE of beech wood samples before and after treatments with four PF resin types with different M_w values at four different concentrations (5, 10, 15, and 20%). At all concentrations, PF resin treatments led to slightly lower MOE values compared with those in the control samples. The MOE was reduced by approximately 7.9% when the PF resin treatment had M_w of 403 g/mol. However, the MOE did not differ markedly among sample groups modified with PF resins with different M_w values at the same concentration. Thermoset resins are usually used to modify wood. The impregnated wood then forms a rigid resin cell wall network in the wood structure after it is pressed in a thermal mold. A reduced MOE value can be attributed to the fracturing of the rigid network of the PF resin, which was precured in a thermal press mold [16]. The decrease in MOE could also be due to the thermal degradation of wood cell wall polymers, especially hemicelluloses, which are affected by high temperature. High temperature exposure can change the structure of fibers under thermal treatment [17,18]. The present results indicated that PF resin with a low M_w in the wood structure serves as a plasticizer and leads to substantial softening of the cell wall, inducing a reduction in the MOE of the cell wall in the cross fiber direction [19].

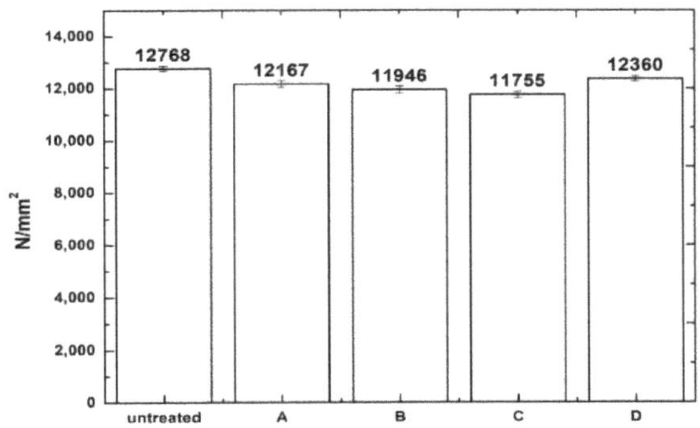

Figure 4. MOE of untreated and PF resin-treated beech wood samples. (A) M_w values of 237 g/mol, (B) M_w values of 305 g/mol, (C) M_w values of 403 g/mol, (D) M_w values of 520 g/mol.

3.4. IBS

Impact strength indicates the ability of a material to achieve a rapid transfer of stress/strain into a bulk material without critical stress peaks [5]. The impact strengths of PF resin-modified and -unmodified wood samples are presented in Figure 5. The IBS of the treated wood samples was lower than that of the unmodified specimens. However, there were no differences in IBS among the modified wood samples at all concentration levels. The loss in IBS was approximately equivalent among concentrations and ranged from 59.62 to 63.9% compared with that in the untreated samples. IBS can be reduced following chemical modification for several reasons. The IBS of the modified beech wood samples may have been reduced because PF resin impregnation reduced the pliability of the wood. It is also possible that the IBS was affected by the acidic hydrolysis of polysaccharides due to the use of magnesium chloride as a catalyst. The movement limitation of the cell wall due to the crosslinking with the PF resin could also reduce the IBS value [20–24]. The cell wall polymers in the inner wood contribute to IBS. The large decrease in IBS in the modified wood samples could also be attributed to reduced pliability due to a rigid, tree-dimensional, corset of PF resin in the beech wood matrix [25].

3.5. Morphology Analysis

The distribution of PF resin in untreated and PF resin-treated beech wood samples was characterized using SEM and light microscopy (Figure 6). As shown in the SEM images, the unmodified beech wood samples exhibited an obvious pore structure. After impregnation with the PF resin, the transverse section of the modified beech samples contained a remarkable amount of PF resin. Moreover, the unmodified beech wood sample had a red color from the adsorbed safranin in the wood. In the modified beech wood samples, this red color became pale yellow. This color change can be attributed to the impregnation of PF resin into the beech wood structure [13,26]. In the modified wood samples, wood cells near the vessel were filled with resin, and the PF resin was located in the lumen of vessels [27]. More bubbles in the resin films were formed in the radical section of the treated wood samples (Figure 6d,e). Additionally, the PF resins were uniformly distributed in the wood samples and existed at high levels in the fiber lumens.

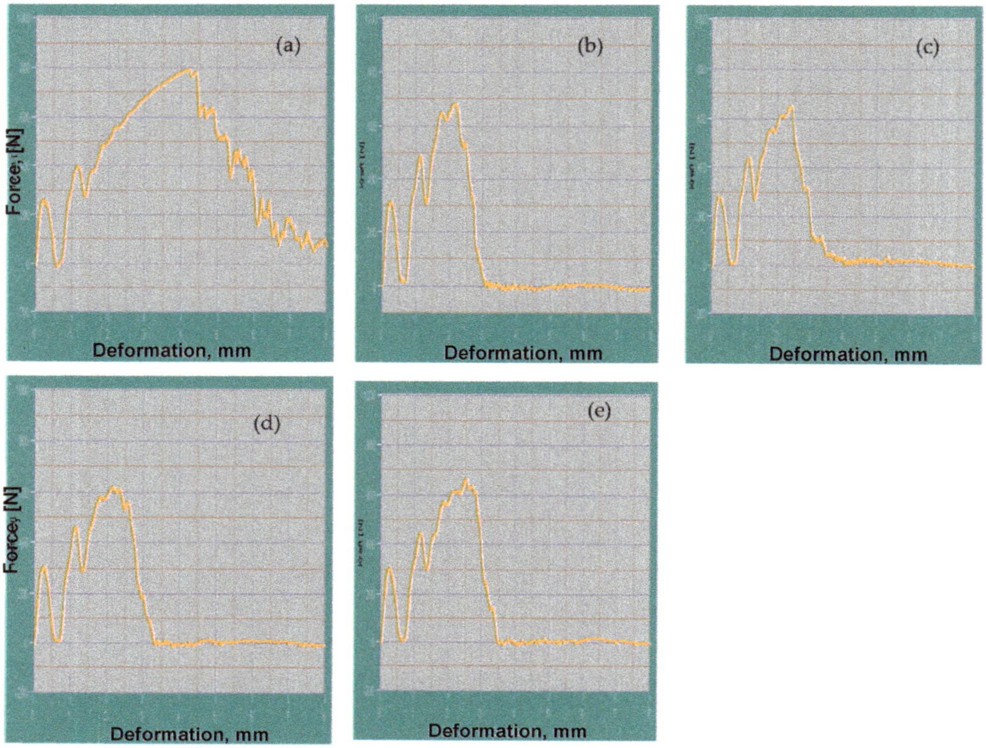

Figure 5. Impact bending strength of untreated and PF resin-treated wood samples. (**a**) Untreated control, and wood treated with PF resins with M_w values of (**b**) 237, (**c**) 305, (**d**) 403, and (**e**) 520 g/mol.

3.6. Mechanism Underlying the Modification of Beech Wood via PF Resin

The structures in beech wood that favor resin impregnation are the lumen and cell wall [28]. When a PF resin with a low M_w is impregnated into the wood, the resin is mainly distributed in small amounts in the lumen through the cell wall. Moreover, the molecule size of the PF resin is small, illustrating that a small molecule resin has good permeability, which easily penetrates the cell wall. When a PF resin with a high M_w was impregnated into beech wood, the resin was mainly distributed in the lumen [29–31], (see Figure 7).

Beech wood has many porous structures. The major types of internal void are large voids and cell wall microvoids [32,33]. The porous structure of wood provides many places into which PF resin can impregnate. As a water solution with polymer molecules, no chemical reaction usually takes place between the PF resin and functional groups in the wood structure. Therefore, resin and wood are mutually selective in the modification process. PF resins have different properties based on their M_w and shape. As passive impregnation agents, it is important to evaluate the relationships among the penetration depths of PF resins, capacity of PF resins, and chemical and physical characteristics of the wood structure [34–36].

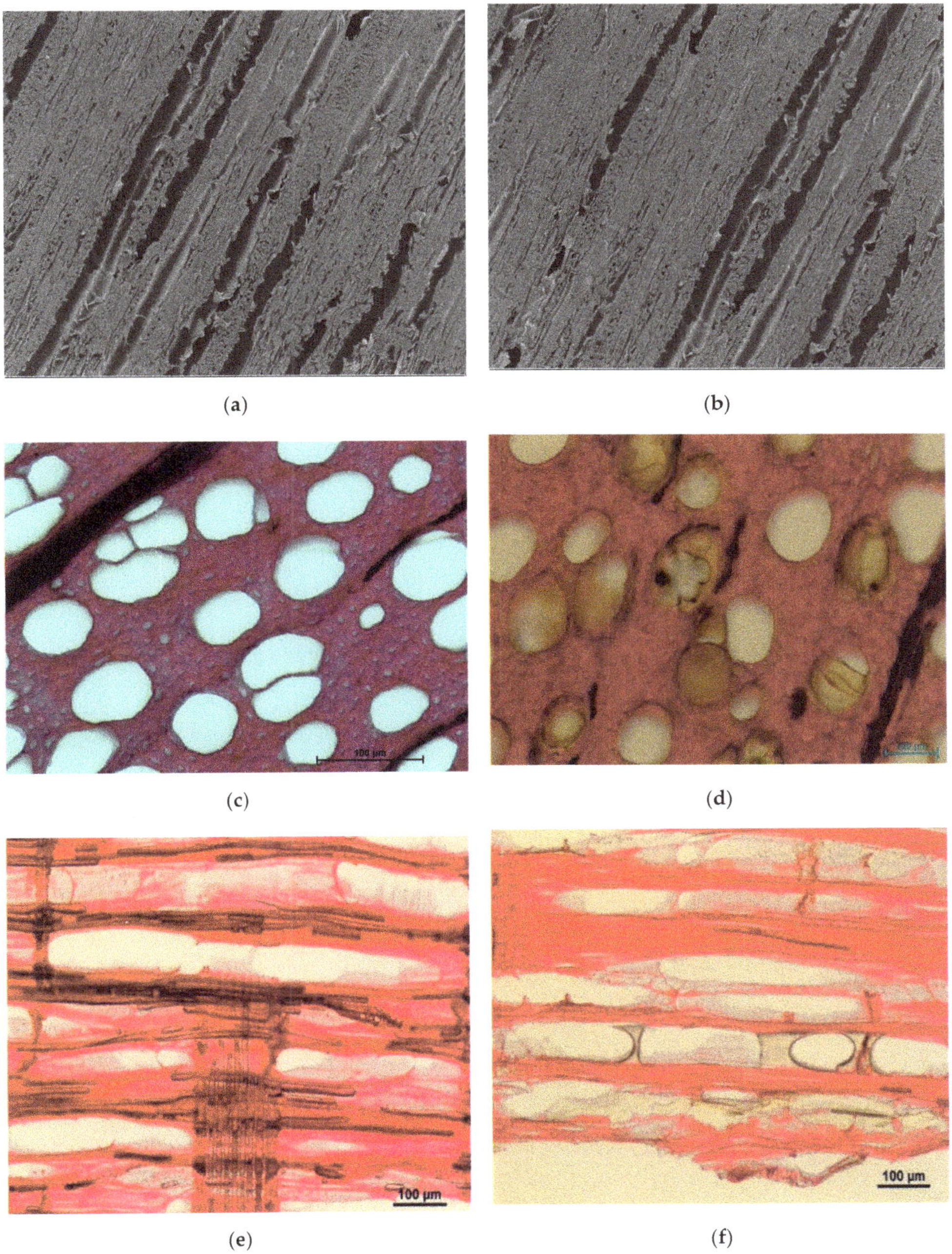

Figure 6. Morphology of untreated and PF resin-treated beech wood. SEM images of (**a**) untreated wood samples and (**b**) PF resin-treated wood samples. Light microscopy images of the transverse sections of (**c**) untreated wood samples and (**d**) PF resin-treated wood samples. Light microscopy images of the radical sections of (**e**) untreated wood samples and (**f**) PF resin-treated wood samples.

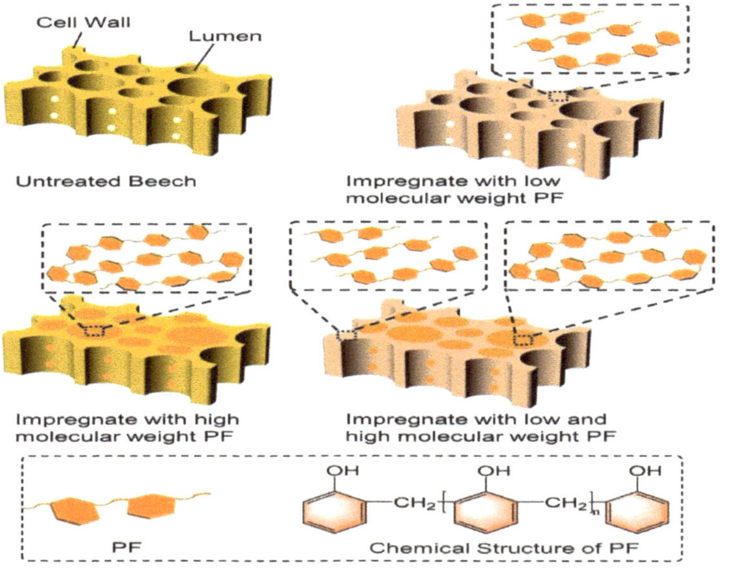

Figure 7. Mechanism underlying the modification of PF resin.

4. Conclusions

PF resins with the four different M_W values at four concentrations (5, 10, 15, and 25%) in water were impregnated into beech wood. The average WPG of the modified beech wood increased linearly as the concentration of PF resin increased. Beech wood treated with a PF resin with a low M_W (237 g/mol) exhibited a higher WPG. The bulking effect values following treatment with PF resin with a M_W of 305 g/mol were higher than those following treatments with PF resins with M_W values of 237, 403, and 520 g/mol. The MOE and IBS results showed that the M_W of a resin is important for the impregnation of resin oligomers into the beech wood cell walls. The relationship between the PF resin and beech wood cell structure was investigated using SEM and light microscopy, and the PF resins were more uniformly distributed in the modified beech wood samples and mainly existed in the fiber lumens. The physical properties of modified beech wood were improved by PF resin treatments with different M_W values. A low average M_W oligomer and small polydispersity index are conductive to the penetration of a PF resin.

Author Contributions: Conceptualization, Q.L. and V.B.; methodology, Q.L. and V.B.; software, Q.L. and V.B.; validation, H.M.; formal analysis, H.M; investigation, Q.L. and V.B.; resources, H.M.; data curation, H.M.; writing—original draft preparation, Q.L.; writing—review and editing, Q.L.; visualization, Q.L. and V.B.; supervision, H.M.; project administration, Q.L.; funding acquisition, Q.L. and V.B. All authors have read and agreed to the published version of the manuscript.

Funding: The research was funded by the China Postdoctoral Science Foundation (2017M621234), Heilongjiang Province Natural Science Foundation (LH2020C002), Heilongjiang Province Postdoctoral Science Foundation (LBH-Z17029) and High-end Foreign Experts Introduction Program(G2022011017L).

Institutional Review Board Statement: Not applicable.

Data Availability Statement: All data included in this study are available upon request by contact with the corresponding author.

Acknowledgments: The authors thank colleagues in Northeast Agricultural University for giving good suggestions regarding experiments and the manuscript.

Conflicts of Interest: The authors declare no conflict of interest.

References

1. Donath, S.; Militz, H.; Mai, C. Wood modification with alkoxysilanes. *Wood Sci. Technol.* **2004**, *38*, 555–566. [CrossRef]
2. Klüppel, A.; Mai, C. The influence of curing conditions on the chemical distribution in wood modified with thermosetting resins. *Wood Sci. Technol.* **2013**, *47*, 643–658. [CrossRef]
3. Hill, C.A. *Wood Modification: Chemical, Thermal and Other Processes*; John Wiley and Sons: Hoboken, NJ, USA, 2007; Volume 5.
4. Xie, Y.; Hill, C.A.; Xiao, Z.; Mai, C.; Militz, H. Dynamic water vapour sorption properties of wood treated with glutaraldehyde. *Wood Sci. Technol.* **2011**, *45*, 49–61. [CrossRef]
5. Müller, M.; Radovanovic, I.; Grüneberg, T.; Militz, H.; Krause, A. Influence of various wood modifications on the properties of polyvinyl chloride/wood flour composites. *J. Appl. Polym. Sci.* **2012**, *125*, 308–312. [CrossRef]
6. Furuno, T.; Imamura, Y.; Kajita, H. The modification of wood by treatment with low molecular weight phenol-formaldehyde resin: A properties enhancement with neutralized phenolic-resin and resin penetration into wood cell walls. *Wood Sci. Technol.* **2004**, *37*, 349–361.
7. Gabrielli, C.; Kamke, F.A. Treatment of chemically modified wood with VTC process to improve dimensional stability. *Forest Prod. J.* **2008**, *58*, 82–86.
8. Wang, X.; Chen, X.; Xie, X.; Cai, S.; Yuan, Z.; Li, Y. Multi-Scale Evaluation of the Effect of Phenol Formaldehyde Resin Impregnation on the Dimensional Stability and Mechanical Properties of Pinus Massoniana Lamb. *Forests* **2019**, *10*, 646. [CrossRef]
9. Anwar, U.M.K.; Paridah, M.T.; Hamdan, H.; Sapuan, S.M.; Bakar, E.S. Effect of curing time on physical and mechanical properties of phenolic-treated bamboo strips. *Ind. Crop Prod.* **2009**, *29*, 214–219. [CrossRef]
10. Persson, P.V.; Hafren, J.; Fogden, A.; Daniel, G.; Iversen, T. Silica Nanocasts of Wood Fibers: A Study of Cell-Wall Accessibility and Structure. *Biomacromolecules* **2004**, *5*, 1097–1101. [CrossRef]
11. Epmeier, H.; Westin, M.; Rapp, A. Differently modified wood: Comparison of some selected properties. *Scand. J. Forest Res.* **2004**, *19*, 31–37. [CrossRef]
12. Biziks, V.; Bicke, S.; Militz, H. Penetration depth of phenol-formaldehyde (PF) resin into beech wood studied by light microscopy. *Wood Sci. Technol.* **2019**, *53*, 165–176. [CrossRef]
13. Gabrielli, C.P.; Kamke, F.A. Phenol–formaldehyde impregnation of densified wood for improved dimensional stability. *Wood Sci. Technol.* **2010**, *44*, 95–104. [CrossRef]
14. Ryu, J.Y.; Imamura, Y.; Takahashi, M.; Kajita, H. Effect of molecular weight and some other properties of resins on the biological resistance of phenolic resin treated wood. *Mokuzai Gakkaishi* **1993**, *39*, 486–492.
15. Imamura, Y.; Kajita, H.; Higuchi, N. Modification of wood by treatment with low molecular phenol-formaldehyde resin (1). Influence of neutral and alkaline resins (in Japanese). In Proceedings of the Abstracts of the 48th Annual Meeting of Japan Wood Research Society, Shizuoka, Japan, 3–5 April 1998.
16. Xie, Y.; Fu, Q.; Wang, Q.; Xiao, Z.; Militz, H. Effects of chemical modification on the mechanical properties of wood. *Eur. J. Wood Wood Prod.* **2013**, *71*, 401–416. [CrossRef]
17. Lopes, D.B.; Mai, C.; Militz, H. Mechanical properties of chemically modified portuguese pinewood. *Maderas Cienc. Tecnol.* **2015**, *17*, 179–194. [CrossRef]
18. Hillis, W.E. High temperature and chemical effects on wood stability. *Wood Sci. Technol.* **1984**, *18*, 281–293. [CrossRef]
19. Hermoso, E.; Fernández-Golfín, J.; Conde, M.; Troya, M.T.; Mateo, R.; Cabrero, J. Characterization of thermally modified Pinus radiata timber. *Maderas Cienc. Technol.* **2015**, *17*, 493–504.
20. Kielmann, B.C.; Militz, H.; Mai, C.; Adamopoulos, S. Strength changes in ash, beech and maple wood modified with a N-methylol melamine compound and a metal-complex dye. *Wood Res.* **2013**, *58*, 343–350.
21. Evans, P.D.; Schmalzl, K.J. A Quantitative Weathering Study of Wood Surfaces Modified by Chromium VI and Iron III Compounds. Part 1. Loss in Zero-Span Tensile Strength and Weight of Thin Wood Veneers. *Holzforschung* **1989**, *43*, 289–292. [CrossRef]
22. Mai, C.; Xie, Y.; Xiao, Z.; Bollmus, S.; Vetter, G.; Krause, A.; Militz, H. Influence of the modification with different aldehydebased agents on the tensile strength of wood. In *European Conference on Wood Modification*; Bangor University: Cardiff, Wales, 2007; pp. 49–56.
23. Dieste, A.; Krause, A.; Bollmus, S.; Militz, H. Physical and mechanical properties of plywood produced with 1.3-dimethylol-4.5-dihydroxyethyleneurea (DMDHEU)-modified veneers of Betula sp. and Fagus sylvatica. *Holz Als Roh-Und Werkst.* **2008**, *66*, 281. [CrossRef]
24. Xie, Y.; Krause, A.; Militz, H.; Turkulin, H.; Richter, K.; Mai, C. Effect of treatments with 1.3-dimethylol-4, 5-dihydroxy-ethyleneurea (DMDHEU) on the tensile properties of wood. *Holzforschung* **2007**, *61*, 43–50. [CrossRef]
25. Pittman, C.U., Jr.; Kim, M.G.; Nicholas, D.D.; Wang, L.; Kabir, F.A.; Schultz, T.P.; Ingram, L.L., Jr. Wood enhancement treatments I. Impregnation of southern yellow pine with melamine-formaldehyde and melamine-ammeline-formaldehyde resins. *J. Wood Chem. Technol.* **1994**, *14*, 577–603. [CrossRef]
26. Modzel, G.; Kamke, F.A.; Carlo, D.F. Comparative analysis of wood: Adhesive bondline. *Wood Sci. Technol.* **2011**, *45*, 147–158. [CrossRef]
27. Biziks, V.; Bicke, S.; Koch, G.; Militz, H. Effect of phenol-formaldehyde (PF) resin oligomer size on the decay resistance of beech wood. *Holzforschung* **2021**, *75*, 574–583. [CrossRef]

28. Nishida, M.; Tanaka, T.; Miki, T.; Hayakawa, Y.; Kanayama, K. Integrated analysis of solid-state NMR spectra and nuclear magnetic relaxation times for the phenol formaldehyde (PF) resin impregnation process into soft wood. *RSC Adv.* **2017**, *7*, 54532–54541. [CrossRef]
29. Yi, T.; Guo, C.; Zhao, S.; Zhan, K.; Gao, W.; Yang, L.; Du, G. The simultaneous preparation of nano cupric oxide (CuO) and phenol formaldehyde (PF) resin in one system: Aimed to apply as wood adhesives. *Eur. J. Wood Wood Prod.* **2020**, *78*, 471–482. [CrossRef]
30. Kordkheili, H.Y.; Pizzi, A. Improving properties of phenol-lignin-glyoxal resin as a wood adhesive by an epoxy resin. *Eur. J. Wood Wood Prod.* **2021**, *79*, 199–205. [CrossRef]
31. Vitas, S.; Segmehl, J.S.; Burgert, I.; Cabane, E. Porosity and pore size distribution of native and delignified beech wood determined by mercury intrusion porosimetry. *Materials* **2019**, *12*, 416. [CrossRef] [PubMed]
32. Hass, P.; Wittel, F.K.; McDonald, S.A.; Marone, F.; Stampanoni, M.; Herrmann, H.J.; Niemz, P. Pore Space Analysis of Beech Wood: The Vessel Network. *Holzforschung* **2010**, *64*, 639–644. [CrossRef]
33. Wang, X.; Chen, X.; Xie, X.; Yuan, Z.; Cai, S.; Li, Y. Effect of phenol formaldehyde resin penetration on the quasi-static and dynamic mechanics of wood cell walls using nanoindentation. *Nanomaterials* **2019**, *9*, 1409. [CrossRef] [PubMed]
34. Pečnik, J.G.; Kutnar, A.; Militz, H.; Schwarzkopf, M.; Schwager, H. Fatigue behavior of beech and pine wood modified with low molecular weight phenol-formaldehyde resin. *Holzforschung* **2021**, *75*, 37–47. [CrossRef]
35. Li, W.; Zhang, Z.; Yang, K.; Mei, C.; Van den Bulcke, J.; Van Acker, J. Understanding the effect of combined thermal treatment and phenol–formaldehyde resin impregnation on the compressive stress of wood. *Wood Sci. Technol.* **2022**, *56*, 1071–1086. [CrossRef]
36. Hong, S.; Gu, Z.; Chen, L.; Zhu, P.; Lian, H. Synthesis of phenol formaldehyde (PF) resin for fast manufacturing laminated veneer lumber (LVL). *Holzforschung* **2018**, *72*, 745–752. [CrossRef]

forests

MDPI

Article

Water Vapor Sorption Kinetics of Beech Wood Modified with Phenol Formaldehyde Resin Oligomers

Qian Lang [1,*], Vladimirs Biziks [2] and Holger Militz [2]

[1] Faculty of Engineering, Northeast Agricultural University, Harbin 150030, China
[2] Wood Biology and Wood Product, Burckhardt-Institute, Georg-August University of Göttingen, Büsgenweg 4, 37077 Goettingen, Germany; vbiziks@gwdg.de (V.B.); hmilitz@gwdg.de (H.M.)
* Correspondence: langqian4521@126.com; Tel.: +86-451-5519-1606

Abstract: Beech is an important tree species in Europe. This study aimed to elucidate the influence of four molecular weights of phenol formaldehyde (PF) resin (266, 286, 387, and 410 g/mol) on the sorption behavior of unmodified and modified beech wood samples using a dynamic vapor sorption (DVS) apparatus. The variations in the environmental relative humidity and moisture content (MC) of the samples were recorded, and the DVS isotherms were plotted from the equipment. During the sorption process, the MC of the modified samples decreased in comparison to that of the unmodified samples, and both apparently decreased with the increasing molecular weight of the PF resin. The DVS isotherm hysteresis plot illustrated a reduction in sorption hysteresis for the modified wood with varying PF resins compared to the unmodified samples. Based on the DVS isotherm adsorption and desorption plots, the decrease in the equilibrium of the MC can be attributed to there being fewer sorption sites in the modified samples, which exhibited the lowest hygroscopicity. Overall, the moisture sorption mechanism for both types of samples was clarified, highlighting a clear correlation between the molecular weight of the applied PF resin and its influence on moisture sorption behavior.

Keywords: phenol formaldehyde resin; dynamic vapor sorption; wood modification; beech wood

1. Introduction

Wood, a porous natural polymer composite material, typically exhibits ample water sorption sites, specifically hydroxyl groups, when the saturated vapor partial pressure on the surface of the water contained in the pore is less than the environmental pressure [1]. In response to fluctuations in the external humidity, wood adjusts its moisture content (MC) through absorption or desorption to attain equilibrium, and this affects the quality of wood products. Thus, the viscoelastic properties of unmodified and modified wood samples have been assessed through dynamic vapor sorption (DVS).

The moisture absorption characteristics of wood are contingent on the temperature and relative humidity of the external environment [2], and moisture expedites surface photodegradation during outdoor exposure [3]. Adsorption isotherms have been used to evaluate these characteristics, illustrating the correlation between the relative humidity (RH) and equilibrium moisture content (EMC) of the wood environment under a steady temperature and pressure. This correlation is pivotal in modifying the service lifespan of outdoor exposed wood.

Phenol formaldehyde (PF) resin, a water-soluble, low-molecular-weight polymer, bolsters the dimensional stability of wood materials by infiltrating and inflating their cell wall [4]. Upon modification, a nonbonding network is established between wood and resin, reducing the hygroscopicity of wood. The hydrophilic hydroxyl groups in cell walls are partially obstructed, and the cell wall nanopores (nanocapillaries) become filled with PF resin. This process reduces the number of sorption sites for water and minimizes the space available for external moisture. The hygroscopic structure of wood is, thus, impeded by PF resin, yielding drier cell walls and an improved dimensional stability [5].

Beech wood is classified as a softwood [6]. In response to environmental conditions such as wind, softwoods exhibit adaptive growth, bending, and tilting of the trunk or branches under pressure, a phenomenon known as pressure wood; the non-stressed part is referred to as the corresponding wood [7]. Compared with the corresponding or normal wood, pressure wood demonstrates a thicker tracheid wall, reduced cellulose content, and increased lignin content. However, the propensity of beech toward cracking and warping in practical processing, which are deemed as defects, hinders the efficient use of plantation wood [8]. The water adsorption characteristics of wood correlate with the content of chemical components and the pore size of the cell wall. Consequently, understanding the water adsorption characteristics of pressure wood has both theoretical and practical implications for promoting the development and utilization of plantation wood.

Various theoretical models have been proposed to interpret the complex water adsorption process, including the Hailwood–Horrobin model, the Brunauer–Emmett–Teller (BET) model, the Guggenheim–Anderson–de Boer (GAB) model, and the Halsey model [9–11]. Hills et al. [12] used the parallel exponential kinetics model to analyze the water vapor sorption behavior of flax fibers, while Krabbenhoft et al. [13] proposed a model for non-Fickian moisture transfer in wood. Moreover, Xie et al. [14] investigated the water vapor sorption behavior of three celluloses using the Kelvin–Voigt viscoelastic model. Each model, however, exhibits deviations between the actual and measured values under varying RH conditions.

This study aimed to analyze the sorption behavior of untreated and treated beech samples, using PF resin of differing molecular weights at various concentrations. A DVS apparatus was used to record the sorption dynamics, sorption rate, and hysteresis in both unmodified and modified beech specimens.

2. Materials and Methods

2.1. Wood Materials

Samples of European beech wood (Fagus sylvatica), with an initial MC of 14.5%, were sourced from Germany. Samples with an average density ranging between 620 and 660 kg/m^3 were selected for the experiment. Samples intended for impregnation were cut into dimensions of 25 mm × 25 mm × 10 mm (R × T × L) and 10 mm × 10 mm × 150 mm (R × T × L). After drying at 103 ± 2 °C for 28 h, the oven-dry weights of these samples were determined. All samples were free of apparent defects.

2.2. Wood Modifier

Four types of PF resin were employed to impregnate the beech wood samples. Ten samples were impregnated in aqueous solutions with the PF resin solid contents of 5%, 10%, 15%, and 25% (*w*/*w*) for each of the four resins under assessment. The molecular weights of the four PF resins were 266, 286, 387, and 410 g/mol. The PF resin (M_W = 410 g/mol) has the following characteristics. Viscosity: 186 mPa s, density: 1.56 g/cm^3, gel time: 130–150 °C. The physical–chemical properties of the four resins are presented in Table 1. The aqueous solutions were prepared using these four types of resin (supplied by Hexion Oy, Helsinki, Finland); these resins had a purity grade of 99%, similar to that of the catalyst.

Table 1. Properties of the PF resins for the modified beech samples.

Resin	Molecular Weight Mn [g/mol]	Solid Content [%]	Catalyst	Amount of Formaldehyde [%]	Free Phenol [%]
A	266	49.5	NaOH	<1	<4
B	286	49.8	NaOH	<1	<4
C	387	49.0	NaOH	<1	<4
D	410	58.4	NaOH	1.88	0.24

2.3. Wood Treatment

Beech samples, dried in an oven, were impregnated with an aqueous solution of 5%, 10%, 15%, and 20% (*w*/*w*) PF solid content using an autoclave, with ten samples impregnated at each concentration. Before autoclave impregnation, the beech wood samples were subjected to a vacuum for approximately 0.5 h at 0.008 kPa gauge pressure. All samples were then immersed in the PF resin solution and exposed to a gauge pressure of 101.32 kPa for 60 min at standard room temperature (20 °C). After impregnation, all the wood samples were removed from the PF resin solution and placed in an air-circulating environment for 24 h to achieve moisture equilibrium.

2.4. Dynamic Vapor Sorption (DVS)

The sorption behavior was assessed using a DVS intrinsic device (DVS Advantage, Surface Measurement Systems, London, UK). Both the untreated and treated beech samples were milled and sieved through a 2 mm mesh. Approximately 20 mg of wood powder was placed on the sample holder, which was connected to a microbalance via a hanging wire and situated in a thermostatically controlled cabinet. The sorption processes were conducted at a steady temperature of 25 °C. The DVS schedule was set to 20 levels of RH (0%, 5%, 10%, 15%, 20%, 25%, 30%, 35%, 40%, 45%, 50%, 55%, 60%, 65%, 70%, 75%, 80%, 85%, 90%, and 95% in a reverse sequence for the desorption isotherm). The instrument maintained each target RH until the mass change of the sample (dm/dt) was <0.002% min^{-1} over a 10-min span. The target RH, actual RH, sample mass, and running time were recorded every 60 s during the sorption process. The weight at equilibrium condition at each RH was recorded using the microbalance, based on which the EMC was calculated. The EMC of both the unmodified and modified beech samples was calculated based on the following equation:

$$MC = (m_2 - m_1)/m_1 \times 100, \quad (1)$$

where m_1 represents the dry mass of wood samples, and m_2 represents the equilibrium mass of wood at a given RH.

3. Results

3.1. Dynamic Vapor Sorption (DVS) Behavior of Beech Wood

The DVS characteristics of both the unmodified and modified beech samples were investigated to determine the impact of PF resin modification. The response of a wood sample to a change in the set RH produces an asymptotic curve that approaches the EMC after an infinite time of exposure at a given RH. As depicted in Figure 1, the EMC of all the wood samples increased with the RH in the adsorption stage, with the EMC first increasing rapidly before increasing slowly to equilibrium at each humidity level. In the desorption stage, the EMC gradually decreased with diminishing humidity for all samples, with the EMC decreasing rapidly at first and then slowly dropping until reaching equilibrium at each RH level. At the 95% stage, the EMC of the sample increases rapidly and cannot be completely balanced. As evident from Figure 1, the PF resin modifications induced a decline in the measured MC during the sorption process, with the MC decreasing more noticeably with the increasing molecular weight of the PF resin. Compared with the EMC (23.7% at RH 95%) of the unmodified sample, that of the modified beech wood at M_w = 286 g/mol and M_w = 410 g/mol demonstrated a reduced MC (MCR) of 11.73 and 14.26%, respectively. The results suggest that PF resin modification considerably reduces moisture adsorption [15]. Reportedly, low-molecular-weight resins penetrate the wood cell wall more easily, and, ultimately, bulk the wood material to a greater extent, thereby offering a superior dimensional stability; however, high-molecular-weight resins fill only the cell lumina [16,17], which diminishes the adsorption of hydrate water in low RH ranges. Moreover, PF resin can crosslink microfibrils, thereby constraining swelling when modified samples are exposed to humidity. In moist environments, water diffusion into the crosslinked cell wall creates a compressive stress because it cannot swell, contrary to its unmodified counterpart. Furthermore, the incorporation of PF resin into the nanopores of the cell wall pre-swells the cell wall, consequently decreasing the size (free space) of the nanopores as well as the cell wall swelling caused by water adsorption. Eventually, the accommodation for water (primarily dissolved and condensed water) diminishes in the high RH range [18,19].

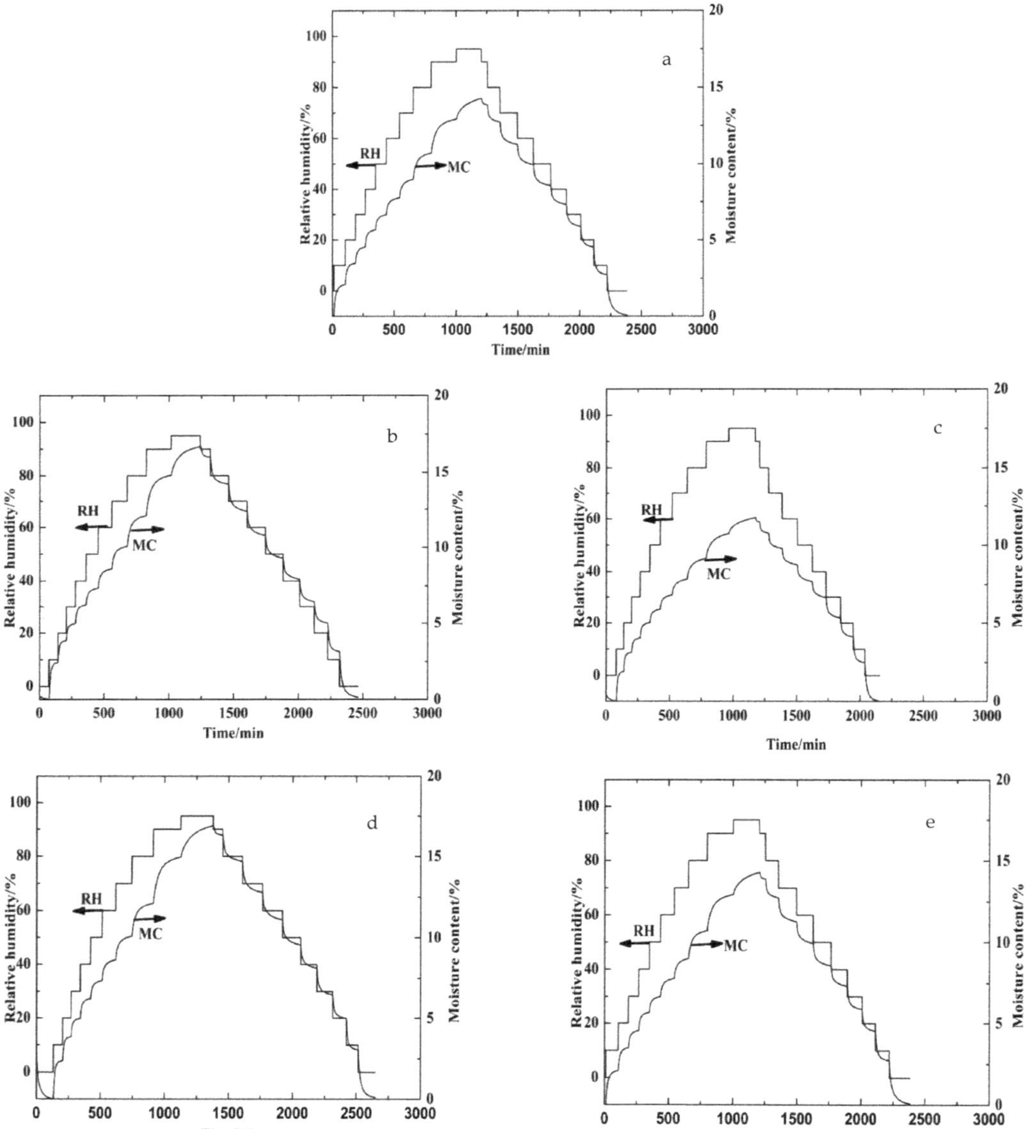

Figure 1. The measured moisture content (MC) of untreated and treated wood samples during sorption under stepwise target relative humidity (RH). (**a**) Untreated beech samples, (**b**) treated with M_w = 266 g/mol, (**c**) treated with M_w = 286 g/mol, (**d**) treated with M_w = 387 g/mol, and (**e**) treated with M_w = 410 g/mol.

3.2. DVS Isotherm Hysteresis Plot

Hygroscopic hysteresis, a common phenomenon observed in the water adsorption process of wood materials, is driven by a reduction in the number of accessible hydroxyl groups within the wood material. The number of available hydroxyl groups in wood that can bind to water molecules decreases with increasing water content, thereby diminishing the EMC value attained through moisture absorption. The modification of wood with the same exposure history with four PF resins of different molecular weights reduced the sorption hysteresis compared to that of unmodified samples (Figure 2). Modified beech samples with a higher molecular weight (M_w = 410 g/mol) displayed less hysteresis at RH levels of 80% and 90% compared with unmodified beech samples. At a lower molecular weight (M_w = 286 g/mol), the sorption hysteresis of the modified beech was reduced across the full hygroscopic range, particularly in the high RH range, and the maximum hysteresis point shifted from 80% RH in the unmodified wood to 70% RH in the modified wood.

The reduction in hysteresis in beech woods can be attributed to the changes in the cell wall structure caused by PF impregnation. The hysteresis effect has been previously interpreted based on sorption phenomena on a glassy solid below the glass transition temperature [20]. In the adsorption stage, the adsorption of the water molecules onto the sorption sites of cell walls is an exothermic process. The thermal motion of incoming water molecules causes the expansion of cell wall nanopores, and thereby creates a new internal surface. During desorption, the relaxation of the cell wall matrices to their state during adsorption is kinetically hindered [21].

Modification with PF resin leads to crosslinking and bulking in the cell walls due to chemical incorporation [22,23]. Crosslinking reduces the degree of freedom of cell wall polymers, and bulking pre-swells the cell wall. Both effects contribute to an increase in wood stiffness after modification, leading to less deformation during the sorption process.

3.3. DVS Isotherm Adsorption and Desorption Plot

During the adsorption process, water molecules sever the hydrogen bonds between cellulose macromolecular chains, producing new free hydroxyl groups, which continue to interact with water molecules to form new hydrogen bonds. Moreover, the distance between molecular chains increases during the adsorption process. In the desorption process, hydrogen bonds between adjacent molecular chains break, and the distance between molecular chains decreases due to the elimination of water molecules. These phenomena result in changes to the physico-mechanical properties of beech samples, ranging from the cellular to the macroscopic level.

Figure 3 shows the adsorption and desorption plots of untreated and treated beech samples. When the RH exceeds 60%–70%, the water adsorption isotherm exhibits an upward-curving trend, displaying an S-shaped curve, indicative of a class II isotherm with multimolecular layer suction characteristics [23]. The EMC of all wood samples increased with the increasing RH. Moreover, the adsorption curve of both the unmodified and modified wood samples steepened with the increasing RH. This phenomenon can be attributed to the adsorption of water vapor molecules onto the adsorption sites on the sample surface, causing the EMC value to increase during the early stage of adsorption. As adsorption persists, a monolayer forms, and a multimolecular layer gradually materializes, causing a swift increase in the equilibrium water content. In the later stage of adsorption, the adsorption gradually saturates, capillary condensation occurs, and the water content increases further. Compared with the modified wood, the unmodified wood samples exhibited a higher EMC upon adsorption and desorption.

The reduction in the EMC can be attributed to the increased quantity of PF resin deposited in beech samples, confirming that the EMC is considerably affected by the presence of PF resin in wood [24–27]. Typically, PF resin does not interact with wood components and deposits as a coating on the internal cell wall structure, as visualized via scanning electron microscopy. The impregnated PF resin can form irreversible hydrogen bonds within hemicellulose and amorphous cellulose, which are dispersed in cell wall voids. The decrease in the EMC of modified beech wood can be primarily attributed to the reduction in sorption sites. In conclusion, the modified beech wood samples exhibited the lowest hygroscopicity.

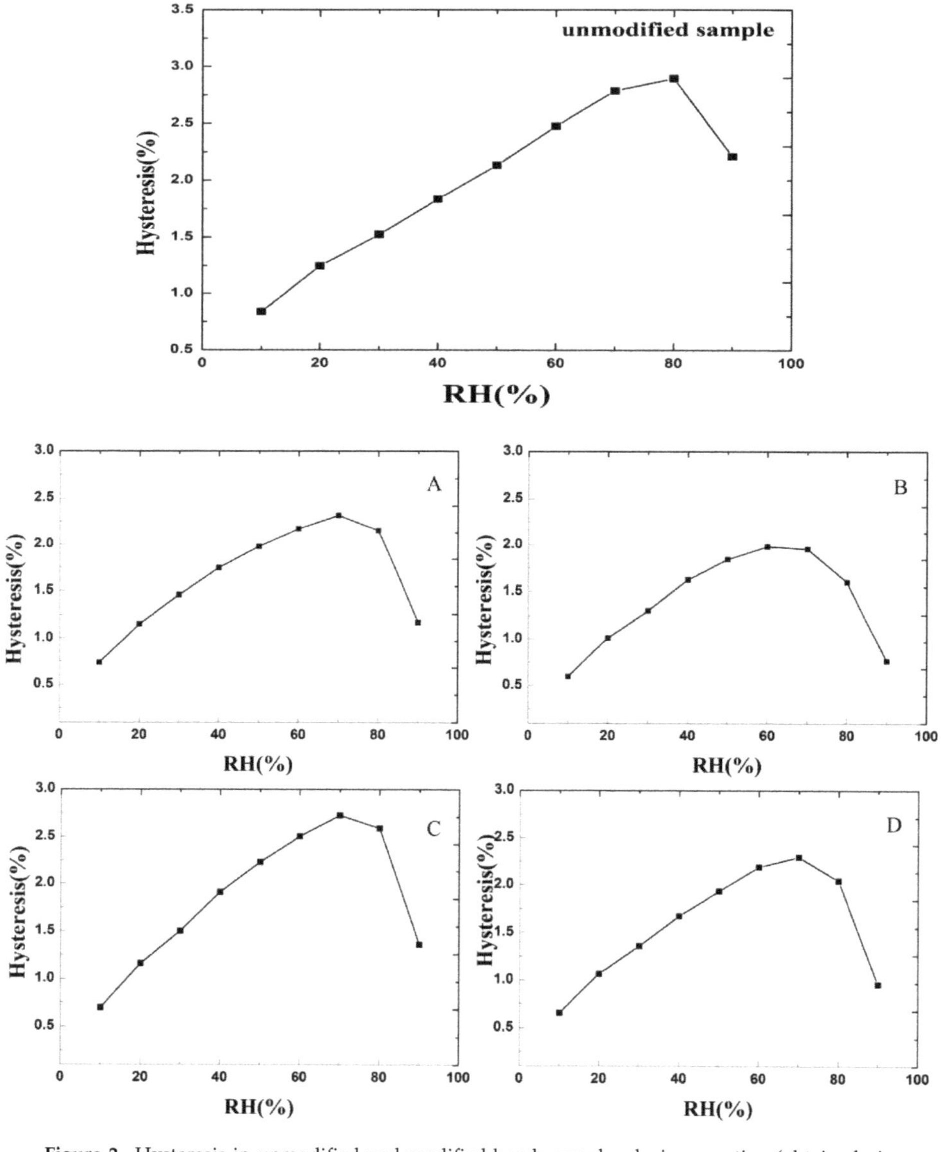

Figure 2. Hysteresis in unmodified and modified beech samples during sorption (obtained via subtracting equilibrium moisture contents between desorption and adsorption at the same RH level). (**A**) Treated with M_w = 266 g/mol, (**B**) treated with M_w = 286 g/mol, (**C**) treated with M_w = 387 g/mol, and (**D**) treated with M_w = 410 g/mol.

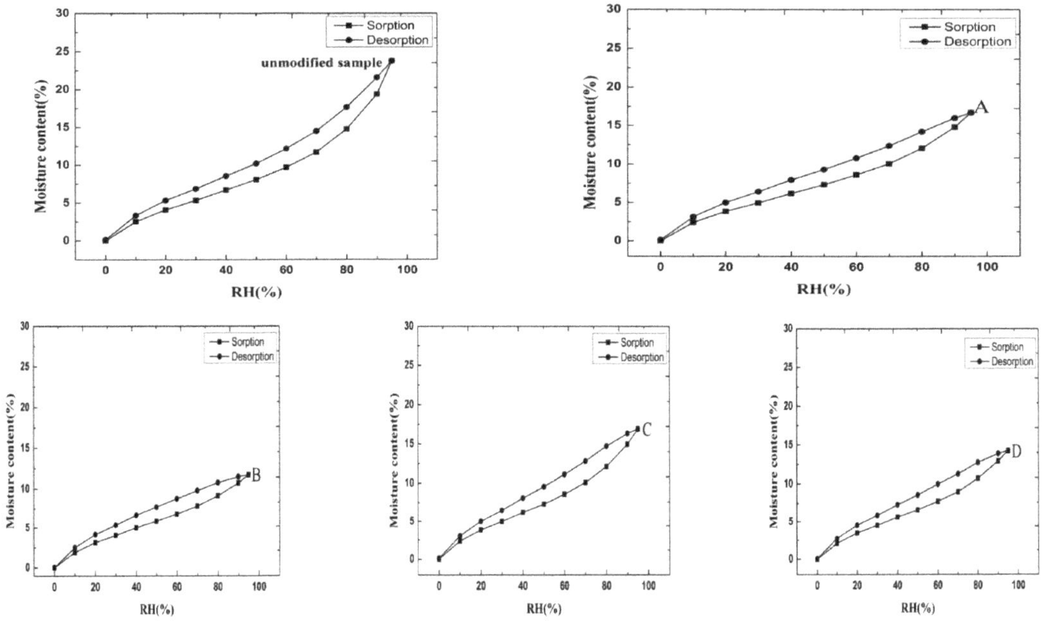

Figure 3. The moisture adsorption and desorption behavior of unmodified wood and modified wood. (**A**) Treated to M_W = 266 g/mol, (**B**) treated to M_W = 286 g/mol, (**C**) treated to M_W = 387 g/mol, and (**D**) treated to M_W = 410 g/mol.

3.4. Mechanism of Adsorption and Desorption of Beech Wood Modified with PF

The process of wood production is extensive and subject to fluctuating ambient temperature and humidity, resulting in wood products undergoing varying moisture absorption or desorption processes and dynamic water-content changes. Figure 4 depicts the mechanism of adsorption and desorption in beech wood modified using PF. Beech is a crucial broadleaf wood type. In broadleaf wood, a higher density corresponds to thicker wood fiber cell walls, smaller cell cavities, and a lower total porosity, resulting in slower water movement. The moisture absorption rate serves as a vital parameter to characterize the moisture absorption capacity of the wood. The measurement and investigation of this rate is critical to understanding the wood adsorption mechanism.

The moisture expansion rate of wood increases with the increasing RH of its environment. This can be attributed to the minimal variations in MC during the hygroscopic stage at a low RH, resulting in a thin hygroscopic water layer between the wood cell wall microfibers and minor alterations in the distance between microfibers. However, when the RH exceeds 95%, a substantial quantity of water infiltrates the space between the wood microfibrils, increasing the thickness of the water layer and resulting in a sharp increase in the distance between microfibrils. Therefore, the wet expansion rate at a high RH is substantially higher than that at a low RH, and large variations in water content remarkably increase the wet expansion rate and reduce the dimensional stability. At each stage of moisture absorption, the partial pressure of the water vapor inside the container is higher than that on the wood surface, inducing rapid water vapor absorption by the wood and a swift increase in the MC. During the hygroscopic process, the partial pressure difference of water vapor decreases, and the hygroscopic rate of wood gradually decreases until dynamic equilibrium is achieved. In all cases, the maximum adsorption rate occurs in the initial stage.

In the desorption process, the partial pressure of water vapor on the wood surface exceeds that inside the container, enabling wood to rapidly expel water, which results in an increased desorption rate. As the desorption process continues, the partial pressure difference in the water vapor decreases, gradually reducing the desorption rate of the wood.

Initially, water molecules form a single layer of molecular water adsorption by interacting with polar groups at the adsorption sites. However, the number of adsorption sites on cellulose and hemicelluloses inside wood is limited, with most binding sites occupied at a low RH. When the RH increases, the single-layer molecular adsorption curve remains relatively gentle owing to the limited adsorption capacity. The single-layer molecular adsorbed water then interacts with water molecules again to form multilayer molecular adsorbed water, which increases with the increasing RH until it reaches equilibrium with the single-layer molecular adsorbed water. As the RH increases, the adsorption capacity of multilayer molecules increases rapidly, causing a swift ascent in the multilayer adsorption curve. Therefore, studying the dynamic change in the MC of beech wood under different environmental temperatures, humidities, and moisture absorption processes is vital to guide the use, storage, and drying of wood products.

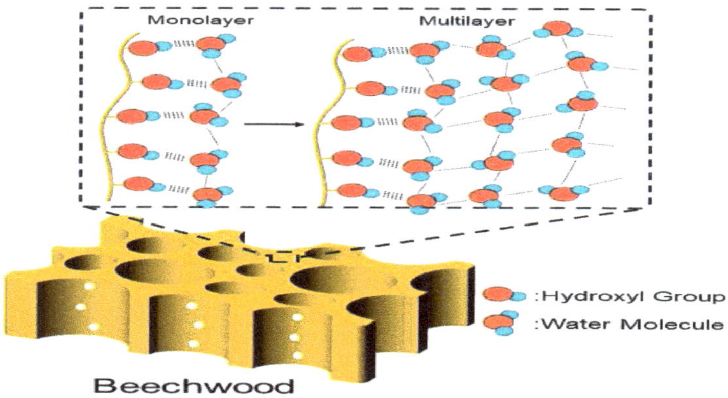

Figure 4. The mechanism of adsorption and desorption in PF-treated beech samples.

4. Conclusions

The modification of beech wood using PF resins of various molecular weights led to a decrease in the EMC. The results demonstrated that, compared with unmodified beech samples, all PF resins of different molecular weights exhibited a reduced sorption hysteresis. Furthermore, all sorption isotherms for unmodified wood and modified beech wood with PF resins of different molecular weights displayed characteristic sigmoidal shapes. The decrease in the EMC value of the modified beech samples can be attributed to the deposition of PF resin in the nanopores of the beech wood cell walls, which reduces sorption sites.

Furthermore, the results revealed that the water content absorbed by the sample increases in stages with increasing humidity. In each RH stage, the water content of the sample initially increases rapidly, then decreases gradually until equilibrium is attained. In the desorption stage, the MC of the sample decreases in stages with decreasing humidity. In each RH stage, the MC of the sample rapidly declines, then gradually decreases until equilibrium, in contrast to the adsorption stage. Impregnating beech samples with PF resin improves their hygroscopicity. DVS can provide valuable insights into the structure and cell wall composition of beech wood fibers during their development.

Author Contributions: Conceptualization, Q.L. and V.B.; methodology, Q.L. and V.B.; software, Q.L. and V.B.; validation, H.M.; formal analysis, H.M.; investigation, Q.L. and V.B.; resources, H.M.; data curation, H.M.; writing—original draft preparation, Q.L.; writing—review and editing, Q.L.; visualization, Q.L. and V.B.; supervision, H.M.; project administration, Q.L.; funding acquisition, Q.L. All authors have read and agreed to the published version of the manuscript.

Funding: The research was funded by Heilongjiang Province Natural Science Foundation (LH2020C002), Heilongjiang Province Postdoctoral Science Foundation (LBH-Z17029), China Scholarship Council (CSC) and the High-end Foreign Experts Introduction Program (G2022011017L).

Institutional Review Board Statement: Not applicable.

Data Availability Statement: All data included in this study are available upon request via contact with the corresponding author.

Acknowledgments: The authors thank colleagues in Georg-August University of Göttingen for giving good suggestions regarding experiments and the manuscript.

Conflicts of Interest: The authors declare no conflict of interest.

References

1. Xie, Y.; Hill, C.A.; Xiao, Z.; Jalaludin, Z.; Militz, H.; Mai, C. Water vapor sorption kinetics of wood modified with glutaraldehyde. *J. Appl. Polym. Sci.* **2010**, *117*, 1674–1682. [CrossRef]
2. Beck, G.; Strohbusch, S.; Larnøy, E.; Militz, H.; Hill, C. Accessibility of hydroxyl groups in anhydride modified wood as measured by deuterium exchange and saponification. *Holzforschung* **2018**, *72*, 17–23. [CrossRef]
3. Xie, Y.; Hill, C.A.; Xiao, Z.; Mai, C.; Militz, H. Dynamic water vapour sorption properties of wood treated with glutaraldehyde. *Wood Sci. Technol.* **2011**, *45*, 49–61. [CrossRef]
4. Gabrielli, C.P.; Kamke, F.A. Phenol–formaldehyde impregnation of densified wood for improved dimensional stability. *Wood Sci. Technol.* **2010**, *44*, 95–104. [CrossRef]
5. Ghavidel, A.; Hosseinpourpia, R.; Militz, H.; Vasilache, V.; Sandu, I. Characterization of Archaeological European White Elm (*Ulmus laevis* P.) and Black Poplar (*Populus nigra* L.). *Forests* **2020**, *11*, 1329. [CrossRef]
6. Aicher, S.; Zachary, C.; Maren, H. Rolling shear modulus and strength of beech wood laminations. *Holzforschung* **2018**, *70*, 773–781. [CrossRef]
7. Gardiner, B.; Barnett, J.; Saranpää, P. *The Biology of Reaction Wood*; Springer: Berlin/Heidelberg, Germany, 2014.
8. Hill, C.; Moore, J.; Jalaludin, Z.; Leveneu, M.; Mahrdt, E. Influence of earlywood/latewood and ring position upon water vapour sorption properties of Sitka spruce. *Int. Wood Prod. J.* **2011**, *2*, 12–19. [CrossRef]
9. Emmerich, L.; Altgen, M.; Rautkari, L.; Militz, H. Sorption behavior and hydroxyl accessibility of wood treated with different cyclic N-methylol compounds. *J. Mater. Sci.* **2020**, *55*, 16561–16575. [CrossRef]
10. Kurkowiak, K.; Emmerich, L.; Militz, H. Sorption behavior and swelling of citric acid and sorbitol (SorCA) treated wood. *Holzforschung* **2021**, *75*, 1136–1149. [CrossRef]
11. Hailwood, A.J.; Horrobin, S. Absorption of water by polymers: Analysis in terms of a simple model. *Trans. Faraday Soc.* **1946**, *42*, B084–B092. [CrossRef]
12. Hill, C.A.; Norton, A.; Newman, G. The water vapor sorption behavior of flax fibers—Analysis using the parallel exponential kinetics model and determination of the activation energies of sorption. *J. Appl. Polym. Sci.* **2010**, *116*, 2166–2173. [CrossRef]
13. Krabbenhoft, K.; Damkilde, L. A model for non-Fickian moisture transfer in wood. *Mater. Struct.* **2004**, *37*, 615–622. [CrossRef]
14. Xie, Y.; Hill, C.A.; Jalaludin, Z.; Sun, D. The water vapour sorption behaviour of three celluloses: Analysis using parallel exponential kinetics and interpretation using the Kelvin-Voigt viscoelastic model. *Cellulose* **2011**, *18*, 517–530. [CrossRef]
15. Murakami, S.; Maeda, H.; Imamura, S. Fuzzy decision analysis on the development of centralized regional energy control system. *IFAC Proc. Vol.* **1983**, *16*, 363–368. [CrossRef]
16. Rodwell, M.J.; Rowell, D.P.; Folland, C.K. Oceanic forcing of the wintertime North Atlantic Oscillation and European climate. *Nature* **1999**, *398*, 320–323. [CrossRef]
17. Kajita, H.; Imamura, Y. Improvement of physical and biological properties of particleboards by impregnation with phenolic resin. *Wood Sci. Technol.* **1991**, *26*, 63–70. [CrossRef]
18. Shams, M.I.; Yano, H.; Endou, K. Compressive deformation of wood impregnated with low molecular weight phenol formaldehyde (PF) resin I: Effects of pressing pressure and pressure holding. *J. Wood Sci.* **2004**, *50*, 337–342. [CrossRef]
19. Xiao, Z.; Xie, Y.; Militz, H.; Mai, C. Modification of wood with glutaraldehyde. In Proceedings of the 4th European Conference on Wood Modification, Stockholm, Sweden, 13 May 2009.
20. Hill, C.A.; Norton, A.; Newman, G. The water vapor sorption behavior of natural fibers. *J. Appl. Polym. Sci.* **2009**, *112*, 1524–1537. [CrossRef]
21. Lu, Y.; Pignatello, J.J. Demonstration of the "conditioning effect" in soil organic matter in support of a pore deformation mechanism for sorption hysteresis. *Environ. Sci. Technol.* **2002**, *36*, 4553–4561. [CrossRef]

22. Yasuda, R.; Minato, K. Chemical modification of wood by non-formaldehyde cross-linking reagents: Part 3. Mechanism of dimensional stabilization by glyoxal treatment and effect of the addition of glycol. *Wood Sci. Technol.* **1995**, *29*, 243–251. [CrossRef]
23. Hosseinpourpia, R.; Adamopoulos, S.; Holstein, N.; Mai, C. Dynamic vapour sorption and water-related properties of thermally modified Scots pine (*Pinus sylvestris* L.) wood pre-treated with proton acid. *Polym. Degrad. Stab.* **2017**, *138*, 161–168. [CrossRef]
24. Brunauer, S.; Deming, L.S.; Deming, W.E.; Teller, E. On a theory of the van der Waals adsorption of gases. *J. Am. Chem. Soc.* **1940**, *62*, 1723–1732. [CrossRef]
25. Liu, M.; Guo, F.; Wang, H.; Ren, W.; Cao, M.; Yu, Y. Highly stable wood material with low resin consumption via vapor phase furfurylation in cell walls. *ACS Sustain. Chem. Eng.* **2020**, *8*, 13924–13933. [CrossRef]
26. Shen, X.; Guo, D.; Jiang, P.; Yang, S.; Li, G.; Chu, F. Water vapor sorption mechanism of furfurylated wood. *J. Mater. Sci.* **2021**, *56*, 11324–11334. [CrossRef]
27. Lang, Q.; Biziks, V.; Militz, H. Influence of Phenol–Formaldehyde Resin Oligomer Molecular Weight on the Strength Properties of Beech Wood. *Forests* **2022**, *13*, 1980. [CrossRef]

forests

MDPI

Communication

Timber Properties and Cellulose Crystallites Size in Pine Wood Cut in Different Sawing Patterns after Pretreatment with CH_3COOH and H_2O_2 and Densification

Andi Detti Yunianti *, S Suhasman, A Agussalim, Musrizal Muin and Heru Arisandi

Faculty of Forestry, Hasanuddin University, South Sulawesi, Makassar 90245, Indonesia;
suhasman@yahoo.com (S.S.); agussalim.madjid@yahoo.co.id (A.A.); musrizal@yahoo.com (M.M.);
heru.arisandi@yahoo.com (H.A.)
* Correspondence: dettiyunianti70@yahoo.com

Abstract: One process to improve wood quality is densification or wood surface compression. Our study analyzed the changes in some basic properties of pine wood, including its anatomical structure, density, modulus of elasticity (MOE), and dimensions of cellulose crystallites, after densification following soaking pretreatment in CH_3COOH and H_2O_2 at a concentration of 20%. Samples were sawn in radial and tangential directions for analysis of the wood. The results showed a change in the shape of tracheid cells from hexagonal to oval, as well as damage to the ray cell constituents on the tangential surface. The thickness decrease of the samples was in accordance with the target, which meant that spring-back was short. In general, the tangential boards had a higher density than the radial boards, with a lower MOE and crystallite dimensions. Our findings showed that the densified tangential board was stronger than the radial board.

Keywords: densification; soaking; CH_3COOH; H_2O_2; radial; tangential

1. Introduction

The availability of wood from natural forests is becoming limited, which has affected forest product industries and led to increased use of wood from plantation forests. However, wood from plantation forests is known to have lower quality than wood from natural forests because it is generally harvested at a young age. Consequently, it has a high portion of juvenile wood and thus low durability.

According to Panshin and de Zeeuw [1] and Bowyer et al. [2], juvenile wood is characterized by low specific gravity and density, short fibers, a large amount of longitudinal shrinkage, and a mostly flatter microfibril angle. It also has less dimensional stability and lower amounts of the chemical components that compose cellulose, but it has a higher lignin content than mature wood. Owing to these characteristics, the processing is needed to improve the quality of wood from fast-growing wood species from plantation forests.

One of the processes for improving the quality of wood is densification or wood surface compression. According to Neyses [3], densification is the process of compressing the wood surface, which results in increased density and hardness of the wood surface in part or all of the densified material. One of the parts of the densification process is the pre-treatment before the densification process.

Neyses et al. [4] reported that pretreatment with a solution of sodium silicate, sodium hydroxide, methacrylate resin, and an ionic liquid increased the surface hardness of pine wood. Previous research by Yunianti et al. [5] showed that pretreating wood by soaking it in a mixture of CH_3COOH and H_2O_2 increased wood density. They also found that modifying temperature and time parameters within the densification process resulted in different responses in pine and gmelina wood. Pretreatments using high-temperature immersion [6] and thermal compression under several conditions [7] also influenced wood modification.

In addition to the increase in density, densification changes the anatomical structure of the wood. In particular, it alters the shape of vessels and causes tracheids to become more oval [8–10], which can affect wood properties. However, it is important to consider that wood is an anisotropic material that generally has different behaviors and responses based on the three directions of the wood grain. Grain direction refers to the longitudinal/axial direction, which is parallel to the tree trunk axis, and the radial and tangential directions, which are perpendicular to the tree trunk axis. The differences in the behaviors and responses of wood are caused by the arrangement of cells wood grain. Cell arrangement affects the anatomical structure of wood, which in turn affects the other wood properties. The anisotropy of wood is particularly notable in light of the sawing patterns used to cut logs into wood boards. Previous studies on wood boards used as samples in testing did not report the sawing pattern, and this omission does not allow for determining the different responses of a wood type receiving the same pre-treatment before the densification process.

In addition, the increase of wood density in the densification process is thought to be influenced by the cell wall ultrastructure, including the microfibril angle and cellulose crystallites. According to Stuart and Evans [11], Butterfield [12], Saranpää [13], Peura et al. [14], Lachenbruch et al. [15], Yin et al. [16], and Tabet and Fauziah [17], the microfibril angle and cellulose crystallites are the main factors affecting the physical properties of wood (especially density and shrinkage), its mechanical properties (especially tensile strength and stiffness), and its chemical content.

Knowledge is needed on the changes in the characteristics of cellulose crystallites after the densification process. Such knowledge could help determine the success of the densification process. It could also explain different responses, especially in wood that has the same density or the same cell wall thickness. In addition, the sawing pattern, along with different cell types, will affect how wood responds to treatment before and during the densification process. Therefore, this study assessed the changes in some basic properties of wood, including anatomical structure, density, modulus of elasticity (MOE), and dimensions of cellulose crystallites, after the densification process. Wood was cut with different sawing patterns and underwent a pre-soaking treatment in a solution of CH_3COOH and H_2O_2, following previous research by Yunianti et al. [5].

2. Materials and Methods

2.1. Wood Materials

This study used pinewood (*Pinus merkusii*) that was cut into 1-m-long logs, which were subsequently cut into tangential and radial boards. The size of each board was 100 cm (L) × 25 cm (W) × 2 cm (H). Boards were then cut into defect-free samples that were 30 cm (L) × 25 cm (W) × 2 cm (H) in size. Ten samples were prepared for each tangential and radial board, of which five samples were densified and did not treat the others.

2.2. Densification Process

Densification was done following the process reported by Yunianti et al. [5]. The wood samples were pre-treated by soaking in a solution of CH_3COOH and H_2O_2 at a concentration of 20%. Afterward, the pre-treated samples were densified under a pressure of 35 kg/cm^2 at 150 °C for 30 min to obtain a thickness reduction of about 30%. All the densified wood samples were conditioned for 24 h in a desiccator and stored until use in the following experiments. Untreated wood samples were also prepared for use as a control group.

2.3. Sample Preparation

The untreated and densified treated wood samples were cut into various sizes depending on the experiments in which they were to be used. For microscopic observation and measurement of density and set recovery, samples were cut to 2 cm × 2 cm ×

1.4 cm. The samples for MOE testing were 2 cm × 1.4 cm × 30 cm. In addition, for the measurement of cellulose crystallites dimensions, samples were ground to pass 40 mesh.

2.4. Microscopic Observation

Microscopic observation was undertaken to determine changes in cell shape following the same preparations method as in the analysis of wood anatomy. The observations were made with regard to the three directions of the wood grain (axial, tangential, and radial) from the radial and tangential boards.

2.5. Recovery of Set

The measurement of set recovery was carried out on densified wood samples with a thickness reduction target of 30%. The set recovery (SR) is defined as:

$$\text{SR}\ (\%) = \frac{R\prime_C - R_C}{R_0 - R_C}$$

where

R_0 = Oven-dry dimension of the specimen in the densification direction before densification (mm),
R_C = Dimension of the specimen after densification (mm),
$R\prime_C$ = Dimension of the densified sample after the wet-dry cycling (mm).

2.6. Wood Density Measurement

The density was measured under air-dry conditions according to British Standard Methods No. 373 [18] with the modification of sample size due to the changes in thickness that occurred after the densification process. The measurement of wood used the dip method. The density (ρ) was calculated as follows:

$$\rho = \frac{m}{v}$$

where

m = weigth of sample under air-dry conditions (g),
v = volume of sample under air-dry conditions (cm^3).

2.7. Modulus of Elasticity Measurement

The MOE was determined under air-dry conditions according to ASTM D 143-94 [19] with a modification of sample size due to changes in thickness that occurred after the densification process. The equation for calculating MOE was as follows:

$$\text{MOE} = \frac{\Delta PL^3}{4\Delta\delta \text{bd}^3}$$

where

ΔP = loading increment in the proportional section of the load deflection diagram (elastic zone) (kg);
L = span (cm);
$\Delta\delta$ = deflection measured at mid-span (cm); b = width of sample (cm); d = thickness of sample (cm).

2.8. Measurement of Dimensional Crystallite

The measurement of dimensional crystallites used X-ray, with CuK (λ = 0.154 nm) as the radiation source at 40 kV and 30 mA with a diffraction angle (2θ) and measured between 20° until 40° with a speed at 2°/min. The crystallite width was measured at reflection 200 (2θ = 24.4°) and the length was measured at reflection 004 (2θ = 34.4°).

The average width and length of crystallites (D_{hkl}) were calculated according to the Scherrer formula [14]:

$$D_{hkl} = \frac{K\lambda}{\beta \cos \theta hkl}$$

where D_{hkl} = dimensions of the crystallites (nm); λ = wavelength targets (copper = 0.154 nm); β = FWHM in radians; θ = half of the scattering angle $2\theta_{hkl}$; K = Constant for graphite sheets (0.9 to for reflection 200 and 1.0 reflection 004).

3. Results and Discussion

Results of this study indicated changes in the anatomical structure, thickness (set of cover), density, and MOE of radial and tangential boards as an effect of the densification treatment. Different sawing patterns were associated with different changes in several characteristics of the basic properties of pinewood.

3.1. Changes in Anatomical Structure

According to Frans [20], a solution of H_2O_2 and CH_3COOH at a certain concentration can soften the wood and separate the individual cells of wood fibers. In our study, samples soaked in a solution of H_2O_2 and CH_3COOH at a concentration of 20% for 24 h at 80 °C prior to a densification process that used a pressure of 35 kg/cm^2 at a temperature of 150 °C for 30 min, causing changes in the pine wood constituent cells. The hexagonal tracheids became oval shaped on the axial, radial, and tangential surfaces (Figure 1), and the ray cells were damaged on the tangential surface (Figure 2). According to Brauns and Rocens [21] and Yunianti and Suhasman [10], during the densification process tracheid/vessel cells become oval shaped because of the compression of the cell wall.

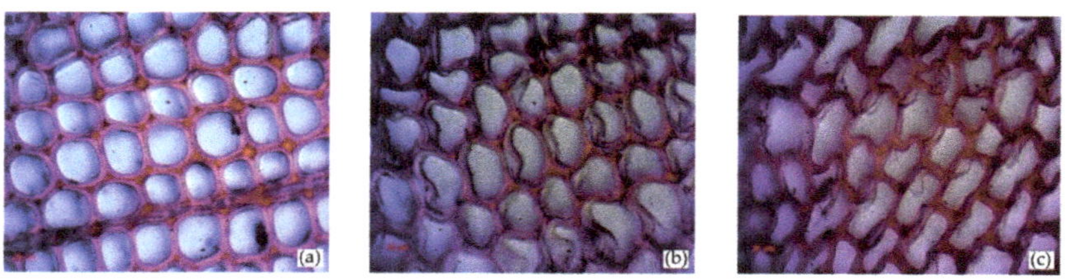

Figure 1. The changes of tracheid cell, (**a**) control, (**b**) radial surface, and (**c**) tangential surface, under high magnification (×40).

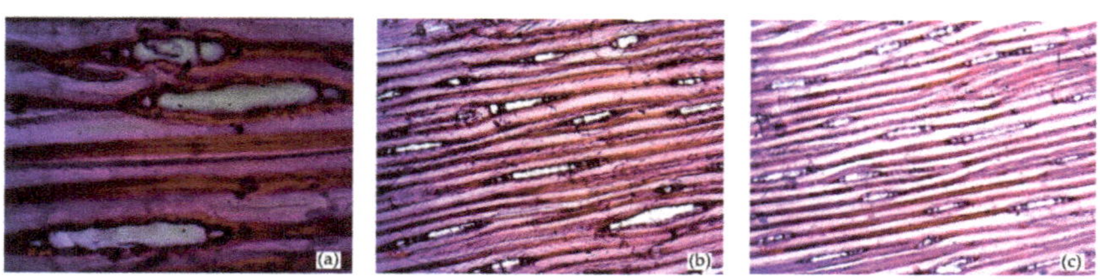

Figure 2. The damage of the ray cells on the tangential surface of the radial and tangential boards (**a**) under high magnification (×40), and (**b**,**c**) under low magnification (×10).

In the tangential board, the pressure was applied in the radial direction, where the transverse ray cells are present. The damage mostly affected the tracheid cells, which have a larger diameter than the ray cells. Owing to the larger diameter, the tracheids have more space and therefore become denser when pressed. In the radial boards, the pressure was applied in the tangential direction. In general, the damage or flattening was related to the ray cells, which have a small diameter. Therefore, less free space was lost when the board was pressed.

When the pressure occurs in the radial direction, the strength is in the ray cell, and when the pressure occurs in the tangential direction, the strength is in the sapwood [22]. Consequently, wood in the radial and tangential boards responded differently during the densification process. In addition, the damage to cells and the changes to wood properties were also different. According to Nairn [23], during the pressing process, the response of the wood depends on the density, percentage of sapwood, the volume of the ray cells, and the direction of the pressure surface. In the current study, the first damage occurred in the sapwood and continued to the whole layer.

3.2. Thickness Changes (Set of Recovery)

According to Neyses et al. [4], set of recovery is a dimensional change in the thickness of densified wood, measured from the initial dimensions and the dimensions after removal of the deformation. The set of recovery can be reduced by chemical modification. In our study, pre-treatment was based on soaking the wood samples in a 20% solution of H_2O_2 and CH_3COOH before the densification process softened the wood due to the separation of individual cells. Based on previous research by Yunianti et al. [5], the 20% concentration was associated with the best quality of densified wood compared with the other concentrations.

In our study, pre-treatment based on soaking the wood samples in a 20% solution of H_2O_2 and CH_3COOH before the densification process softened the wood due to the separation of individual cells. Based on previous research Yunianti et al. [5], the 20% concentration was associated with the best quality of densified wood compared with the other concentrations.

The thickness changes of the radial boards and the tangential boards of pinewood after the densification process were in accordance with the target of thickness reduction of 30%. The decreased thickness of the tangential board was 32%, whereas the radial board thickness decreased by 31%. This outcome indicates that the densified pinewood did not have much spring back and tended to be stable. According to Rautkari [24], Hartono et al. [25], and Báder and Németh [26], some wood component cells sustain shape changes during the densification process. If the cells return to their original shape when the pressure is removed, spring back has occurred.

3.3. Density

Figure 3 shows that wood density increased from 11.76% to 18.91% in the radial and tangential boards, respectively. This research result is in line with a previous report by Mania et al. [27], who found that the density of poplar and birch wood increased after the densification process.

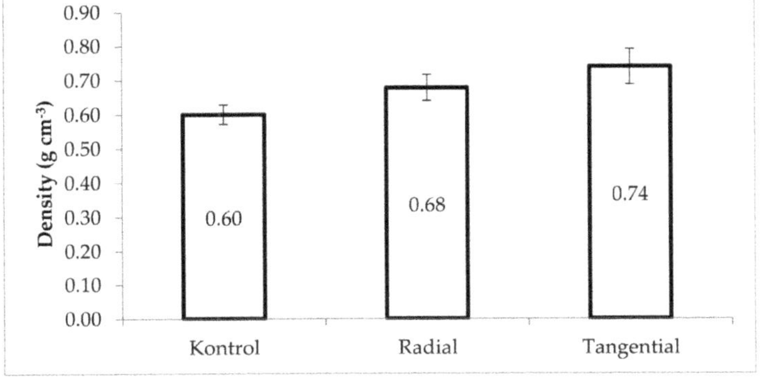

Figure 3. The Density of Pine Wood from Radial Board and Tangential Board.

According to Rautkari [24], a shorter time in the pressing process along with a lower temperature will cause the density on the wood surface to increase. In the current study, the densification process at a temperature of 150 °C for 30 min generated a change in the density of radial and tangential boards.

The increased density was associated with the different sawing patterns, which were closely related to the changes in the densified wood cells. The increased number of cells per unit of the same area indicated the cause of the density change in the densified wood. The damage to the tracheid cells caused the increased number of cells per unit of the same area, and therefore, the tangential board was found to have a higher density than the radial board. Meanwhile, since the ray cells became the most damaged cell on the radial board the area with these cells was smaller than in the tangential board. The increase in density is influenced by the anisotropic properties of wood [22].

3.4. Changes in MOE

Figure 4 shows that MOE of the radial and the tangential boards increased after pinewood was pretreated with 20% H_2O_2 and CH_3COOH and densified.

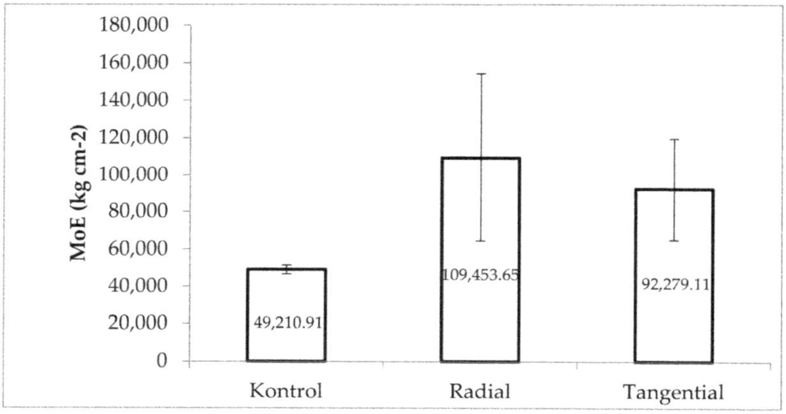

Figure 4. MOE of densified Pinewood on The Radial Board and The Tangential Board.

According to Brauns and Rocens [21], the different relations between wood elasticity and densified wood occurs in longitudinal, radial, and tangential directions. The strength increases as the wood elasticity increases. The relation varies depending on the type of wood and the method of densification used. In our study, the MOE value increased in the radial and tangential boards, which was in line with the increase in the wood density. According to Blomberg [28], semi-isostatic densified wood resulted in low cracking rates and high strength, even in an irregular shape. The deformation is determined by the anisotropic properties of wood.

3.5. Changes in Dimensions of Crystallites

Cellulose crystallites affect the physical, mechanical, and chemical properties of lignocellulosic materials. The crystalline part of the wood is the fraction of the densely crystalline material of cellulose microfibrils, and it is affected by the length and width of the crystallites [29]. The crystallite dimensions have been reported for some conifers by Andersson et al. [29], Anderson [30], and Peura et al. [14], and their results showed that the crystallite length ranges from 6.50 to 36.40 nm, and the width ranges from 2.5 to 3.60 nm. Our study showed that the width and length of the crystallites differed between radial boards and tangential boards following pre-treatment and densification (Table 1).

Table 1. Changes in crystallite dimensions in radial and tangential boards after densification.

	Crystallite Width (nm)	
Sawing Pattern	**Before Densification**	**After Densification**
Radial	2.22 ± 1.25	1.30 ± 0.16
Tangential		1.31 ± 0.19
	Crystallite Length (nm)	
Sawing Pattern	**Before Densification**	**After Densification**
Radial	6.14 ± 3.88	3.78 ± 0.26
Tangential		3.26 ± 0.18

Changes in the dimensions of the crystallites (i.e., length and width) can especially indicate changes in the density of the wood. The smaller the cellulose crystallite dimensions, the higher the density. These changes may be related to the chemical components of cell walls, especially cellulose. According to Panshin and de Zeeuw [1], crystallites are part of the crystalline microfibrils formed from cellulose chains. In addition, according to Kutnar and Sernek [31], chemical densification causes changes in the chemical components of cell walls and is formed as covalent bonds. The soaking treatment of H_2O_2 and CH_3COOH caused the composition of the chemical components in cells to change, which led to changes in the dimensions of the crystallites in the cell walls. In this research, the change in the dimensions of the cellulose crystallites was associated with the increased density of the radial and tangential boards.

4. Conclusions

In association with pre-treatment using a solution of CH_3COOH and H_2O_2 at a concentration of 20% before the densification process, the radial and tangential boards showed changes in the shape of tracheid cells from hexagonal to oval and damage to the ray cell constituents on the tangential surface. The thickness decreased in accordance with the target amount (30%), indicating the occurrence of short spring back. In general, tangential boards have a higher density (0.74 g/cm^3) than radial boards (0.68 g/cm^3), with a lower MOE and smaller crystallite dimensions. Therefore, a tangential board is stronger than a radial board after the densification process.

Author Contributions: Conceptualization, A.D.Y. and S.S.; methodology, A.D.Y. and S.S.; formal analysis, A.A.; investigation, A.A. and H.A.; resources, H.A.; data curation, A.A. and H.A.; writing—original draft preparation, A.D.Y.; writing—review and editing, M.M. and S.S.; visualization, A.A.; supervision, M.M.; project administration, A.D.Y. All authors have read and agreed to the published version of the manuscript.

Funding: This research was funded by Deputy for Strengthening Research and Development, Ministry of Research and Technology/National Research and Innovation Agency grant number 752/UN4.22/PT.02.00/2021.

Institutional Review Board Statement: Not applicable.

Informed Consent Statement: Not applicable.

Data Availability Statement: Not applicable.

Acknowledgments: The authors sincerely thank all the parties who have helped and supported this research. Special thanks to the Deputy for Strengthening Research and Development, Ministry of Research and Technology/National Research and Innovation Agency, which provided funding through the University Priorities Basic Research Grant.

Conflicts of Interest: The authors declare no conflict of interest.

References

1. Panshin, A.J.; de Zeeuw, C. *Textbook of Wood Technology*; McGraw Hill Book, Co.: New York, NY, USA, 1980; Volume I.
2. Bowyer, J.L.; Shmulsky, R.; Haygreen, J.G. *Forest Products and Wood Science: An Introduction*, 4th ed.; Iowa State Press: Ames, IA, USA, 2003.
3. Neyses, B. Surface Densification of Solid Wood Paving the Way towards Industrial Implementation. Ph.D. Thesis, Division of Wood Technology, Luleå University of Technology, LTU, Skellefteå, Sweden, 2019.
4. Neyses, B.; Rautkari, L.; Yamamoto, A.; Sandberg, D. Pre-treatment with Sodium Silicate, Sodium Hydroxide, Ionic Liquids or Methacrylate Resin to Reduce the Set-Recovery and Increase the Hardness of Surface-Densified Scots Pine. *iForest Biogeosci. Forest.* **2017**, *10*, 857–864. [CrossRef]
5. Yunianti, A.D.; Tirtayasa, P.K.; Suhasman, S.; Taskirawati, I.; Agussalim, A.; Muin, M. Modified Densification Process for Increasing Strength Properties of Pine and Gmelina Wood from Community Forests. *J. Korean Wood Sci. Technol.* **2019**, *47*, 418–424.
6. Phebryanti, S. Coconut Wood as Raw Material for Furniture at House in Public Area. *J. Intra* **2015**, *3*, 53–56. (In Indonesian)
7. Lee, J.M.; Lee, W.H. Dimensional Stabilization through Heat Treatment of Thermally Compressed Wood of Korean Pine. *J. Korean Wood Sci. Technol.* **2018**, *46*, 471–485.
8. Darwis, A. Fiksasi Kayu Agathis dan Gmelina Terpadatkan pada Arah Radial serta Observasi Struktur Anatominya. Master's Thesis, Institut Pertanian Bogor, Bogor, Indonesia, 2008.
9. Hadiyane, A. Perubahan Sifat-sifat Komponen Penyusun Kayu, Struktur Sel Kayu dan Sifat-sifat Dasar Kayu Terdensifikasi secara Parsial. Ph.D. Thesis, Sekolah Pascasarjana, Institut Pertanian Bogor, Bogor, Indonesia, 2011. Unpublished work.
10. Yunianti, A.D.S. Perubahan mikroskopis kayu sengon dan kayu agathis hasil proses densifikasi dengan modifikasi awal. In *Prosiding Seminar Nasional Teknologi & Inovasi Industri*; Kementerian Perindustrian, Badan Penelitian dan Pengembangan Industri, BRSI: Banjar Baru, Indonesia, 2018.
11. Stuart, S.A.; Evans, R. X-ray diffraction estimation of microfibril angle variation in eucalypt increment cores. In *The CRC for Hardwood Fibre & Paper Science*; Research Report; University of Melbourne: Melbourne, Australia, 1994.
12. Butterfield, B.G. Wood Anatomy in Relation to Wood Quality. In *Wood Quality and Its Biological Basis*; Barnett, J.R., Jeronimidis, G., Eds.; Blackwell Publishing: Oxford, UK, 2003; pp. 30–52.
13. Saranpää, P. Wood Density and Growth. In *Wood Quality and Its Biological Basis*; Barnett, J.R., Jeronimidis, G., Eds.; Blackwell Publishing: Oxford, UK, 2003; Chapter 4.
14. Peura, M.; Müller, M.; Vainio, U.; Sarén, M.; Saranpää, P.; Serimaa, R. X-ray Microdiffraction Reveals the Orientation of Cellulose Microfibrils and the Size of Cellulose Crystallites in Single Norway Spruce Tracheids. *Trees* **2008**, *22*, 49–61. [CrossRef]
15. Lachenbrunch, B.; Johnson, G.R.; Downes, G.M.; Evans, R. Relationships of Density, Microfibril Angle and Sound Velocity with Stiffness and Strength in Mature Wood of Douglas-fir. *Can. J. For. Res.* **2010**, *40*, 55–64. [CrossRef]
16. Yin, Y.; Xiaomei, J.; Kunlin, S. Influence of Microfibril Angle on Within-Tree Variations in the Mechanical Properties of Chinese Fir (*Cunninghamia lanceolata*). *IAWA J.* **2011**, *32*, 431–442. [CrossRef]
17. Tabet, T.A.; Fauziah, A.A. Cellulose Microfibril Angle in Wood and its Dynamic Mechanical Significance. In *Cellulose, Fundamental Aspects*; van de Ven, T.G.M., Ed.; Intech: London, UK, 2013; pp. 113–142.
18. British Standard Institution. *Methods of Testing Small Clear Specimens of Timber*; Serial BS 373; British Standard Institution: London, UK, 1957.
19. ASTM International. Wood. D 143. Standard Methods Test of Testing Small Clear Specimens of Wood. In *Annual Book of ASTM Standards*; ASTM International: West Conshohocken, PA, USA, 2005; Volume 04.10.
20. Frans, A.W. Adequate Pandanus Fiber (Tectorius Lamk.) Boiling Time Using Forest Product Laboratory Method. Bachelor's Thesis, Faculty of Forestry in Papua, Manokwari, Indonesia, 2013. Unpublished work. (In Indonesian)
21. Brauns, J.; Rocens, K. Modification of Wood: Mechanical Properties and Application. In *Encyclopedia of Materials: Science and Technology*; Elsevier: Amsterdam, The Netherlands, 2007; pp. 1–9. [CrossRef]
22. Blomberg, J.; Persson, B.; Blomberg, A. Effects of Semi-isostatic Densification of Wood on the Variation in Strength Properties with Density. *Wood Sci. Technol.* **2005**, *39*, 339–350. [CrossRef]
23. Nairn, J.A. Numerical Simulations of Transverse Compression and Densification in Wood. *Wood Fiber Sci.* **2006**, *38*, 576–591.
24. Rautkari, L. *Thermo-Hydro-Mechanical Modification*; Puu-28.4002 Wood Modification; Aalto University: Helsinki, Finland, 2011.
25. Hartono, R.; Wahyudi, I.; Febrianto, F.; Dwianto, W.; Hidayat, W.; Jang, J.H.; Lee, S.H.; Park, S.H.; Kim, N.H. Quality Improvement of Oil Palm Trunk Properties by Close System Compression Method. *J. Korean Wood Sci. Technol.* **2016**, *44*, 172–183. [CrossRef]
26. Báder, M.; Németh, R. Spring-back of Wood after Longitudinal Compression. In *IOP Conf. Series: Earth and Environmental Science 505.012018 Proceedings of the 6th International Conference on Environment and Renewable Energy, Hanoi, Vietnam, 24-26 February 2020*; IOP Publishing: Bristol, UK, 2020. [CrossRef]
27. Mania, P.; Wróblewski, M.; Wójciak, A.; Roszyk, E.; Molinski, W. Hardness of Densified Wood in Relation to Changed Chemical Composition. *For. J.* **2020**, *11*, 506. [CrossRef]
28. Blomberg, J. Mechanical and Physical Properties of Semi-isostatically Densified Wood. Ph.D. Thesis, Division of Wood Technology, Luleå University of Technology, Skellefteå, Sweden, 2006.
29. Andersson, S.; Serimaa, R.; Paakari, T.; Saranpää, P.; Pesonen, E. Crystallinity of Wood and the Size of Cellulose Crystallites in Norway Spruce (*Picea abies*). *J. Wood Sci.* **2003**, *49*, 531–537. [CrossRef]

30. Andersson, S. A Study of the Nanostructure of the Cell Wall of the Tracheids of Conifer Xylem by X-ray Scattering. Ph.D. Thesis, University of Helsinki, Helsinki, Finland, 2006.
31. Kutnar, A.; Sernek, M. Densification of Wood. *Zbornik Gozdarstva in Lesarstva* **2007**, *82*, 53–62.

forests

MDPI

Article

Influence of Thermo-Mechanical Densification (TMD) on the Properties of Structural Sawn Timber (*Pinus sylvestris* L.)

Marek Grześkiewicz, Sławomir Krzosek, Izabela Burawska *, Piotr Borysiuk and Piotr Mańkowski

Institute of Wood Sciences and Furniture, Warsaw University of Life Sciences-SGGW, Nowoursynowska 159, 02-776 Warsaw, Poland

* Correspondence: izabela_burawska@sggw.edu.pl

Abstract: The article presents the results of thermo-mechanical densification tests conducted on Scots pine timber. The densification process was carried out in industrial conditions with a high-pressure press, which allowed flat compression of boards that were up to 2.5 m long. A phenomenon of elastic redeformations was observed in the densified boards after each pulse of compression. As a result of thermo-mechanical compression, the average timber moisture content dropped to 9%, and the average density increased by 13.5%, from the level of 547 to 621 kg/m^3. As a result of thermo-mechanical densification, the strength class C of most Scots pine timber pieces improved. Most timber pieces that were subjected to thermo-mechanical densification have improved their strength class, C, by one (72.7% of the tested batch) or two C classes (3.6% of the batch under study).

Keywords: density; MOE; strength grading; structural timber; TMD

1. Introduction

The growing demand for wood and engineered wood products for construction and the limited availability of high-quality wooden resources have stimulated the search for new, alternative materials to use in construction. One of the possible directions to solve the problem of raw materials is to develop treatments and processes to modify conventional materials in order to complement or replace the domestic forest species [1]. One of the factors of an effective economy that takes into account the need to simultaneously mitigate climate change is an ecologically sustainable wood treatment that consists of exposing it to appropriate temperature and pressure at the same time, without the need to add any chemical substances [2,3]. Thermo-mechanical treatment (TMD) of wood is a process of densification that has been known for decades and results in the increase of material density through compression perpendicular to the grain [4]. The density of wood and engineered wood products (EWP) is one of the criteria for conditioning the technical quality of a material. In general, wood species and EWP with greater density are preferred for many applications in the construction industry, due to their higher resistance and natural durability [5]. However, the availability of high-density wood is limited, and as a result, their purchase requires significant financial spending. Therefore, the use of wood with low or medium density subjected to a process of densification can improve its technical parameters, while at the same time taking into account the aspect of affordability.

There are many premises that justify the application of TMD as a structural timber modification process. For example, exposing wood to high temperatures increases its resistance to biological factors [6–9] and its dimensional stability [10,11], and reduces its hygroscopicity [12–15]. By adding the compression (densification) process, we can improve the physical and mechanical properties of wood [16–20]. As a result of the surface densification of wood, it is possible to increase its hardness even twice [21] and significantly improve the durability of the material. These features translate into the potential for the application of densified wood, especially in external layers of layered floor materials [22–24]. The concept of using a material that was initially worse, and whose resistance properties improve after modification, is especially important in the case of wood with low initial density, such as poplar [25,26], coniferous species [27,28], or eucalyptus [29,30]. One of the areas where wood with originally worse quality can be used is construction. As a result of the densification of spruce wood, it is possible to improve both MOE and MOR by up to 200%–300% in comparison with native (not densified) wood [31,32]. In the case of pine sapwood, the MOE of wood subjected to thermo-mechanical treatment amounted to 14.4 GPa (a 150% increase compared to native wood), while its density incremented from 385 kg/m^3 to 1040 kg/m^3 [33]. Similar results have been obtained for pine wood by Esteves et al. [34], who densified wood between press shelves at temperatures of 150 °C, 180 °C, and 200 °C. Wood with an initial density of 615 kg/m^3, 614 kg/m^3, and 512 kg/m^3 after densification achieved the following values: 1048 kg/m^3, 1031 kg/m^3 and 1041 kg/m^3, respectively. The MOE of densified wood was higher than native wood by 42%, 40%, and 71%, respectively. In the case of spruce wood, at a compression rate (CR) equal to 22%, 50%, 60%, and 67%, the density increase of native wood was in the range between 33% and 78%, while the elasticity modulus was between 79% and 114% [35]. Excessively elevated temperatures in the densification process affect the mechanical properties of wood negatively. This can be caused by the growing chemical degradation, which exacerbates together with rising temperatures. The most optimal temperature level for densification is considered to be 120 °C. For pine wood, a densification process performed at a temperature of 120 °C resulted in an 85% density increase, leading to an improvement of bending strength by 42%, shear strength by 20%, and compressive strength by 47% [35]. In the case of poplar wood, depending on the treatment temperature (120–200 °C), wood density increased by 144%–188%, which translated into an MOE that was 73%–130% higher and MOR higher by 3%–72% [36]. As to sugi wood, the densification process allows us to achieve an MOE increase by even 309%, and MOR by up to 183% [37]. TM wood processing is a scientific field that has been well-studied, which is confirmed by multiple literature publications on this topic, focusing on analysing the influence of CR, steam temperature, pressure, species tested, the direction of compression, etc. Nonetheless, it should be taken into account that most studies were carried out on small samples, at a laboratory scale, which—of course—does allow us to observe the influence of a given factor on the densification effect or lack thereof. However, study results obtained at a laboratory scale, looking at the physical and mechanical properties of wooden samples, can be affected by the so-called scale effect [38], which is a very important aspect when it comes to considerations of wood applications in engineering. In the stochastic aspect, the higher the volume of a wooden element (laboratory/full-scale technical sample), the higher the probability of defects in the structure of the wood.

Nevertheless, the high effectiveness of thermo-mechanical treatment, confirmed by the improvement of physical and mechanical properties of wood, allows us to conclude that with appropriate scaling and adjustment of treatment parameters, the process of wood densification can be used to enhance wood quality, not only in timber from plantations but also in traditional, domestic species, whose average density values are low, which will allow for their use in construction applications (as independent structural elements or for the production of layered composites). One of the necessary conditions for safe building with timber is the use of wood with adequate strength parameters. Sawn timber used in construction for structural purposes must be subjected to strength grading. There are two methods for strength grading of structural sawn timber: visual and machine. Machine strength grading results in much higher efficiency and wood pieces are classified into higher classes, with fewer timber pieces rejected in comparison to the results of visual strength grading of the same batch of wood [39]. As a result of machine strength grading, the timber pieces are classified into C strength classes in accordance with the EN 338 standard [40].

The study described herein refers to the analysis of the effects of thermo-mechanical timber densification in industrial conditions on the selected physical and mechanical properties of wood. The technical scale of the study permits us to directly apply the obtained results to the characteristic resistance of wood, corresponding to a given sorting class and

C strength grade [40]. It is also possible to specify the structure of C strength classes of a given batch of timber before and after densification, which has both an economical and ecological dimension.

2. Material and Methods

The study was conducted on a batch of boards made of pine wood (*Pinus silvestris* L.) that contained 400 pieces with an average moisture content of 13%. The timber originated from the Masovian Forestry Region (central Poland). The nominal board dimensions were: 24 × 90 × 2400 mm. The boards had a flatsawn growth ring layout (tangential sections on the wide board surfaces). The average growth ring width amounted to 1.50 mm, and timber was obtained from sapwood, with a small share of resin substances and a low amount of knots.

The scope of the research included: specifying the values of density (DEN), dynamic modulus of elasticity before thermo-mechanical modification (MOE), the performance of thermo-mechanical timber modification in industrial conditions, as well as specifying the density and modulus of elasticity of modified timber (DEN_{TMD}, MOE_{TMD}). The density of the timber under examination was determined with the stereometric method. The determination of the modulus of elasticity was carried out with the MTG device (Brookhuis Microelectronics BV, Holland). The device uses the impact method to determine MOE_{dyn}. The machine was also used for the strength grading of the timber batch under research before and after thermo-mechanical modification. The density profile of sample planks was determined with the use of a profile meter: Laboratory Density Analyser DAX GreCon (Fagus-Grecon Greten GmbH & Co. KG, Alfeld, Germany). A density measurement was made every 0.02 mm at the measurement speed of 0.05 mm/s. Before the test, the samples used to determine the density profile were planed on both sides to avoid any concave or convex areas on the wide surfaces of the boards. The samples used to specify the density profile had the dimensions of 50 × 50 mm and thickness remaining after planing on both sides and were cut from the middle part of the compressed boards.

In order to carry out the densification process in industrial conditions at the Wersal company, a high-pressure press was used, produced by Italpress, model GL/260-PS, which allowed flat compression of boards that were up to 2.5 m long. Before densification, timber, MC, 11.5%, was heated between the press shelves at a temperature of 90 °C, over the course of 20 min, in order to soften the lignin, and then it was densified in three steps. In the first step, the timber was flattened from 24 mm to 22 mm (4 mm, press shelf move), and the press was opened (the top shelf was moved upwards from the compressed timber piece); in the second pulse, timber was compressed to 20 mm (6 mm press shelf move), and the shelf was opened once again, separating it from the compressed board, to allow for stress relaxation. In the third pulse, the timber was compressed to 18 mm (8 mm press shelf move). In the last pulse of compression, the timber remained closed tightly in the press for 30 s, after which the press shelves were opened.

Statistical analysis of the results was carried out using Statistica version 13 (TIBCO Software Inc., Palo Alto, CA, USA). The analysis of variance (ANOVA) was used to test ($\alpha = 0.05$) for significant differences between factors. A comparison of the means was performed using Tukey's test, with $\alpha = 0.05$.

3. Results and Discussion

3.1. Density of Wood before and after Thermo-Mechanical Densification

A phenomenon of elastic redeformations was observed in the densified boards after each pulse of compression, and after the last pulse, there was a redeformation back to the average thickness of ca. 20 mm. Timber pieces, whose initial nominal thickness amounted to 24 mm, were compressed to 20 mm (CR compression ratio 20%), while the initial board width, 90 mm, increased to 92 mm. The obtained cross-sections of the boards after compression were not flat. As a result of the thermo-mechanical compression, the average timber moisture content dropped to 9% and the average density increased by 13.5%, from of 547 to 621 kg/m^3.

Figure 1 presents the relation between timber density before and after TMD. Both sets have a normal distribution. It has been observed that as a result of thermo-mechanical densification, timber density increased in a statistically significant manner ($p < 0.05$), compared to the initial density before TMD. The average density of the entire timber batch (400 pcs.) increased by 14%. The relation between timber density before and after TMD is high (determination coefficient equals 0.78). A high determination coefficient reflects a quite uniform augmentation of the density of individual timber pieces as a result of TMD. Of course, the increase in density of a single piece of timber depended on its initial density. In the case of timber pieces with high initial density (and, at the same time, low porosity), the increase in density caused by thermo-mechanical densification was smaller (Table 1).

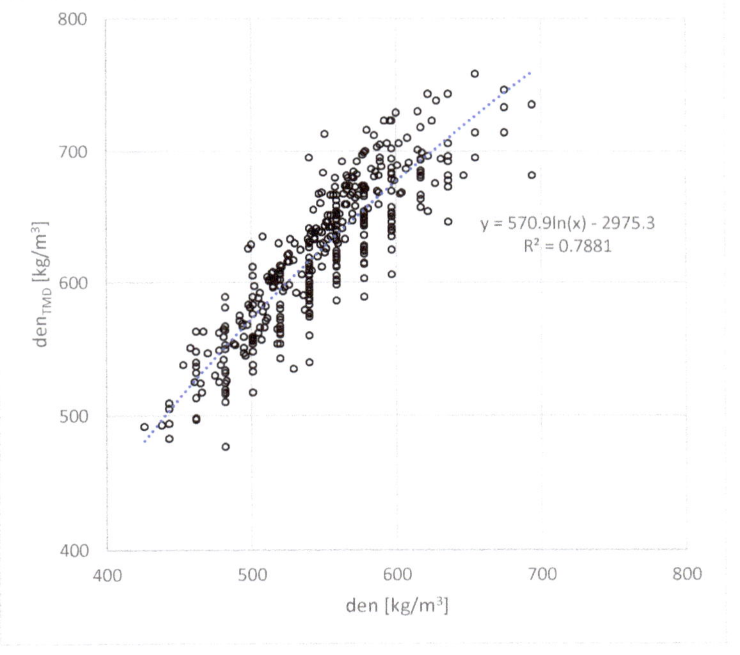

Figure 1. Relation between timber density before and after densification.

Table 1. Timber density before and after thermo-mechanical densification for timber classified in different strength grades before thermo-mechanical densification.

C Class of Graded Sawn Timber Determined before Densification	Average Density before Densification	Average Density after Densification	Density Increase
	[kg/m^3]	[kg/m^3]	[%]
C18	518 (54)	583 (62)	13
C24	506 (31)	573 (40)	13
C30	559 (25)	639 (33)	14
C35	600 (28)	679 (30)	13
C40	675 (19)	720 (30)	7
ALL	547 (48)	621 (48)	14

The density of wood before and after thermo-mechanical densification, taking into account the results of strength grading before and after thermo-mechanical densification has been presented in Tables 1 and 2. The results refer to the same batches of boards classified in a given class before densification.

Table 2. Timber density after thermo-mechanical densification for timber classified in different strength grades after densification.

C Class of Graded Sawn Timber Determined after Densification	Average Density after Densifiction
	[kg/m^3]
C18	652
C24	550
C30	586
C35	639
C40	652
ALL	621

By analysing the results presented in Table 1, it can be concluded that the highest density increase took place in the case of boards classified into class C30 after TMD (14.3%), and the smallest in the case of boards classified in the C40 class (6.7%). The density increase for all the boards amounted to 13.5%. In the case of densified timber from class C18, it stands out for its high density in comparison with the remaining C classes of densified timber (Table 2) and is 25.9% higher than the density of C18 timber before thermo-mechanical densification.

Figure 2 presents density distribution curves for the timber under research before and after modification. The curves show the changes in density that happened as a result of thermo-mechanical modification. TMD caused an increase in the density of timber under research by about 14%. Both before and after TMD modification, there is a negative kurtosis, which indicates a left-skewed asymmetry of distribution. The obliqueness, in both cases, is close to zero (distribution similar to a symmetrical one). After the modification, the obliqueness is skewed to the left (domination of lower-density values). The distribution is close to the normal distribution (kurtosis value = 0). Kurtosis is negative, which means there are fewer positive outliers than in a normal distribution.

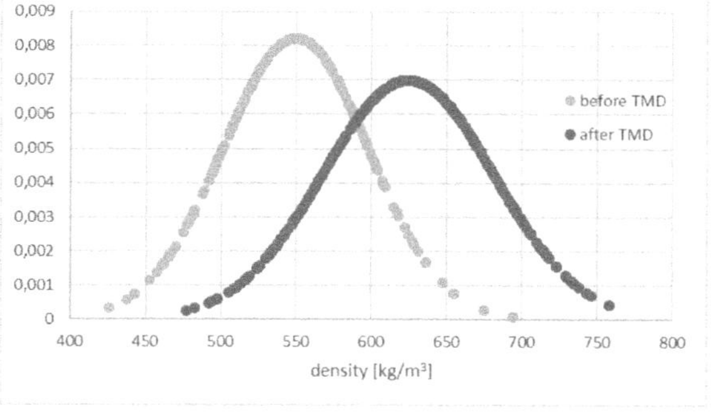

Figure 2. Timber density distribution before and after thermo-mechanical modification.

Examples of density profiles before and after thermo-mechanical densification have been presented in Figure 3 and in Figure 4. The thickness of pieces after TMD for which the density profile has been determined was smaller than the thickness of timber pieces for which strength class was determined, which resulted from their planing in order to achieve flatness. The obtained density profiles confirm that timber has been densified in a uniform manner throughout its entire volume. As a result of the thermo-mechanical densification of pine wood, the differences between earlywood and latewood become smaller. Before thermo-mechanical modification, the density of the entire batch of tested timber was between 426 kg/m^3 and 694 kg/m^3. After thermo-mechanical modification, the density of the tested batch of timber was between 477 kg/m^3 and 758 kg/m^3.

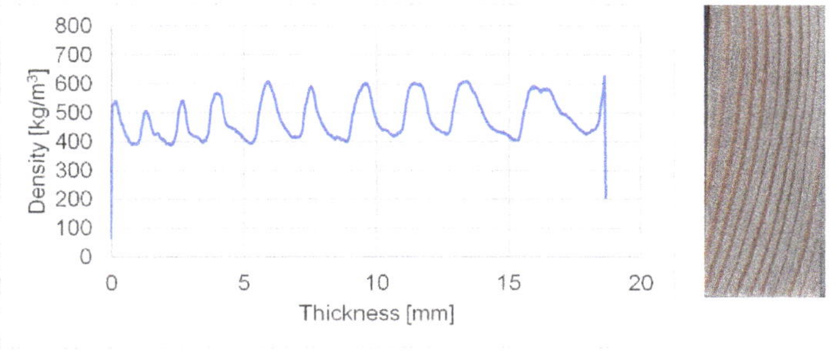

Figure 3. Density profile and cross-section of sawn timber before thermo-mechanical densification, average density of sample 482 kg/m^3. Sawn timber samples before modification were planed on both sides to achieve flatness.

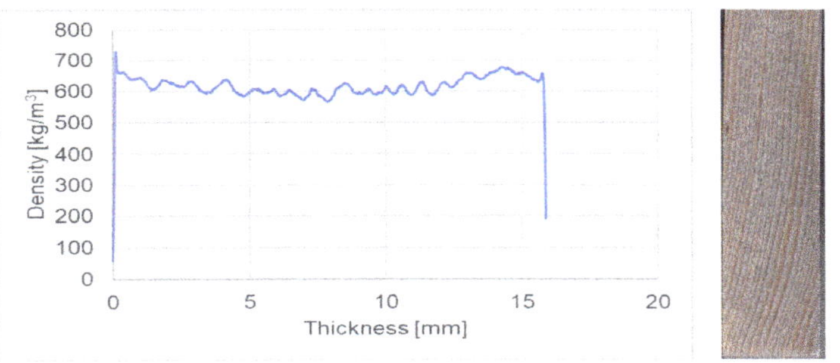

Figure 4. Density profile and cross-section of sawn timber after thermo-mechanical densification, average density of sample 615 kg/m^3. Sawn timber after densification was planed on both sides to achieve flatness.

The thermo-mechanical modification of pine wood influences both its internal structure and its properties. The density profile of pine boards became more homogenous. Before densification, there were visible differences between the density of earlywood and latewood (Figure 3). After the modification, the density of pine wood in the element's thickness became more uniform (Figure 4), and more intense material densification occurred in earlywood areas. The same has also been confirmed by the observations of Nilsson et al. (2011) [41]. This is due to the fact that in coniferous wood, earlywood cells have bigger cross-section dimensions and thinner cell walls than latewood. A similar effect has been reported in the literature in reference to densified spruce wood [13,42,43]. In turn, the way cells are densified has a very significant influence on the mechanical and physical properties of the material under modification [43].

3.2. Dynamic Modulus of Elasticity before and after Thermo-Mechanical Densification

The relation between the modulus of elasticity before and after TMD modification, for the entire batch of timber under study, has been presented in Figures 5 and 6. For both sets, departures from normality were observed.

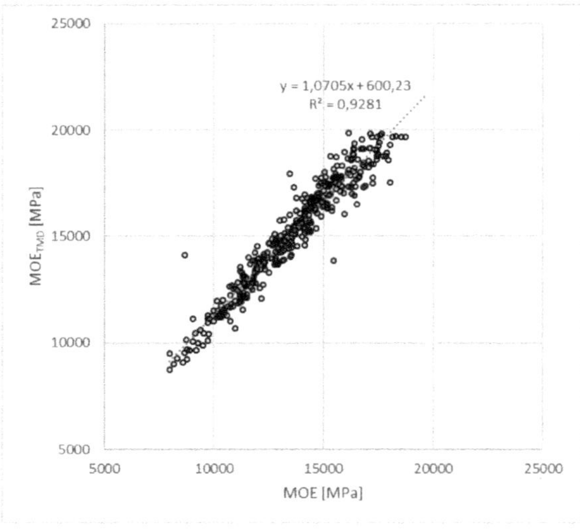

Figure 5. Relation between MOE before and after densification.

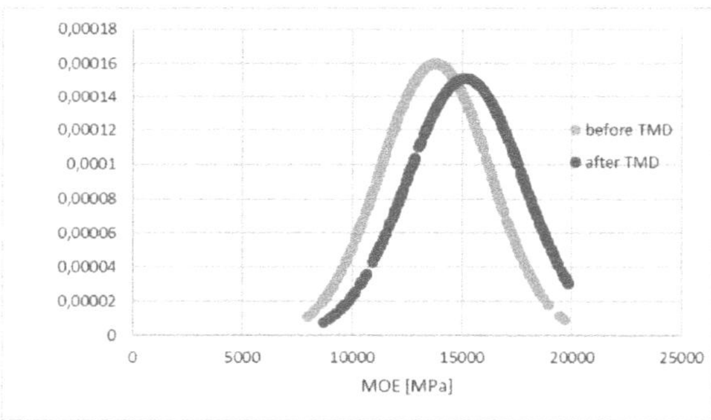

Figure 6. MOE distribution before and after thermo-mechanical modification in the timber under study.

Figure 6 shows an increase in MOE as a result of TMD modification. The main value of MOE incremented by ca. 10%. Both before and after TMD modification, there is a negative kurtosis, which indicates a left-skewed asymmetry of distribution. The obliqueness, in both cases, is close to zero (distribution similar to a symmetrical one). After the modification, the obliqueness is skewed to the left (domination of lower MOE values). The distribution is close to the normal distribution (kurtosis value = 0). Kurtosis is negative, which means there are fewer positive outliers than in a normal distribution.

The dynamic modulus of elasticity of wood (MOE_dyn) before and after thermo-mechanical densification, for the various C classes of strength-graded timber determined before densification, has been presented in Table 3.

By analysing the results presented in Table 3, it can be concluded that the largest increase of MOE_dyn took place in the case of boards classified into the C18 strength grade before densification (13.6%) and the smallest in the case of boards classified as C40 (1.2%). For the remaining classes, the changes in MOE values were the following: C24 (12.1%), C30 (12.0%) and C35 (9.9%). The increase of dynamic modulus of elasticity for all the boards amounted to 11.4%. In the case of the group of boards that, before densification, was classified as C40, the increase of the modulus of elasticity was the smallest. For a part of the boards, 17 pieces, including 4 pieces from the C40 class and 13 pieces from the C35 class (graded before densification), whose density increased, as a result of compression, from 613 kg/m^3 to 705 kg/m^3, we were not able to specify the strength class using the MTG device. C classes and average MOE_dyn board after wood densification are presented in Table 4.

Table 3. Dynamic modulus of elasticity before and after thermo-mechanical densification for timber classified in different strength grades before thermo-mechanical densification.

C Class of Graded Sawn Timber Determined before Densification	Average Dynamic Modulus of Elasticity before Densification	Average Dynamic Modulus of Elasticity after Densification	Modulus of Elasticity Increase
	[MPa]	[MPa]	[%]
C18	8913 (575)	10,128 (1179)	14
C24	11,626 (990)	13,030 (1355)	12
C30	14,286 (871)	16,000 (1268)	12
C35	16,690 (872)	18,346 (952)	10
C40	18,385 (693)	18,602 (1523)	1
ALL	13,586 (2495)	15,131 (2494)	11

Table 4. Dynamic modulus of elasticity after thermo-mechanical densification for timber classified in different strength grades specified after thermo-mechanical densification.

C Class of Graded Sawn Timber Determined after Densification	Average Dynamic Modulus of Elasticity MOE_dyn after Densification
	[MPa]
C18	9127
C24	11,304
C30	13,570
C35	16,187
C40	18,555
ALL	15,131

Figure 7 presents the relation between the density and MOE of timber, taking into account the results of strength grading.

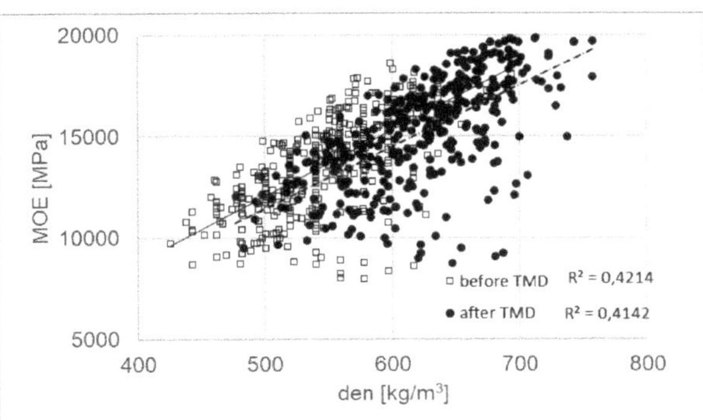

Figure 7. Relation between density and MOE (before and after TMD modification).

3.3. Structure of C Strength Grades before and after Thermo-Mechanical Densification

As a result of the thermo-mechanical densification of pine-sawn timber, there have been changes in the structure of share of strength classes in the batch of timber under study

(Figure 8). In the case of the set of timber pieces strength graded with the machine method before thermo-mechanical densification, the dominant timber classes were C24 and C30. As a result of thermo-mechanical densification, C35 became the most dominant timber class. The second largest share corresponded to class C30, while C40 came third.

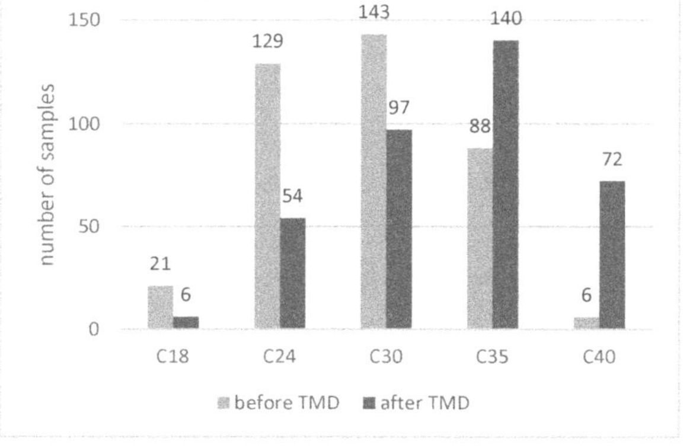

Figure 8. Structure of C classes before and after thermo-mechanical densification.

The obtained results permit us to conclude that, as a result of thermo-mechanical densification of timber, its density grows, which translates, as a consequence, into an increase of the dynamic modulus of elasticity specified with a non-destructive method, using the MTG device (Mobile Timber Grader). The strength grading of timber before and after densification results in a significant increase in the share of pieces in higher C classes.

The effectiveness of board densification (density increase) fell within the range of 8%–14%. The highest densification (12%–14%) has been observed for timber from classes C18, C24, C30, and C35. In each case, the observed differences in density values before and after densification were statistically significant (different homogeneous groups in Table 5). In the case of modification of wood from the C40 class, the density increase we observed (8%) was visible but statistically insignificant (the same homogeneous group "f" in Table 5). It is worth noting that the highest percentual influence on material densification (54.1%—Table 6) resulted from the initial class of wood, which depended on the material's initial parameters. Both the densification process itself and the interaction between the densification process and the initial class of timber were characterised by a lower influence on the density increase than other factors that have not been taken into account in the current study (error = 36.2%).

Together with the densification increase, we also observed an increase in MOE for the timber under research. It fell within the range of 9%–14% for timber from classes C18, C24, C30, and C35. In each case, the increase of MOE values was statistically significant (different homogeneous groups in Table 5). The highest improvement of this parameter (14%) was observed for the lowest initial class, C18. In the case of the initial class C40, the modification process did not cause an improvement in MOE values (the same homogeneous group "H" in Table 1). It is worth noting that, similarly to density, the highest percentual influence on MOE values increase (81.5%—Table 6) resulted from the initial class of wood, which depended on the material's initial parameters. Both the densification process itself and the interaction between the densification process and the initial class of timber were characterised by a lower influence on the MOE values increase than other factors that have not been taken into account in the current study (error = 17.5%).

Table 5. Properties (homogeneous groups).

Modification	Class (before Densification)	Density [kg/m^3]	Homogeneous Groups	MOE [MPa]	Homogeneous Groups
NM	C18	518	a	8913	A
TMD		583	d, c, d	10,128	B
NM	C24	506	a	11,626	C
TMD		573	c	13,030	D
NM	C30	559	b	14,286	E
TMD		639	e	16,000	F
NM	C35	599	d	16,863	G
TMD		681	f	18,346	H
NM	C40	675	e, f	18,602	G, H
TMD		727	f	18,864	H

NM—not modified; TMD—modified.

Table 6. ANOVA for selected factors affecting density and MOE of tested wood.

Source of Variance	Density		MOE	
	p	P (%)	P	P (%)
densification	0.000000	9.2	0.000000	0.7
initial class	0.000000	54.1	0.000000	81.5
Densification × initial class	0.047054	0.5	0.101145	0.2
Error		36.2		17.5

p—probability of error, P—percentage of contribution.

4. Conclusions

The thermo-mechanical densification of flatsawn pine timber boards increases the values of its density by 13.5% on average, which translates into the average increase of the dynamic modulus of elasticity by 9% and causes changes in the number of timber pieces classified into the individual C strength grades.

As a result of thermo-mechanical densification in industrial conditions, the strength class C of most Scots pine timber pieces improves. Most timber pieces that were subjected to thermo-mechanical densification have improved their strength class, C, by one (72.7% of the tested batch) or two C classes (3.6% of the batch under study).

Thermo-mechanical densification causes an increase in the quality (expressed by average values of density and dynamic modulus of elasticity) of timber for construction applications.

Author Contributions: Conceptualization, I.B. and S.K.; methodology, I.B. and S.K.; software, I.B. and P.B., validation, I.B. and S.K.; formal analysis, I.B., P.M., P.B. and S.K.; investigation, I.B., P.M. and S.K.; resources, I.B., M.G. and S.K.; data curation, I.B, P.M. and P.B.; writing—original draft preparation, I.B., P.M., M.G., P.B. and S.K.; writing—review and editing, I.B.; visualization, I.B.; supervision, S.K.; project administration, I.B. and S.K.; funding acquisition, S.K. All authors have read and agreed to the published version of the manuscript.

Funding: The authors are grateful for the support of the National Centre for Research and Development, Poland, under "Environment, agriculture and forestry" – BIOSTRATEG strategic R&D programme, Agreement No. BIOSTRATEG3/344303/14/NCBR/2018.

Informed Consent Statement: Not applicable.

Data Availability Statement: Not applicable.

Conflicts of Interest: The authors declare no conflict of interest.

References

1. Payn, T.; Carnus, J.-M.; Freer-Smith, P.; Kimberley, M.; Kollert, W.; Liu, S.; Orazio, C.; Rodriguez, L.; Silva, L.N.; Wingfield, M.J. Changes in planted forests and future global implications. *For. Ecol. Manag.* **2015**, *352*, 57–67. [CrossRef]
2. Canadell, J.G.; Schulze, E.D. Global potential of biosphericcarbon management for climate mitigation. *Nat. Commun.* **2014**, *5*, 5282. [CrossRef]
3. Sandberg, D.; Haller, P.; Navi, P. Thermo-hydro and thermo-hydro-mechanical wood processing: An opportunity for future environmentally friendly wood products. *Wood Mater. Sci. Eng.* **2013**, *8*, 64–88. [CrossRef]
4. Hajihassani, R.; Mohebby, B.; Najafi, S.K.; Navi, P.; Hajihassani, R.; Mohebby, B.; Najafi, S.K.; Navi, P. Influence of combined hygro-thermo-mechanical treatment on technical characteristics of poplar wood. *Maderas-Cienc. Tecnol.* **2018**, *20*, 117–128. [CrossRef]
5. Knapic, S.; Santos, J.; Santos, J.A.D.; Pereira, H. Natural durability assessment of thermo-modified young wood of eucalyptus. *Maderas-Cienc. Tecnol.* **2018**, *20*, 489–498. [CrossRef]
6. Kamperidou, V. The Biological Durability of Thermally- and Chemically-Modified Black Pine and Poplar Wood Against Basidiomycetes and Mold Action. *Forests* **2019**, *10*, 1111. [CrossRef]
7. Dubey, M.K.; Pang, S.S.; Chauhan, S.; Walker, J. Dimensional Stability, Fungal Resistance and Mechanical Properties of Radiata Pine after Combined Thermo-Mechanical Compression and Oil Heat-Treatment. *Holzforschung* **2016**, *70*, 793–800. [CrossRef]
8. Sandberg, D.; Kutnar, A.; Mantanis, G. Wood modification technologies—A review. *IForest* **2017**, *10*, 895–908. [CrossRef]
9. Lesar, B.; Humar, M.; Kamke, F.A.; Kutnar, A. Influence of the thermo-hydro-mechanical treatments of wood on the performance against wood-degrading fungi. *Wood Sci. Technol.* **2013**, *47*, 977–992. [CrossRef]
10. Fang, C.H.; Cloutier, A.; Blanchet, P.; Koubaa, A. Densification of veneers combined with oil-heat treatment. Part I: Dimensional stability. *BioResources* **2011**, *6*, 373–385. [CrossRef]
11. Navi, P.; Girardet, F. Effects of thermo-hydro-mechanical treatment on the structure and properties of wood. *Holzforschung* **2000**, *54*, 287–293. [CrossRef]
12. Fang, C.H.; Cloutier, A.; Blanchet, P.; Koubaa, A. Densification of veneers combined with oil-heat treatment. Part II: Hygroscopicity and mechanical properties. *BioResources* **2012**, *7*, 925–935. [CrossRef]
13. Heger, F.; Groux, M.; Girardet, F.; Welzbacher, C.; Rapp, A.O.; Navi, P. Mechanical and Durability Performance of THM-Densified Wood. In Proceedings of the Final Conference COST Action E22 Environmental Optimization of Wood Protection, Estoril, Portugal, 22–23 March 2004; pp. 1–10.
14. Pelaez-Samaniego, M.R.; Yadama, V.; Lowell, E.; Espinoza-Herrera, R. A review of wood thermal pretreatments to improve wood composite properties. *Wood Sci. Technol.* **2013**, *47*, 1285–1319. [CrossRef]
15. Kubovský, I.; Kačíková, D.; Kačík, F. Structural changes of oak wood main components caused by thermal modification. *Polymers* **2020**, *12*, 485. [CrossRef] [PubMed]
16. Grześkiewicz, M.; Poddębski, K. Thermal properties and density profile of poplar wood (*Populus nigra* L.) thermally and thermo-mechanically modified. In Proceedings of the 9th European Conference on Wood Modification, Arnhem, The Netherlands, 17–18 September 2018; pp. 1–6.
17. Kozakiewicz, P.; Drożdżek, M.; Laskowska, A.; Grześkiewicz, M.; Bytner, O.; Radomski, A.; Zawadzki, J. Effects of thermal modification on selected physical properties of sapwood and heartwood of black poplar (*Populus nigra* L.). *BioResources* **2019**, *14*, 8391–8404. [CrossRef]
18. Kozakiewicz, P.; Drożdżek, M.; Laskowska, A.; Grześkiewicz, M.; Bytner, O.; Radomski, A.; Mróz, A.; Betlej, I.; Zawadzki, J. Chemical Composition as a Factor Affecting the Mechanical Properties of Thermally Modified Black Poplar (*Populus nigra* L.). *BioResources* **2020**, *15*, 3915–3929. [CrossRef]
19. Anshari, B.; Guan, Z.; Kitamori, A.; Jung, K.; Hassel, I.; Komatsu, K. Mechanical and moisture-dependent swelling properties of compressed Japanese cedar. *Constr. Build. Mater.* **2011**, *25*, 1718–1725. [CrossRef]
20. Huang, C.; Chui, Y.; Gong, M.; Chana, F. Mechanical behaviour of wood compressed in radial direction: Part II. Influence of temperature and moisture content. *J. Bioresour. Bioprod.* **2020**, *5*, 266–275. [CrossRef]
21. Rautkari, L.; Laine, K.; Kutnar, A.; Medved, S.; Hughes, M. Hardness and density profile of surface densifies and thermally modified Scots pine in relation to degree of densification. *J. Mater. Sci.* **2013**, *48*, 2370–2375. [CrossRef]
22. Zhou, Q.; Chen, C.; Tu, D.; Zhu, Z.; Li, K. Surface densification of poplar solid wood: Efects of the process parameters on the density profile and hardness. *BioResources* **2019**, *14*, 4814–4831. [CrossRef]
23. Welzbacher, C.R.; Wehsener, J.; Rapp, A.O.; Haller, P. Thermo-mechanical densification combined with thermal modification of Norway spruce (Picea abies Karst) in industrial scale—Dimensional stability and durability aspects. *Holz Roh. Werkst.* **2008**, *66*, 39–49. [CrossRef]
24. Yoshihara, H.; Tsunematsu, S. Bending and shear properties of compressed Sitka spruce. *Wood Sci. Technol.* **2007**, *41*, 117–131. [CrossRef]
25. Bao, M.; Huang, X.; Jiang, M.; Yu, W.; Yu, Y. Effect of thermo-hydro-mechanical densification on microstructure and properties of poplar wood. *J. Wood Sci.* **2017**, *63*, 591–605. [CrossRef]
26. He, Z.; Qi, Y.; Zhang, G.; Zhao, Y.; Dai, Y.; Liu, B.; Lian, C.; Dong, X.; Li, Y. Mechanical Properties and Dimensional Stability of Poplar Wood Modified by Pre-Compression and Post-Vacuum-Thermo Treatments. *Polymers* **2022**, *14*, 1571. [CrossRef] [PubMed]

27. Şenol, S.; Budakçı, M. Effect of Thermo-Vibro-Mechanic Densification Process on the Gloss and Hardness Values of Some Wood Materials. *BioResources* **2019**, *14*, 9611–9627. [CrossRef]
28. Cruz, N.; Avila, C.; Aguayo, M.G.; Cloutier, A.; Castillo, R. Impact of the Chemical Composition of Pinus radiata Wood on its Physical and Mechanical Properties Following Thermo-Hygromechanical Densification. *BioResources* **2018**, *13*, 2268–2282. [CrossRef]
29. Balasso, M.; Kutnar, A.; Niemelä, E.P.; Mikuljan, M.; Nolan, G.; Kotlarewski, N.; Hunt, M.; Jacobs, A.; O'Reilly-Wapstra, J. Wood Properties Characterisation of Thermo-Hydro Mechanical Treated Plantation and Native Tasmanian Timber Species. *Forests* **2020**, *11*, 1189. [CrossRef]
30. Pertuzzatti, A.; Missio, A.L.; Cademartori, P.H.G.; Santini, E.J.; Haselein, C.R.; Berger, C.; Gatto, D.A.; Tondi, G. Effect of Process Parameters in the Thermomechanical Densification of *Pinus elliottii* and *Eucalyptus grandis* Fast-growing Wood. *BioResources* **2018**, *13*, 1576–1590. [CrossRef]
31. Carbal, J.P.; Kafle, B.; Subhani, M.; Reiner, J.; Ashraf, M. Densification of timber: A review on the process, material properties and application. *J. Wood Sci.* **2022**, *68*, 24. [CrossRef]
32. Kutnar, A.; Kamke, F.A. Compression of wood under saturated steam, superheated steam, and transient conditions at 150 °C, 160 °C, and 170 °C. *Wood Sci. Technol.* **2012**, *46*, 73–88. [CrossRef]
33. Kamke, F.A. Densified radiata pine for structural composites. Maderas. *Cienc. Tecnol.* **2006**, *8*, 83–92. [CrossRef]
34. Esteves, B.; Ribeiro, F.; Gruz-Lopes, L.; Domingos, J.F.I. Densification and heat treatment of Martine pine wood. *Wood Res.* **2017**, *62*, 373–388.
35. Ulker, O.; Imirzi, O.; Burdurlu, E. The effect of densification temperature on some physical and mechanical properties of Scots pine (*Pinus sylvestris* L.). *BioResources* **2012**, *7*, 5581–5592. [CrossRef]
36. Sözbir, G.D.; Bektaş, İ.; Ak, A.K. Influence of combined heat treatment and densification on mechanical properties of poplar wood. *Maderas Cienc. Tecnol.* **2019**, *21*, 481–492. [CrossRef]
37. Tanaka, K.; Demoto, Y.; Ouchi, J.; Inoue, M. Strength property of densified SUGI adopted as material of connector. In Proceedings of the 11th World Conference on Timber Engineering (WCTE 2010), Trentino, Italy, 20–24 June 2010; pp. 1696–1701.
38. Malaga-Toboła, U.; Łapka, M.; Tabor, S.; Niesłony, A.; Findura, P. Influence of wood anisotropy on its mechanical properties in relation to the scale effect. *Int. Agrophys.* **2019**, *33*, 337–345. [CrossRef]
39. Burawska-Kupniewska, I.; Krzosek, S.; Mańkowski, P. Efficiency of Visual and Machine Strength Grading of Sawn Timber with Respect to Log Type. *Forests* **2021**, *12*, 1467. [CrossRef]
40. *EN 338*; Timber Structures—Strength Classes. European Committee for Standardisation: Brussels, Belgium, 2016.
41. Nilsson, J.; Johansson, J.; Kifetew, G.; Sandberg, D. Shape stability of modified engineering wood product subjected to moisture variation. *Wood Mater. Sci. Eng.* **2011**, *6*, 42–49. [CrossRef]
42. Schrepfer, V.; Schweingruber, F.H. Anatomical Structures in Reshaped Press-Dried Wood. *Holzforschung* **1998**, *52*, 615–622. [CrossRef]
43. Kamke, F.A.; Sizemore, H. Viscoelastic Thermal Compression of Wood. U.S. Patent 7,404,422 B2, 29 July 2008.

forests

MDPI

Article

Experimental Study on Tenon and Mortise Joints of Wood-Structure Houses Reinforced by Innovative Metal Dampers

Shibin Yu [1,2], Wen Pan [1,2,*], Hexian Su [1,2], Liaoyuan Ye [2,3] and Daohang Wang [4]

1 Faculty of Civil Engineering and Mechanics, Kunming University of Science and Technology, Kunming 650500, China; 20191110010@stu.kust.edu.cn (S.Y.); shx870@kust.edu.cn (H.S.)
2 Earthquake Engineering Researching Center of Yunnan, Kunming 650500, China; yly@163.com
3 Communist Party Committee Office, Yunnan Normal University, Kunming 650500, China
4 Faculty of Public Security and Emergency Management, Kunming University of Science and Technology, Kunming 650500, China; wang_daohang@163.com
* Correspondence: panwen@vip.sina.com

Abstract: To improve the seismic performance of Chinese traditional wood-structure houses, this paper proposes to strengthen their mortise and tenon joints by applying an innovative metal damper. According to the dimensions of the "Yikeyin" wood-structure houses in the Tonghai area of Yunnan Province, two groups of six samples of three types of mortise and tenon joints were manufactured, in which one group was mounted with dampers made of Q235 steels. Subsequently, a low-cycle repeated loading test was conducted to examine the overall behavior of these joints. Various characteristics of seismic performance indexes, such as the moment–rotation hysteresis curve, skeleton curve, stiffness degradation, energy dissipation capacity, residual amount of tenon and the removal before and after reinforcements of straight, penetrated and dovetail tenon joints were analyzed. The test results show that these tenons exhibit good deformation capacity, their hysteresis curves became fuller and their "pinch" effects were significantly reduced, all after their joints became strengthened, indicating that their joint slips were reduced during the loading processes and their residual amounts of tenon removals were under effective control. Compared with the blank group, the joint stiffness was substantially improved, and the increase in the reverse stiffness turned greater than that of the positive stiffness at each stage of loading, while the degradation curve of the whole joint stiffness became steeper. After mounting the dampers, the bearing capacity and energy dissipation of the joints were significantly improved, the equivalent viscous damping coefficients of the straight and penetrated tenon joints were increased, but that of the dovetail joint was slightly reduced. These study results can provide a reference for the reinforcement and protection of traditional wood-structure houses.

Keywords: wood joints; dampers; quasi-static test; strengthening; mortise

1. Introduction

Chinese traditional houses are the carriers of national survival wisdom, construction skills and aesthetic consciousness, which also became China's valuable resource of cultural heritage. Some traditional wood-structure houses remain preserved in villages and towns of Southwest China. Due to frequent earthquakes, environmental erosion, lack of regular maintenance and other causes, their main structures have been damaged to various degrees. To preserve these precious traditional wood-structure houses for long enough, it is urgent to study the seismic performance and reinforcement technology of the specific kind for their structures.

According to the survey data on earthquake damages, in some high-intensity areas in Southwest China, the tenon and mortise joints of wood-structure houses were pulled out and removed, and even the wood roof-frame had been tilted as a whole and collapsed [1–5].

This phenomenon is mainly caused by the shrinkage and extrusion deformation of the tenon joint causing serious degradation in the mechanical properties of the joint, which eventually leads to the destruction of the entire structure [6–9].

Joints are the basic and key point of wooden structures, because they carry loads between the various parts of the structure, they maintain the structure and ensure the unity and integrity of the wooden products during its use [10]. Researchers have focused on the strength of mortise and tenon joints, among which Kamperidou et al. and Kamboj G et al. successively used adhesives to strengthen some furniture joints and found that the stiffness of the joints was greatly improved [11,12]. Numerous scholars have put forward some reinforcement measures for the joints and carried out relevant research. Their test results showed that the diagonal brace reinforcement could provide greater stiffness for the joints but is very detrimental to the tenon pull-out resistance [13]. Other reinforcement methods, such as tinplate, carbon fiber cloths, steel members and self-tapping screws are also proven to be capable of certainly improving the stiffness of wood joints and effectively reduce the tenon-pulling; but, the energy dissipation of the joints has not been significantly increased when the structures experienced small deformation [14–19]. Besides, Pan et al. [20] found that the reinforcement of steel members enhanced the stiffness of wood joints and the internal force of some adjacent members, which is prone to causing the tenon failure of adjacent joints, while the dampers could reduce the vibration response of the structure without increasing the stiffness of these joints.

To improve the energy dissipation of wood joints, researchers have consecutively developed a series of damping devices. Zou et al. [21] proposed a new type of angular displacement damper applicable for wood-frame buildings in villages and towns. Their tests verified that the new damper presented good deformation ability, effective fatigue resistance and certain energy dissipation capacity. Lu et al. [22,23] sequentially proposed the technologies of an arc-shaped energy dissipator and mortise-embedded energy dissipator to strengthen the performance of mortise and tenon joints, conducted experimental research on the seismic performance of these joints, and studied the impacts of the above energy dissipators with different parameters (radius, thickness and number of layers) on the seismic performance indexes of mortise and tenon joints. Gao et al. [24] conducted a shaking table test on a bucket type of wood-structure with fan-shaped viscoelastic dampers. Then, Nie et al. [25] investigated the influences of the above fan-shaped viscoelastic dampers on the seismic performance of dovetail joints with different levels of tightness, and summarized that these dampers could improve the energy dissipation ability, stiffness and bearing capacity of the wood-structure. Xue et al. [26–28] sequentially proposed innovative friction dampers and "shape memory alloy" (SMA) steel-wire dampers to strengthen mortise and tenon joints. The pseudo-static test results revealed that these two types of dampers could both reduce the amount of tenon pulling, improve the ultimate bearing capacity and initial stiffness of the joints, and enable the energy dissipation to increase significantly even under small deformations.

Although the above scholars have proposed new damping devices for wood-structure joints and implemented corresponding test research on their performances, the analysis of their test results indicate that most of these reinforcement devices would encounter the following problems: Firstly, the majority of the joint type dampers exhibited low energy dissipation before the joint angle reached 0.05 rad, which only became significant for larger structural deformation. However, the "technical standard for maintenance and strengthening of historic timber building" [29] stipulates that the inter-story displacement angles of wood frames under rare earthquakes must be limited to 1/30 (0.033 rad), which requires the damper to have good energy dissipation capacity even under small angle deformation of the structure. Secondly, some reinforcement devices (such as innovative friction dampers and SMA reinforcements) are mounted at the angles of the wood-structure joints, which achieved a significant damper energy dissipation of the structure even under small deformation, but it certainly compromised the original style of the wood-structure building. In view of the above concerns, this paper puts forward a technology of innovative metal damper reinforcing mortise and tenon joints of wood-structures, which presents two major advantages as follows: First, with the help of extended steel plate, the angular deformation of dampers' energy dissipation section increased (the joint angle shown in Figure 1 is θ and the angle of the energy dissipation section is enlarged to θ′); resultantly, the damper could dissipate as much energy as possible within the limited deformation range of the inter-story displacement angle [30]. Second, the two ends of the damper were fixed to the sides of the beam and column under the effect of self-tapping screws. As a result, the damper could be effectively hidden simply by keeping the color of the damper similar to that of the wood components in reducing its interference to the original architectural style of the wood-structure.

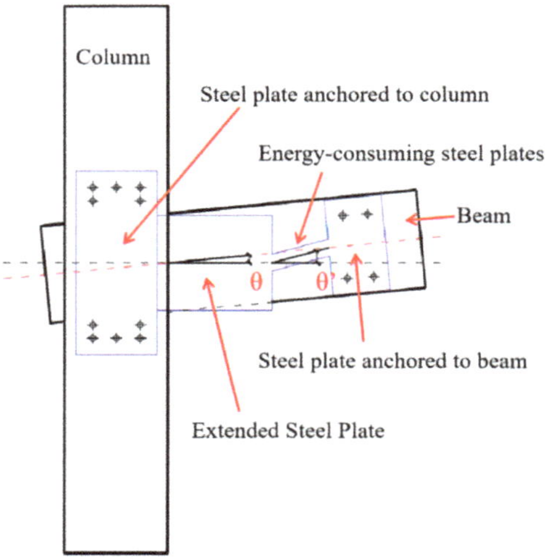

Figure 1. Working principles of the innovative metal dampers.

Through the quasi-static test, the changes to seismic performances of three types of mortise and tenon joints before and after reinforcement were compared, respectively. It was demonstrated that the innovative metal dampers proposed in this paper could not only improve the bearing capacity of the joints, but also greatly elevate their energy dissipation when the structure encountered small deformations, which might provide a reference for the practical reinforcements of traditional wood-structure houses.

The motive and purpose of this paper is to propose a new metal device to enhance the load-bearing capacity and energy dissipation capacity of mortise and tenon joints in wooden houses. This device can effectively reduce the damage of wooden structure or wooden furniture under vibration load and prolong its service life.

2. Materials and Methods

2.1. Designs of Wood Joint Samples and Dampers

Based on the investigation of several wood-structure folk houses in Xingmeng Country, Tonghai County, Yunnan Province, Southwest China, it is found that the wood beam and column members are mostly connected in the form of a straight tenon, penetrated tenon and dovetail tenon. Therefore, two groups of six mortise and tenon joint samples, including two identical straight tenons, two identical penetrated tenons and two identical dovetail tenons, were designed and manufactured with reference to the size of the typical local "Yikeyin" wood-structure of houses. All the tested materials are pine wood, originally from Yunnan Province, and the results of the mechanical properties of the materials are shown in Table 1. The columns and beams are made by multi-functional log lathe, and the mortise on the columns is obtained by grooving machine. See Table 2 for the sample number and corresponding parameters. The detailed structural dimensions of the three mortise and tenon joints are shown in Figure 2.

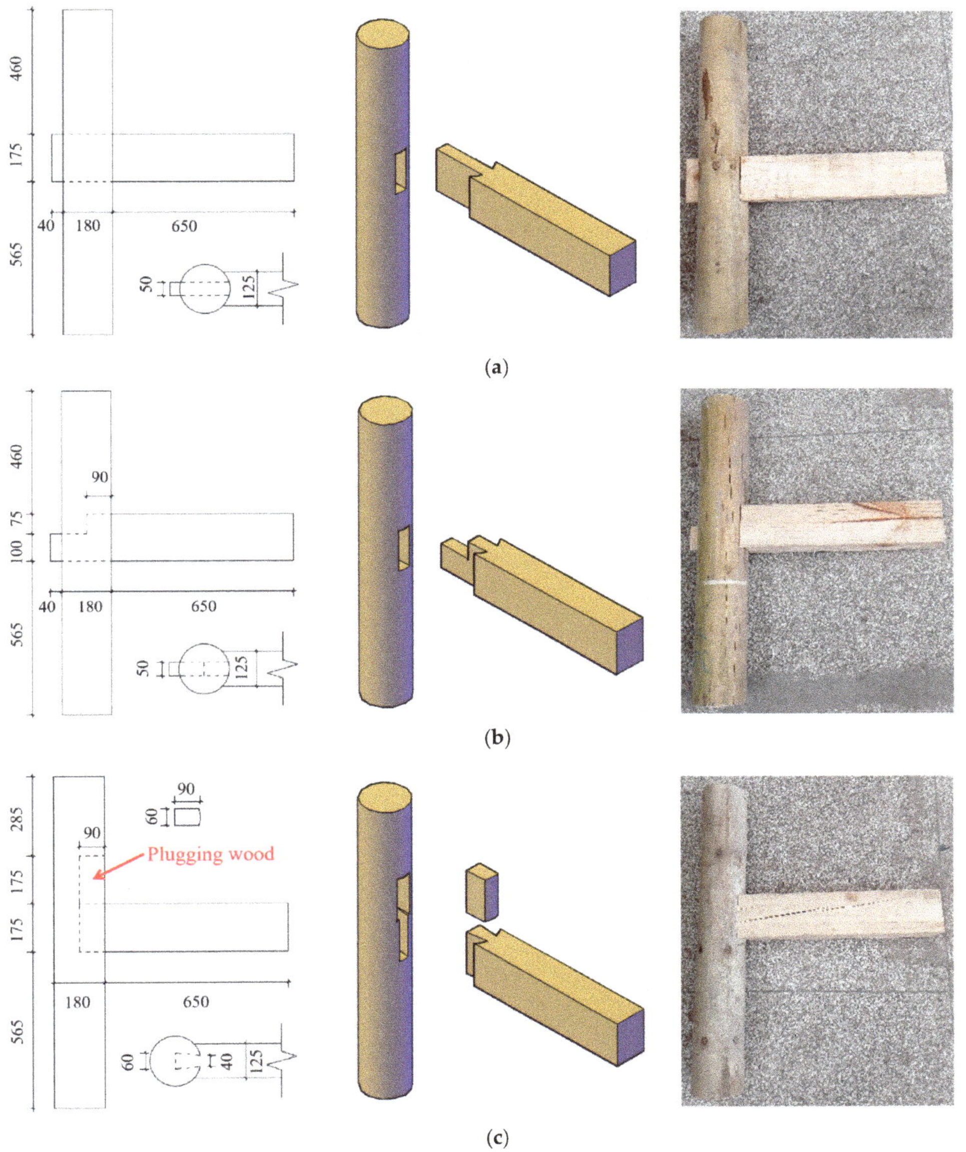

Figure 2. Detailed dimensions of tenon and mortise joints sample (unit: mm): (**a**) Straight tenon; (**b**) Penetrated tenon; (**c**) Dovetail tenon.

Table 1. Mechanical properties of Yunnan pine (unit: Mpa).

E_L	E_R	E_T	C_L	C_R	C_T
26.0	3.7	4.6	10,344.8	375.8	200.4
(2.2)	(0.3)	(0.5)	(1937.8)	(114.4)	(42.3)

Note: The subscript "L", "R" and "T" denote the longitudinal, radial and tangential directions of wood, respectively; the "E" and "C" respectively represent Young's modulus and compressive strength of wood. Each value is the mean of 6 sets of tests, and the values in parentheses are the standard deviations of 6 sets of specimens.

Table 2. Parameters of tenon and mortise joints sample.

Sample No.	Joint Type	Whether to Mount Dampers
ZTS0	Straight tenon	no
TJS0	Penetrated tenon	no
YWS0	Dovetail tenon	no
ZTS1	Straight tenon	yes
TJS1	Penetrated tenon	yes
YWS1	Dovetail tenon	yes

Six metallic dampers are made of Q235 steel plates with a thickness of 5 mm. See Figure 3 for details. A damper is mounted on each side of the mortise and tenon joint of the reinforcement group, and both ends of the damper are respectively connected with the wood column and beam through self-tapping screws. The mounting effect is shown in Figure 4.

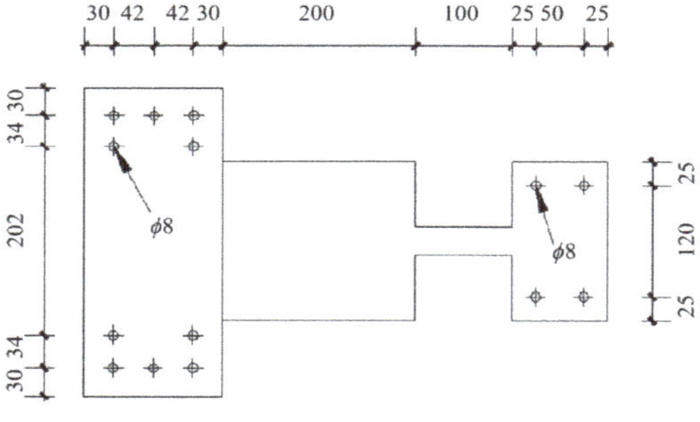

Figure 3. Dimensions of metallic dampers (unit: mm).

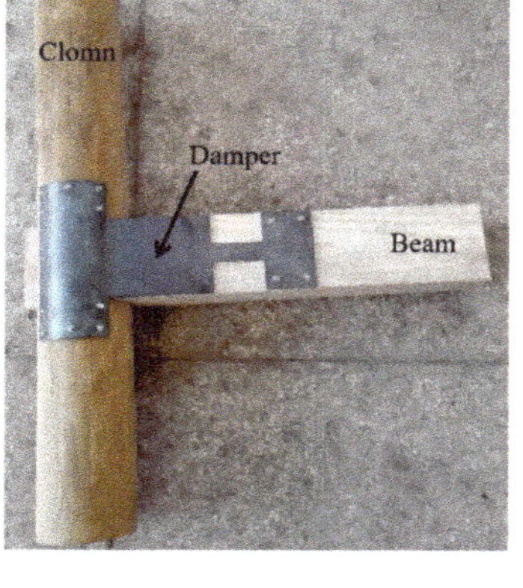

Figure 4. Mounting effect of dampers (unit: mm).

2.2. Test of Loading Scheme

The test was conducted in the Earthquake Engineering Researching Center of Yunnan, which required hydraulic servo mechanism, lifting jack, reaction frame, displacement meter and force sensor. The coordination between each device is shown in Figure 5.

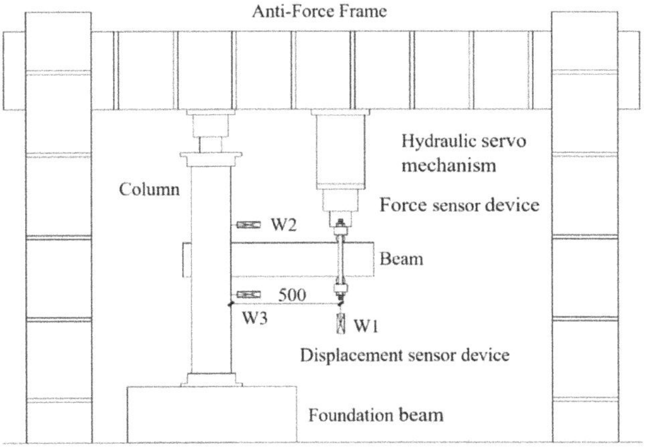

Figure 5. Pseudo-static test on loading devices.

According to the data (axis grid size, etc.) obtained from the field investigation of two-story traditional wood-structure houses in Tonghai area, Yunnan Province, along with the relevant provisions about the dead and live loads in the "Load Code for the Design of Building Structures" [31], the vertical load on the top of the column on the first floor is calculated to be 50 KN. To approach the real stress condition of mortise and tenon joints on the first floor of typical houses, the two ends of the column are restrained by shallow steel sleeves, and the upper end of the column is applied with vertical load by jacks. According to the "Specification for Seismic Test of Buildings" [32], the load at the beam end is stressed by stages under displacement control. The loading point of this test is located on the beam 500 mm away from the column edge. The displacement amplitude of the first stage is 5 mm, and then the displacement amplitude of each stage is increased by 5 mm. The test is completed when the loading amplitude reaches 25 mm (0.05 rad).

2.3. Test Data Measurement

A force sensor is mounted at the end of the hydraulic servo actuator to collect the reaction force at the loading point. A displacement meter (W1, see Figure 5) is mounted at the bottom of the beam directly below the loading point to measure the actual vertical displacement of the loading point. A displacement meter (W2, W3) is placed at the upper and lower sides of the beam to monitor the outward pull-out value of the tenon. The signals measured by each sensor are transmitted to the static strain tester dh3818y, and the collected data are processed to obtain the actual displacement, reaction force and tenon pull-out value at the test loading point.

3. Test Phenomenon

3.1. Test Phenomena before and after Straight Tenons' Reinforcement

For the unreinforced test piece ZTS0, there is a "creak" sound during the whole test loading process. When loading near the amplitude, the sound interval is long and the volume is large, while when unloading, it is continuously compact and the volume is small. At the initial stage of loading (vertical loading displacement of 5 mm and 10 mm), the extrusion deformation of tenon and mortise is not obvious, but the gap between tenon and mortise can be observed to be compacted. As the loading displacement increases step-by-step, not only does mutual extrusion occur at the contact edge between the tenon and the mortise, but relative sliding between them occurs, which is more significant. After the test, the tenon is pulled out about 4 mm from its initial position, and its top and bottom surfaces leave obvious local indentation, and this part becomes very smooth due to sliding friction.

For the strengthened sample ZTS1, the squeezing and sliding friction between the tenon and the mortise, and the friction between the steel plate and the wood beam on both sides, led to the sound in many places during the whole loading, and the "dong dong" sound gradually dominated in the subsequent loading. As for the damper, during the whole loading process, the steel plate of the boom section is firmly connected with the column, and no vertical displacement is observed. There is no relative slip between the beam and the anchor steel plate. The deflection effect of the energy dissipation section is obvious with the increase of the angle (see Figure 6). After the final level of loading (25 mm), a visible transverse crack appeared at the edge of the energy-dissipation steel plate inside the beam (see Figure 7).

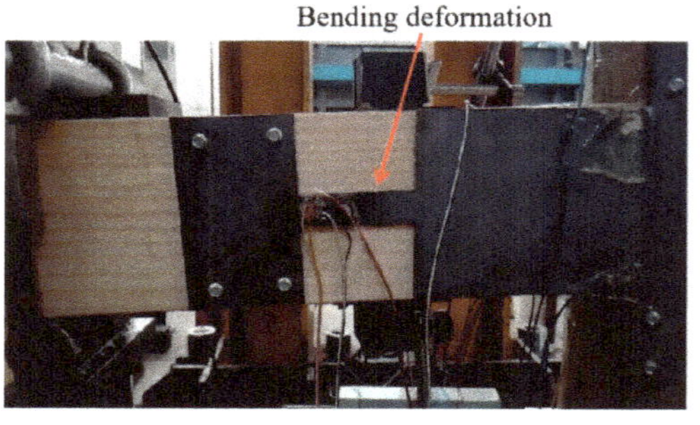

Figure 6. Bending deformation of the energy dissipation section of the metallic damper.

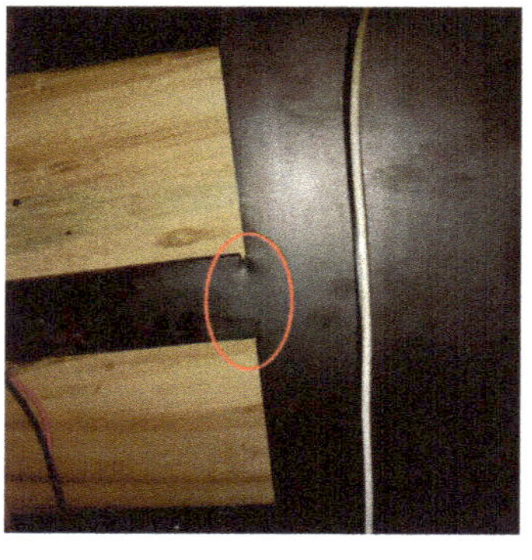

Figure 7. Crack in energy dissipation section of metallic damper.

3.2. Test Phenomena before and after Penetrated Tenons' Reinforcement

The phenomenon in the early test of the unreinforced sample TJS0 is similar to that of the ZTS0. Except, there is a "creaking" sound in each stage of loading and there is a mutual extrusion deformation between the tenon and the mortise; it is also obvious that the tenon can be pulled out step-by-step during the reverse loading due to the existence of steps inside the tenon. The tenon pulled out during the unloading process cannot be completely recovered, and the residual tenon pulling value increases step-by-step.

For the reinforced TJS1, the weak sound of wood tearing can be heard during the first stage loading and the development and extension of dry shrinkage cracks on the column can be observed. The main test phenomenon of the damper is the same as that of ZTS1, except that the upper and lower edges of the energy dissipation section of the inner damper locally bulge outward during the fifth stage (25 mm) loading and are leveled during unloading.

3.3. Test Phenomena before and after Dovetail Tenons' Reinforcement

Due to the shortest dovetail tenon, the embedded compression effect is the weakest under the same angle. In addition, the large joint gap leads to no sound during the first two stages of loading of the unreinforced YWS0. After that, there is a "dong dong" sound during the forward loading and no-reverse loading at each stage. It is observed that the stopper moves outward with the tenon during the downward loading, and the relative sliding and extrusion effect between the top surface of the tenon and the stopper is small.

For the reinforced sample YWS1, the test phenomenon is roughly the same as that of the other two reinforced samples, mainly with the following differences: there are many small dry shrinkage cracks on the wood column, and the width of the dry shrinkage cracks near the wood-fiber tearing-sound sockets increases step-by-step during the loading, and at the same time, it can be observed that ash dust is ejected from the joints. Starting from the third stage (15 mm), the stopper at the upper part of the tenon is repeatedly pulled out and jacked in with the rotation of the beam: the later the stopper is loaded, the more obvious the stopper movement is. The energy dissipation sections of dampers on both sides are slightly bent out of the plane from loading to the end of the test, and small cracks are found at the upper and lower edges.

4. Test Results and Discussion

4.1. Analysis of Moment-Rotation Hysteresis Curve of Joints

This paper specifies that the extension direction of the actuator is positive, and the retraction direction is negative. In the loading test, it is found that only the mortise and mortise are plastic extruded, and the column and beam are almost free of bending deformation. After the test, the beam and column are basically intact, so the sampled joint's bending moment M and rotation angle θ can be calculated from Equations (1) and (2), respectively.

$$M = F \cdot L \tag{1}$$

$$\theta = \delta / L \tag{2}$$

where F is the reaction force measured at the loading point; L is the distance from reaction force to column edge; and δ is the vertical displacement at the loading point.

The moment-rotation hysteresis curve of each sample is shown in Figure 8. The analysis of three types of joints before and after reinforcement is as follows:

(1) Straight tenon. The positive and negative loading hysteretic curves of the joint samples before and after reinforcement are basically symmetrical, and the "pinch" effect of the sample ZTS1 is slower than that of ZTS0, indicating that the metallic damper reduces the relative slip between mortise and mortise during the loading process, and increases the energy dissipation of the joint. The hysteresis loop areas of the second and third cycles of loading at the same level are close to, but less than, the first cycle, which is mainly due to the stiffness degradation caused by the plastic extrusion of the tenon and mortise during the first cycle of loading. It can also be seen from the figure that the curve of ZTS1 is similar to that of ZTS0 under the first and second stage loadings, which indicates that the internal clearance of ZTS1 is large, while the tenon and mortise are not squeezed with each other when the angle deformation is small. At this time, the bending moment is only provided by the damper. From the change of reaction force at the loading point, the bending moments of ZTS0 and ZTS1 grow with the increase in rotation angles. The growth rate of the former is significantly lower than that of the latter. After the reinforcement, the bearing capacity of the joint is significantly improved. Within the rotation range shown in the figure, the maximum bending moment of the ZTS1 joint is 6.4 KN·m, about 2.2 times that of ZTS0.

(2) Penetrated tenon. The unreinforced TJS0's forward and reverse loading hysteresis curves are asymmetric, which is manifested in that the reverse loading's hysteresis loop area and bearing capacity are slightly larger than those of the forward loading, because the step tenon structure is asymmetric, which leads to the difference between the mechanical mechanisms of forward and reverse rotations [33]. The relative symmetry of TJS1 curve is due to the fact that the proportion of the reaction provided by the damper with the same forward and reverse mechanical properties than the wood joint itself means that the difference between the forward and reverse directions of the whole figure is reduced. The curve "pinching" effect of the through mortise joint after reinforcement is greatly reduced, but not as good as that of the straight mortise joint after reinforcement, indicating that the slip of the through mortise joint before and after reinforcement is large. The reaction force at the loading position of TJS1 increases significantly, and the bending moment reaches the maximum value of 5.5 KN·m near the angle of 0.05 rad, while the maximum value of unreinforced TJS0 in the rotation section shown in the figure is 1.9 KN·m.

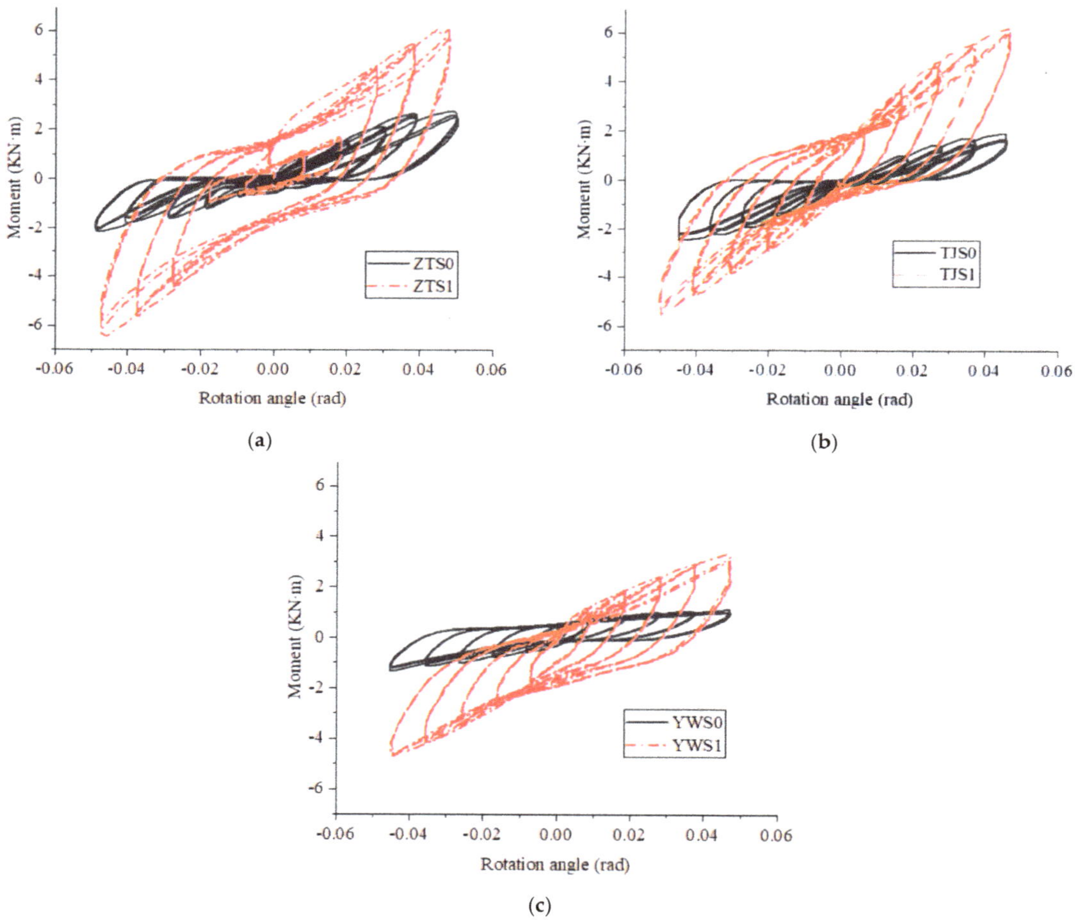

Figure 8. Comparison of M-θ curves of three types of tenon joints before and after reinforcement: (**a**) Straight tenon; (**b**) Penetrated tenon; (**c**) Dovetail tenon.

Dovetail tenon. Among the three types of unreinforced mortise and tenon joints, YWS0 exerted the lightest "pinch" effect. This is because the dovetail tenon enhances the extrusion and friction between its sides and the mortise, effectively limiting the slip of their joint. The specimen was disassembled after the test, and it was found that there was a large amount of wood chips in the mortise, the mortise and the side of the mortise were seriously worn (see Figure 9), and the waste heat after friction was obviously felt. The enhanced YWS1 also exerts a slight "pinch" effect, which is lighter than the other two types of reinforced joints, and the overall inclination of the curve turned out to be the smallest. From the change of reaction force at the loading point, the unreinforced YWS0 bending moment value is small and the growth rate is slow. Under each level of loading, the reaction force of the straight tenon and penetrated tenon is significantly greater than that of the dovetail tenon, because the reaction force is mainly determined by the internal embedded pressure of the joint, while the dovetail tenon is only half the length of the other two types of tenon, and its internal embedded pressure is the smallest. In addition, the processing technology level and wood discreteness also have a certain impact on the reaction force at the loading point. The YWS1 reaction force of the reinforced sample is significantly increased, but still significantly less under the forward loading than under the reverse loading. It is observed that the cork at the upper part of the tenon is pulled out to the outside and the internal gap of the joint becomes larger when loading downward, and the cork is pressed inward when loading upward, so the reaction force increases faster under the reverse loading.

Figure 9. Wear of dovetail tenons and mortises.

4.2. Skeleton Curve Analysis

We connected the extreme points obtained by loading at all levels to obtain the skeleton curve of bending moment and rotation angle of three types of mortise and tenon joints before and after reinforcement, as shown in Figure 10.

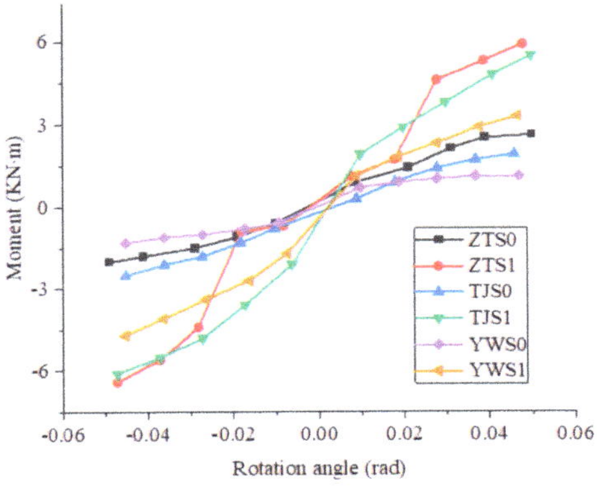

Figure 10. Skeleton curve of each joint sample.

From the analysis of curve change, the skeleton curve of the unreinforced joint sample rises gradually and there is no falling section, indicating that the three types of mortise have good deformation capacity. The bending moments of ZTS0 and TJS0 both grow rapidly with the increase of the rotation angle, while that of YWS0 hardly grows after the rotation angle reaches 0.01 rad. This is because the bearing capacity of straight tenon and penetrated tenon is mainly provided by the embedded pressure inside the joint. The larger the rotation angle is, the stronger the embedded pressure is, while the dovetail tenon is mainly borne by the internal friction, and the friction does not change significantly with the rotation angle. In addition, the slope of the skeleton curve of the unreinforced joint decreases gradually, because the plastic extrusion deformation between the mortise and tenon increases with the increase of the angle, resulting in the attenuation of the joint stiffness, and the slope of the whole curve decreases step-by-step.

The skeleton curve of the strengthened joints ZTS1, TJS1 and YWS1 under forward and reverse loadings is relatively steep on the whole, and the damper has an obvious effect on improving the bearing capacity of each joint. At the initial stage of loading, the slope of the skeleton curve of the damper mounted sample ZTS1 first decreases and then increases sharply, which is the transition from the joint compaction to the mortise and tenon compression, reflecting the large internal clearance of ZTS1: the slope of the enhanced TJS1 and YWS1 skeleton curves decreases gradually from the beginning of loading to the end, indicating that the internal gap of the joints is small.

4.3. Stiffness-Degradation Curve Analysis

The stiffness of the wood joint is closely related to the size, clearance and wood properties of the mortise and tenon. With the increase in the loading displacement, the embedded pressure and friction between the mortise and the mortise will increase, which will cause the stiffness degradation. The positive and negative secant stiffness of the sample under various levels of loading is calculated according to Equation (3).

$$K_i = M_i/\theta_i \quad (3)$$

where M_i is the peak bending moment in the first positive (negative) cycle of the i-stage displacement loading; θ_i is the angle value corresponding to M_i.

We calculated the stiffness of the sample at all angles according to the above formula, and drew the degradation curve into Figure 11:

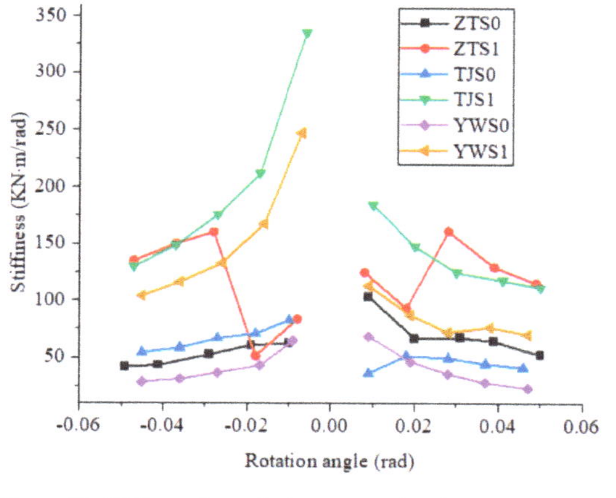

Figure 11. Stiffness-degradation curve of each joint sample.

The average stiffness values of unstiffened joints ZTS0, TJS0 and YWS0 are all as low as 61.7 KN·m/rad, 55.9 KN·m/rad and 40.9 KN·m/rad, respectively. The stiffness decreases gradually, and the degradation rate becomes slow under forward and reverse loadings. From the curve change law, it can be seen that the stiffness of each gradient of the joint under forward loading is: ZTS0 > TJS0 > YWS0, and that under reverse loading is: TJS0 > ZTS0 > YWS0. It can be seen that the stiffness of the TJS0 joint under forward loading is less than that under reverse loading, which is the difference in clockwise and counterclockwise mechanical properties caused by the shape of the tenon ladder.

The joint stiffness of strengthened ZTS1, TJS1 and YWS1 is greatly improved under both forward and reverse loading, and their average values are 120.4 KN·m/rad, 168.7 KN·m/rad and 118.6 KN·m/rad, respectively. In general, the stiffness of each joint deteriorates with the increase in the rotation angle, and the stiffness of ZTS1 decreases first and then increases, which is due to the large internal clearance. The TJS1 stiffness of the sample decreases step-by-step, and the reverse stiffness is slightly greater than the forward stiffness under all levels of loading. The stiffness degradation curve of the YWS1 sample is seriously asymmetric and the reverse stiffness is always greater than the positive one. This is because the cork at the top of the dovetail tenon moves outward during positive loading, resulting in low joint loosening stiffness, while the cork squeezes inward during reverse loading, resulting in high joint stiffness.

4.4. Energy-Dissipation Capacity Analysis

In the low-cycle reciprocating test, the hysteresis loop of the wood joint sample is shown in Figure 12. Through the analysis of the hysteresis loop, the equivalent viscous damping coefficient he of the sample can be obtained, which can reflect the energy dissipation capacity of the joint. Its specific calculative equation [34,35] is as follows:

$$h_e = \frac{S}{2\pi(S_{\Delta CEO} + S_{\Delta DFO})} \tag{4}$$

where the numerator is the area of the hysteresis loop (the shaded part in Figure 12). This paper applies the area of the first cycle of each stage, and the denominator bracket part is the sum areas of the triangles CEO and DFO.

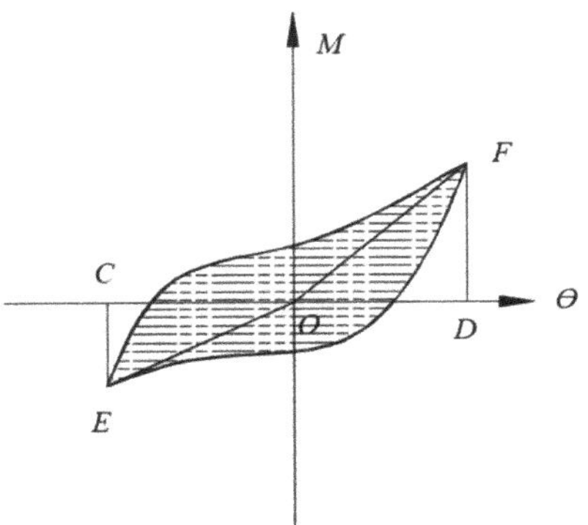

Figure 12. Schematic diagram of calculation method of equivalent viscous damping coefficient.

The variation curve of the equivalent viscous damping coefficient he of each joint with the rotation angle is shown in Figure 13.

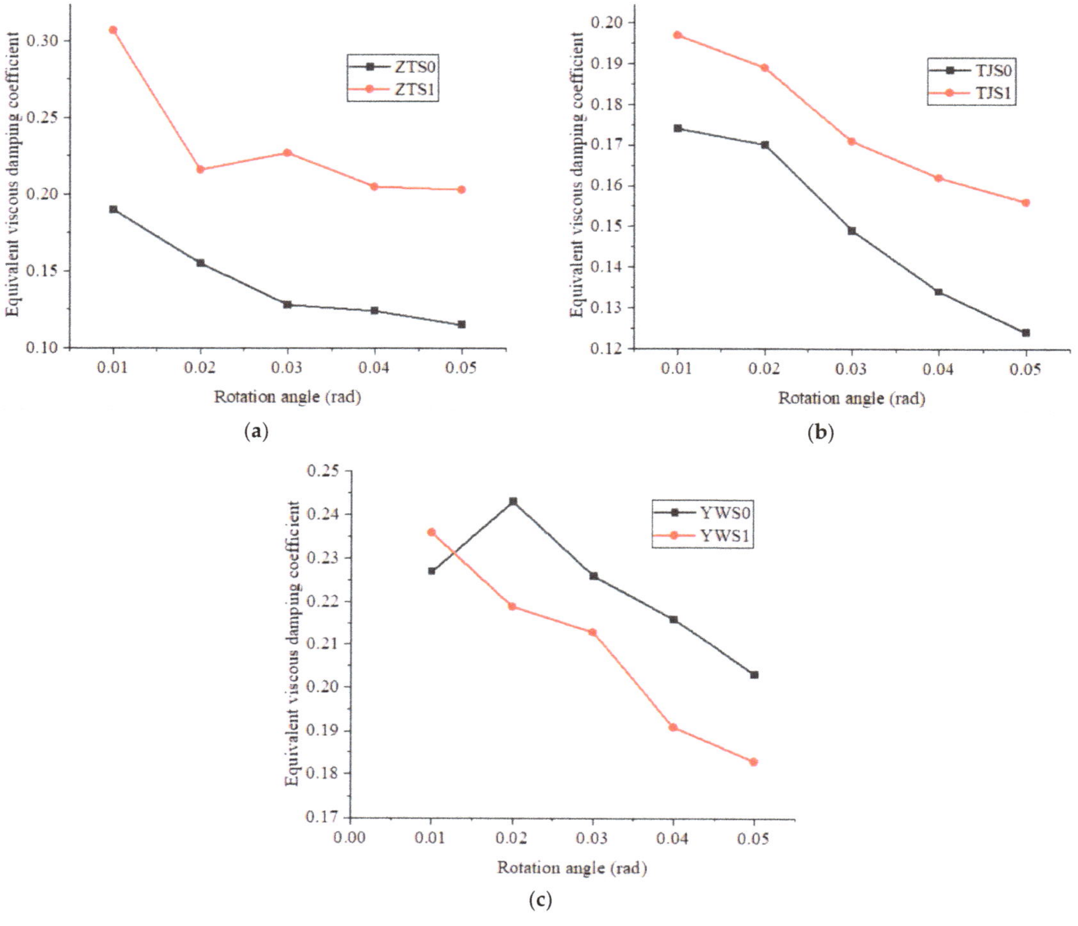

Figure 13. Relationship between equivalent viscous damping coefficients and rotation angles of samples before and after reinforcement: (**a**) Straight tenon; (**b**) Penetrated tenon; (**c**) Dovetail tenon.

In general, the energy dissipation capacity of all unreinforced joints decreases with the increase of the angle, because the energy dissipation of joints mainly depends on the friction between the tenon and the socket and the compression deformation. In the process of repeated loading, the friction surface between the tenon and the socket is gradually smooth, and the plastic deformation caused by the compression is gradually reduced. From the curve comparison, it can be seen that the equivalent viscous damping coefficient of the dovetail joint in the unreinforced joint is the largest, and the reason can be analyzed from the calculative equation: First, as for the molecular term, although the interaction between the top, bottom and the mortise during loading is reduced due to the halving of the length of dovetail joint, the two sides and the mortise will also be squeezed and rubbed to consume energy. In fact, the difference between the total energy dissipation of each stage of the dovetail tenon (numerator in Equation (4)) and the straight tenon and step tenon is small, which can be verified by the calculation results of the hysteretic loop area of each stage of the three joints. Secondly, from the denominator, it can be seen from the previous hysteresis curve that the bearing capacity of dovetail joints with the same angle is relatively small, so its elastic strain energy (denominator in Equation (4)) is the lowest.

The average equivalent viscous damping coefficients of the unreinforced samples ZTS0 and TJS0 are 0.14 and 0.15, respectively, and those of the damper mounted samples ZTS1 and TJS1 are 0.23 and 0.18, respectively, indicating that the energy dissipation capacity of the joints of the mortise and through mortise samples is significantly increased and the lifting range of the mortise is more significant. The equivalent viscous damping coefficient of YWS1 joint is slightly lower than that of unstiffened YWS0, because the growth rate of joint energy dissipation after mounting the damper is less than that of elastic strain energy. This means that to improve the equivalent viscous damping coefficient of dovetail joint, it is necessary to design a damper suitable for the stiffness, and control the increase range of joint strain energy while increasing the joint energy dissipation.

Furthermore, another evaluation index of the energy dissipation capacity of the joint is the staged energy dissipation. In this paper, the average value of the area enclosed by three complete hysteretic curves at each stage is taken as the stage energy dissipation of the joint. The calculation results are shown in the curve in Figure 14. It can be seen from the figure that the energy dissipation curves of the unreinforced straight tenon, step tenon and dovetail tenon almost coincide, and the energy dissipation of three joint stage is low, and the maximum value is close to 0.07 KN·m. After mounting the metallic damper, the energy dissipation of each stage of each joint increases significantly, and the maximum energy dissipation in the ZTS1 stage reaches 0.33 KN·m.

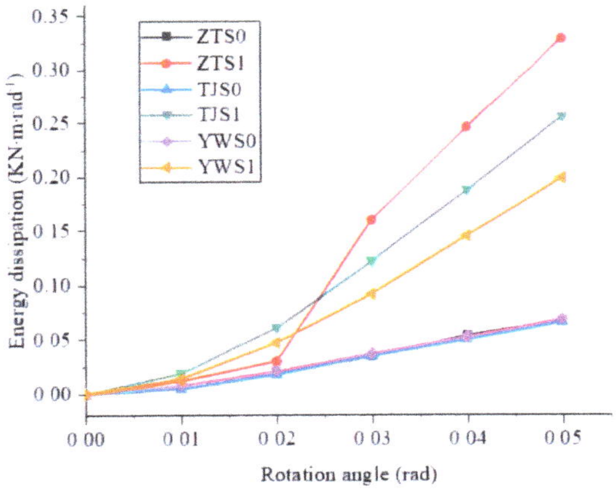

Figure 14. Comparison of energy dissipation of the joint sample at different stages.

4.5. Tenon's Pulling Amount Analysis

Under the low-cycle round-trip load, the tenon is repeatedly pulled out and reset from the mortise, and the horizontal displacement of the tenon relative to the initial position after each level of loading is called the residual tenon pulling amount. Research [36] shows that the tenon pulling amount will not only affect the bearing capacity of the joint, but also lead to the tenon breaking of the joint and the collapse of the building in serious cases. In the test, the residual tenon pulling amount of each test piece measured by the displacement meter is shown in Figure 15. From the figure, it can be seen that the residual tenon pulling amount of each joint test piece increases with the increase of the angle. The tenon pulling amount of the step tenon at the same angle is the largest, followed by the straight tenon, and the dovetail tenon is the smallest. The maximum residual pulling amounts of straight tenon, penetrated tenon and dovetail tenon before reinforcement are 4.5 mm, 8.9 mm and 3.9 mm, respectively. The corresponding tenon pulling values after reinforcement are 0.9 mm, 3.3 mm and 0.54 mm, respectively, indicating that the damper can effectively control the residual tenon pulling of the tenon joint.

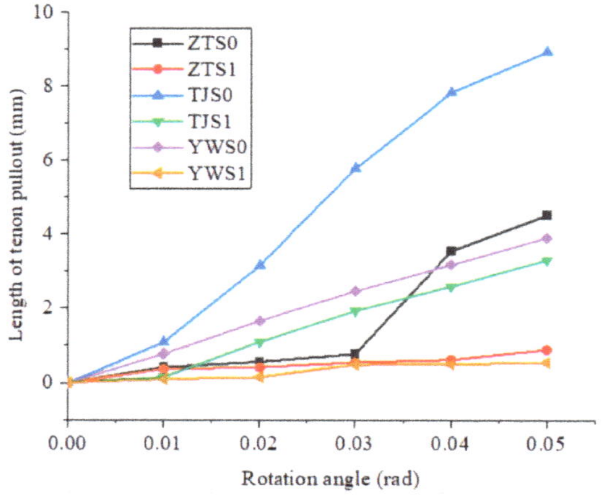

Figure 15. Relationship between tenon's residual pulling amount and angle.

5. Conclusions

In this paper, an innovative metal damper is proposed and applied to the straight tenon, penetrated tenon and dovetail tenon. The seismic performance indexes of the three types of joints before and after reinforcement are compared through low-cycle repeated loading tests, and the following conclusions are obtained:

(1) The moment–rotation hysteretic curves of the samples all have the "pinch" effect, in which the dovetail joint is the lightest and the through joint is the most significant, and the slip of the through joint is the largest during the loading process. After strengthening, the hysteretic curves of the three types of joints are fuller and the "pinch" phenomenon is reduced, the bearing capacity of the joints is significantly increased, and the relative slip is reduced.

(2) The stiffness of ZTS0, TJS0 and YWS0 joints decreases with the increase of the rotation angle and the change rate is relatively slow. The stiffness under the forward load of the same level is: ZTS0 > TJS0 > YWS0, and under the reverse load is: TJS0 > ZTS0 > YWS0; After adding dampers, the stiffness of the joint is greatly increased and the increase in the reverse stiffness is greater than that of the forward stiffness. Compared with the unreinforced joint, the overall stiffness degradation curve becomes steeper.

(3) The energy dissipation of the three types of mortise and tenon joints strengthened with dampers increased significantly, and their equivalent viscous damping coefficients decreased gradually with the increase of the joint angle. Compared with the blank group, the equivalent viscous damping coefficients of the straight mortise and through mortise joints increased, while the dovetail joints decreased slightly.

(4) In this paper, the innovative metal damper can effectively reduce the residual tenon pulling amount of mortise and tenon joints and prevent the premature tenon pulling failure of joints.

(5) Cracks are found in the damper energy dissipation section after the test, and the right-angle transition at both ends is easy to cause stress concentration, so it should be designed as an arc transition section.

Generally, the dampers significantly improve the stiffness and energy dissipation capacity of the mortise and tenon joints of wood-frame products, thereby weakening the adverse effects of horizontal vibration loads on them, and this device can be promoted for use in wood-frame houses and wooden furniture.

Author Contributions: Conceptualization, S.Y. and W.P.; Formal analysis, W.P.; Funding acquisition, L.Y. and D.W.; Investigation, S.Y.; Methodology, S.Y.; Project administration, W.P.; Resources, D.W.; Supervision, H.S.; Validation, H.S.; Writing—original draft, S.Y.; Writing—review & editing, S.Y. All authors have read and agreed to the published version of the manuscript.

Funding: This research was funded by the National Key Research and Development Project of China (No. 2020YFD1100703-04) established by the Ministry of Science and Technology of the People's Republic of China and the Yunnan Provincial Education Department Scientific Research Fund Project (2022J0066) established by the Yunnan Provincial Education Department.

Institutional Review Board Statement: Not applicable.

Informed Consent Statement: Not applicable.

Data Availability Statement: Not applicable.

Acknowledgments: The authors would like to express their gratitude to the teachers and staff employees of the Earthquake Engineering Researching Center of Yunnan for their assistance and support.

Conflicts of Interest: The authors declare no conflict of interest.

References

1. Qu, Z.; Dutu, A.; Zhong, J.; Sun, J. Seismic Damage of Masonry Infilled Timber Houses in the 2013 M7.0 Lushan Earthquake in China. *Earthq. Spectra.* **2015**, *31*, 1859–1874. [CrossRef]
2. Qu, Z.; Zhong, J.; Sun, J. Seismic damages of non-seismic design residential houses in the M7.0 Lushan Earthquake. *J. Build. Struct.* **2014**, *35*, 157–164.
3. Pan, Y.; Chen, J.; Bao, Y.; Peng, X.; Lin, X. Seismic damage investigation and analysis of rural buildings in MS 6.0 Changning earthquake. *J. Build. Struct.* **2020**, *41*, 297–306.
4. Dai, B.; Tao, Z.; Ye, L.; Gao, Y.; Huangfu, S. Investigation and Analysis of Seismic Damage to the Renovated Rural Dilapidated Residences in Mojiang M5.9 Earthquake. *J. Hunan Univ. (Nat. Sci.)* **2021**, *48*, 121–131. (In Chinese)
5. Dai, B.; Tao, Z.; Xu, G.; Li, H.; Wu, L. Investigation of seismic damage to rural houses in the epicenter area of Yangbi MS 6.4 earthquake in Yunnan. *World Earthq. Eng.* **2021**, *7*, 9–18.
6. Zhang, F.; Gao, Z.; Xue, J.; Zhu, W.; Lu, J. Research on failure analysis and reinforcement measures of ancient timber structure under earthquakes. *China Civ. Eng. J.* **2014**, *47*, 29–35.
7. Parisi, M.A.; Cordié, C. Mechanical behavior of double-step timber joints, 2010. *Constr. Build. Mater.* **2010**, *24*, 1364–1371. [CrossRef]
8. Qiao, G.; Li, T.; Chen, Y. Assessment and retrofitting solutions for an historical wooden pavilion in China. *Constr. Build. Mater.* **2016**, *105*, 435–447. [CrossRef]
9. Wang, B.; Huang, Q.; Liu, X. Deterioration in strength of studs based on two-parameter fatigue failure criterion. *Steel Compos. Struct.* **2017**, *23*, 239–250.
10. Efe, H.; Erdil, Y.Z.; Kasal, A.; Imirzi, H.Ö. Withdrawal strength and moment resistance of screwed T-type end-to-side grain furniture joints. *Forest Prod. J.* **2004**, *54*, 91–97.
11. Vassiliou, V.; Barboutis, I.; Kamperidou, V. Strength of corner and middle joints of upholstered furniture frames constructed with black locust and beech wood. *Wood Res.* **2016**, *61*, 495–504.
12. Kamboj, G.; Gaff, M.; Smardzewski, J.; Haviarová, E.; Borůvka, V.; Sethy, A.K. Numerical and experimental investigation on the elastic stiffness of glued dovetail joints. *Constr. Build. Mater.* **2020**, *263*, 120613. [CrossRef]
13. Fu, G. Study on Reinforcement and Transformation Methods for Timber of Traditional Folk Houses. Master's Thesis, Kunming University of Science and Technology, Kunming, China, 2016.
14. Zhou, Q.; Yan, W.; Guan, H. An experimental study on aseismic behavior of Chinese ancient structure by different methods. *China Cult. Herit. Sci. Res.* **2013**, *2*, 56–62.
15. Xiong, H.; Liu, Y.; Yao, Y.; Li, B. Experimental study of reinforcement methods and lateral resistance of glued-laminated timber post and beam structures. *J. Tongji Univ. (Nat. Sci.)* **2016**, *44*, 695–702.
16. Yao, K.; Zhao, H.; Xue, J.; Li, J.; Xie, Q. Experimental studies on seismic characteristic of strengthened mortise-tenon joint in historical timber buildings. *J. Harbin Inst. Technol.* **2009**, *41*, 220–224.
17. Huan, J.; Ma, D.; Guo, X.; Xu, S. Experimental study of aseismic behaviors of flexural tenon joint, through tenon joint and dovetail joint reinforced with flat steel devices. *J. Beijing Univ. Technol.* **2019**, *45*, 763–771.
18. Zhang, F.; Xu, Q.; Zhang, J.; Liu, Q.; Gong, C. Experimental study on seismic behavior of timber frames with mortise-tenon joints under different strengthening methods. *J. Build. Struct.* **2016**, *37*, 307–313.
19. Jin, Y.; Su, H.; Pan, W.; He, Y.; Du, J. Experimental research on seismic performance and reinforcement comparison of mortise-tenon joints in timber structures. *J. Civ. Envir. Eng.* **2022**, *44*, 138–147.
20. Pan, Y.; Wang, C.; Tang, L.; Li, L.; Geng, P. Comparative research on flat steel and damper strengthening of straight type of tenon-mortise joints. *J. Southwest Jiaotong Univ.* **2014**, *49*, 981–986.

21. Zhou, S.; Huo, L.; Li, H. Fundation experiments research of new angle displacement damper for wood frame structures in rural areas. *Eng. Mech.* **2011**, *28*, 62–68. [CrossRef]
22. Lu, W.; Sun, W.; Gu, J.; Deng, D.; Liu, W. Experiment study on seismic performance of timber frames sthengthened with curved steel dampers. *J. Build. Struct.* **2014**, *35*, 151–157.
23. Lu, W.; Wu, W.; Shi, C.; Liu, X. Experiment study on seismic performance of mortise-tenon joints with embedded dampers. *J. Civ. Environ. Eng.* **2022**, *44*, 30–37.
24. Gao, Y.; Tao, Z.; Ye, L.; Yang, M.; Su, H.; Yu, W. Shaking table tests of mortise-tenon joints of a traditional Chuan-Dou wood structure attached with viscoelastic dampers. *China Civ. Eng. J.* **2016**, *49*, 59–68.
25. Nie, Y.; Tao, Z.; Gao, Y. Experimental study on dovetail mortise-tenon joints with viscoelastic dampers in traditional timber structures. *J. Build. Struct.* **2021**, *42*, 125–133.
26. Xue, J.; Wu, C.; Zhang, X.; Qi, Z. Experimental and numerical study of mortise-tenon joints reinforced with innovative friction damper. *Eng. Struct.* **2021**, *230*, 111701. [CrossRef]
27. Xue, J.; Wu, C.; Zhang, X.; Zhang, Y. Experimental study on seismic behavior of mortise-tenon joints reinforced with shape memory alloy. *Eng. Struct.* **2020**, *218*, 110839. [CrossRef]
28. Xue, J.; Wu, C.; Zhang, X.; Zhang, Y. Effect of pre-tension in superelastic shape memory alloy on cyclic behavior of reinforced mortise-tenon joints. *Constr. Build. Mater.* **2020**, *241*, 118136. [CrossRef]
29. *GB/T 50165-2020*; Technical Standard for Maintenance and Strengthening of Historic Timber Building. China Architecture and Industry Press: Beijing, China, 2020.
30. Li, H.; Li, G. Experimental study of structure with "dual function" metallic dampers. *Eng. Struct.* **2007**, *29*, 1917–1928. [CrossRef]
31. *GB 50009-2012*; Load Code for the Design of Building Structures. China Architecture and Industry Press: Beijing, China, 2012.
32. *JGJ/T101—2015*; Specification for Seismic Test of Building. China Architecture and Industry Press: Beijing, China, 2015.
33. Ma, L.; Xue, J.; Dai, W.; Zhang, Y. Moment-rotation relationship of mortise-through-tenon connections in historic timber structures. *Constr. Build. Mater.* **2020**, *232*, 117285. [CrossRef]
34. Shen, Y.; Yan, X.; Yu, P.; Liu, H.; Wu, G.; He, W. Seismic Resistanceof Timber Frames with Mud andStone Infill Walls in a ChineseTraditional Village Dwelling. *Buildings* **2021**, *11*, 580. [CrossRef]
35. Magenes, G.; Calvi, G.M. In-plane seismic response of brick masonry walls. *Earthq. Eng. Struct. Dyn.* **1997**, *26*, 1091–1112. [CrossRef]
36. Xie, Q.; Zhang, L.; Wang, L.; Cui, Y.; Yang, L. Finite element analysis on the cyclic behavior of straight mortise-tenon joints with pullout tenons. *Eng. Mech.* **2019**, *36*, 138–143.

forests

Article

Molecular Dynamics Study on Mechanical Properties of Cellulose with Water Molecules Diffusion Behavior at Different Oxygen Concentrations

Yuanyuan Guo, Wei Wang * and Xuewei Jiang

College of Engineering and Technology, Northeast Forestry University, Harbin 150040, China
* Correspondence: vickywong@nefu.edu.cn; Tel.: +86-133-1361-3588

Abstract: Six groups of cellulose-water-oxygen simulation models with oxygen concentrations of 0%, 2%, 4%, 6%, 8%, and 10% were established by molecular dynamics software to analyze the effect of oxygen concentration on the mechanical properties of wood cellulose during water vapor heat treatment in terms of the number of hydrogen bonds, the diffusion coefficient of water molecules, the mean square displacement of cellulose chains, and mechanical parameters. The results showed that the diffusion coefficient of water molecules increased steadily as oxygen concentration increased, which affected cell size and density to some extent. The mean square displacement of the cellulose chain at a higher oxygen concentration was larger than at a lower oxygen concentration, indicating that the cellulose chain became more unstable at high oxygen concentration. This trend was consistent with the amount of hydrogen bonds inside the cellulose chains. The analysis of mechanical parameters showed that Young's modulus and shear modulus showed a trend of increasing and then decreasing with increasing oxygen concentration, and wood cellulose had good resistance to deformation and rigidity at 2% oxygen concentration. Therefore, during the heat treatment of wood, appropriately increasing the oxygen concentration will potentially improve the rigidity and distortion resistance of wood.

Keywords: water vapor heat treatment; molecular dynamics; oxygen concentration; wood cellulose; mechanical properties

1. Introduction

As a renewable resource, wood provides great convenience for human use because of its light texture and beautiful appearance. Wood can be used as a material for interior decoration and furniture manufacturing, and because of its diversity of design styles, it greatly enriches the choice of wood products. Heat treatment is the most industrially successful and economically efficient method of wood modification [1]. However, the requirements for heat treatment of wood are very high. If care is not taken, deformation or cracking will occur, which affects the use of wood [2]. Therefore, reasonable process optimization of wood has great significance and value.

The hygroscopicity of wood can be effectively reduced by heat treatment technology. Zhou et al. investigated the hygroscopicity of wood by measuring the wood-water related parameters of heat-treated lumber thermally and showed that heat treatment lowered the fiber saturation point of wood, resulting in a decrease in the hygroscopicity of heat-treated wood [3]. Fu et al. investigated the effects of heat treatment on wood moisture absorption and other water-related properties and found that surface wettability was significantly weakened after heat treatment, which was manifested as an increased contact angle and decreased surface free energy [4]. Heat treatment technology also makes the wood more dimensionally stable, more uniform in color, and longer-lasting, but it can also lead to the destruction of the mechanical properties of the wood [5–8]. Bruno et al. found that the dimensional stability of *Paulownia tomentosa (Thunb.) Steud.* wood improved after heat treatment, but its mechanical qualities deteriorated [9]. Lee et al. examined the physical and mechanical properties of particleboards made from heat-treated rubberwood particles and found that heat treatment increased the dimensional stability of particleboards but had a negative effect on their mechanical quality [10]. Lu et al. studied the chemical changes of the heat treatment parameters (temperature and duration) on the surface color and gloss of young eucalyptus lumber. The outcomes showed that the temperature of the heat treatment had a substantial impact on the coloring properties of young eucalyptus lumber [11].

As the most common heat treatment medium in China, superheated steam is more practical, economical, and convenient than other media (inert gas, air, and hot oil). Many water vapor heat treatment processes are carried out at a certain oxygen concentration, which increases the processing efficiency of wood heat treatment to some extent. However, fire incidents occur in the heat treatment kiln. This is likely because the oxygen concentration in the heat treatment kiln is too high. Therefore, the critical value of oxygen is set at 10% in this paper. During superheated steam heat treatment, the oxygen concentration in the kiln has an impact on the mechanical properties of the heat-treated material [12]. In this paper, when the simulation results of each model are compared with the non-oxygen model, it is found that wood cellulose has good resistance to deformation and rigidity at oxygen concentrations below 6%. However, these properties are significantly diminished at oxygen concentrations higher than 6%.

Cellulose makes up a large portion of wood. Natural cellulose is a macromolecular polysaccharide that is composed of interwoven, overlapping crystalline and amorphous regions [13]. The arrangement between cellulose molecules in the crystalline region is very compact and orderly. During water vapor heat treatment, water molecules are limited to the surface of the crystalline area, so they do not interact well with the crystalline area [14]. In the amorphous region, the molecular structure between celluloses is disordered [15], which has a strong adsorption effect on water molecules. This results in the structure between celluloses to become easily destroyed by water molecules. Therefore, this study carefully investigates the effect of the interaction between the amorphous region and water vapor on the heat treatment of wood at different oxygen concentrations.

Nowadays, analog computing techniques have gained prominence. The use of molecular simulations to derive the microscopic properties of molecules and predict the microscopic, mesoscopic, and macroscopic properties of products and materials has become an emerging theoretical development trend. In this paper, simulations are performed using Materials Studio software, which is widely used in materials science and for constructing various 3D molecular models [16]. This software provides greater theoretical support for existing macroscopic experiments and provides a predictive research tool for the implementation of macroscopic experimental protocols, which can reduce the scientific research costs and time costs associated with macroscopic experiments.

Although extensive experimental studies have been conducted to analyze the structure of cellulose, simulation approaches to cellulose studies are relatively rare. This is not surprising because the complexity of cellulose structure makes it difficult to study using traditional molecular simulation methods. A theoretical framework for the study of cellulose polymers was provided by Fukuda et al., who utilized molecular dynamics simulations to examine the diffusion pattern of water molecules in various polymers [17]. In order to model and simulate cellulose, Ftanaka et al. carried out a comparison study of the arrangement of cellulose molecules both with and without moisture [18]. Liao et al. conducted a thorough investigation into the mechanism by which cellulose ages when exposed to moisture. The distribution of hydrogen bonds between molecules was examined under the influence of an increasing temperature and thermal field in a mixed system with a varying moisture distribution. It was found that the diffusion results were unaffected by the distribution of water molecules and that the water molecules would eventually penetrate the amorphous region of cellulose and form hydrogen bonds with other molecules, speeding up the thermal aging of insulating paper [19].

This research examined how varied oxygen concentrations during water vapor heat treatment impacted the mechanical characteristics of wood cellulose. The diffusion coefficient of water molecules, cell volume and density, mean square displacement of cellulose chains, hydrogen bonds, and mechanical parameters were analyzed. Its primary objective is to investigate the changes in macroscopic properties of wood during heat treatment and to give more theoretical support for wood modification.

2. Materials and Methods

2.1. Modeling

Based on the study of Jiang et al. on the effects of different oxygen concentrations on the properties of heat-treated wood, this paper constructs a microscopic model of the main components of wood using molecular dynamics simulation to explain these changes from a microscopic perspective. And according to the settings of oxygen concentration in the macroscopic experiments by Jiang et al., the number of oxygen and water molecules in six samples were set to 0 and 50, 1 and 49, 2 and 48, 3 and 47, 4 and 46, and 5 and 45, corresponding to oxygen concentrations of 0%, 2%, 4%, 6%, 8%, and 10% in this study [12]. The amorphous polymer building process was used to create the amorphous region of cellulose [20]. In actual applications, the chain length of cellulose is up to several thousand, and the degree of polymerization of several thousand is typically not used due to computer performance and calculation speed. The results of the available studies show that the properties of the material are little affected by the degree of cellulose polymerization [21–23]. Therefore, in order to reduce the simulation time and calculation workload, the cellulose chain with a polymerization degree of 20 was chosen for this study [24]. The model was designed using the software's amorphous cell tool, in which the density was established at 1.5 g/cm^3 [25]. Each model contained a wood cellulose molecular chain and a certain number of water and oxygen molecules. The models are shown in Figure 1.

2.2. Dynamic Simulation

In the simulation process, the choice of the ensemble and force field is particularly important. The ensemble is a collection of a large number of independent systems with identical properties and structure, in various states of motion, under certain constraints. The cohesion is a fundamental concept introduced by the statistical method to describe the statistical regularity of a thermodynamic system, which is not the actual object but the system that constitutes the cohesion. The constraints are represented by a set of applied macroscopic parameters. In the application of classical molecular dynamics simulation methods, two states of the system—equilibrium and nonequilibrium—can be simulated. In the equilibrium system state, molecular dynamics simulations can be divided into micro-regular system molecular dynamics (NVE) simulations, regular system molecular dynamics (NVT) simulations, isothermal isobaric system molecular dynamics (NPT) simulations, and isoenthalpic isobaric system molecular dynamics (NPH) simulations [26].

The molecular force field is the key core of molecular dynamics simulation, and the appropriateness of the force field selection directly determines the accuracy of the simulation results [23]. The choice of the force field for different atoms is determined by the type of atom, which is one of the bases of simulation. Many force fields, such as PCFF, UFF, Amber, COMPASS, etc., have been developed according to the nature of different atomic structures, transforming the theoretical study of single atoms at the physical level into a deeper chemical study of atoms of many elements of the periodic table. They can calculate various properties of molecules: bond lengths, bond angle torsions, molecular chain motions, mechanical parameters, thermodynamic properties, etc.

To create the system with the least amount of energy, a preliminary geometry optimization based on a 5000-step smart algorithm was first performed when modeling was complete. The PCFF force field was used throughout the study [27]. After the system's energy was minimized and stabilized, the model was simulated at a temperature of 453.15 K under the NPT ensemble with a random beginning velocity and a total simulation length of 1 ns, which was collected every 5000 steps for analysis. The selection of the simulation step length determines the total simulation time and the accuracy of the results. Too long a step length may cause intense collisions between molecules and lead to data overflow in the system; too short a step length may reduce the ability of the system to search the phase space during the simulation. The temperature control was performed by Nose [28]. The pressure control was carried out by Berendsen [29]. Atom-based analyses were used to determine the Van der Waals force [30]. The Ewald method was employed to calculate the electronic effect [31].

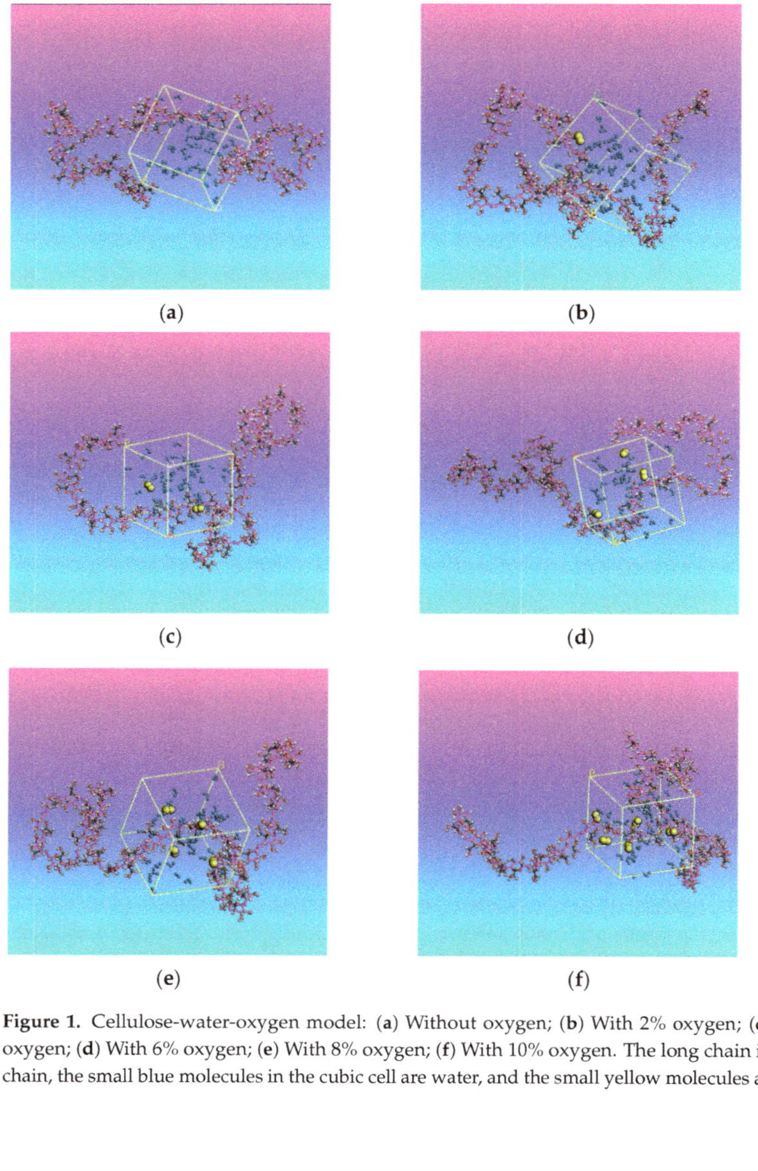

Figure 1. Cellulose-water-oxygen model: (**a**) Without oxygen; (**b**) With 2% oxygen; (**c**) With 4% oxygen; (**d**) With 6% oxygen; (**e**) With 8% oxygen; (**f**) With 10% oxygen. The long chain is cellulose chain, the small blue molecules in the cubic cell are water, and the small yellow molecules are oxygen.

3. Results and Discussion

3.1. Balance of the System

The conditions for determining equilibrium in molecular dynamics simulations are mainly temperature and energy [32,33]. The system is well balanced within a fluctuation interval of 5% to 10%. After simulating the mixed system for 1 ns, a plot of the energy and temperature versus time was obtained for a temperature of 453.15 K, as shown in Figure 2.

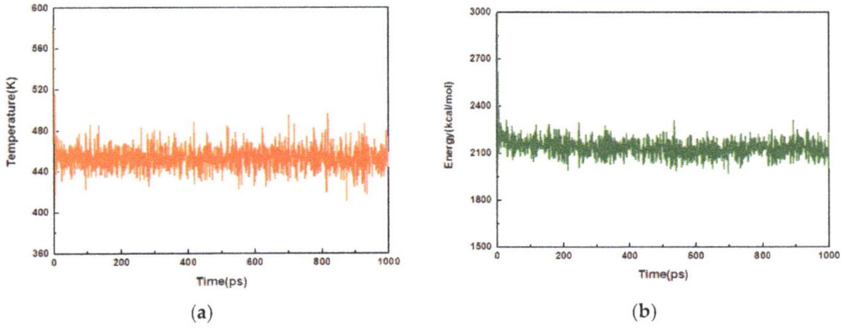

Figure 2. Temperature and energy fluctuations of the system: (**a**) Temperature–Time; (**b**) Energy–Time.

Figure 2a shows the temperature fluctuations of the model, whose values are controlled within ±25 K, which indicates that the system is in equilibrium after energy relaxation. Figure 2b shows the energy fluctuations of the model, which range between 2.22% and 3.10% according to the data in Table 1. In summary, the system maintains a good steady state, which proves the reliability of this study.

Table 1. Energy Fluctuation Values of the Model at Different Oxygen Concentrations.

Oxygen Concentration	0%	2%	4%	6%	8%	10%
Fluctuation Value	2.22%	2.71%	3.10%	2.92%	2.37%	2.79%

3.2. Diffusion Coefficient of Water Molecules

The mean square displacement (MSD) is used to observe the positional deviations of the molecules after kinetic simulations, which can represent the motion behavior and migration paths for small molecules in the model. It refers to the number of masses or moles of a substance diffused vertically through a unit area per unit concentration gradient along the direction of diffusion in unit time. The diffusion coefficient is an important part of the study of mass transfer mechanisms. The magnitude of the diffusion coefficient mainly depends on the temperature, the diffusion medium, and the type of diffusing substance, and the basic data are mainly obtained through experiments at present, but the diffusion coefficients under high temperature, high pressure, and adventitious conditions are difficult to obtain through experiments, and the intuitive process of diffusive motion of microscopic particles in the system cannot be observed experimentally. In molecular dynamics simulation (MD), the simulation calculation of diffusion coefficients not only provides microscopic information on the diffusion coefficients of substances and provides some theoretical support for mechanism analysis but also reduces the experimental cost. The MD simulation not only can observe the visual process of particle diffusion motion but also can analyze the trajectory of the particle and obtain the mean squared displacement (MSD) of the particle as a function of time, as expressed by Equation (1),

$$\mathrm{MSD} = \sum_{i=1}^{n} \left| \vec{r_i}(t) - \vec{r_i}(0) \right|^2 \quad (1)$$

In the above equation, $\vec{r_i}(t)$ and $\vec{r_i}(0)$ denote the coordinates of the molecule at time t and the coordinates of the molecule at the initial time, respectively. <·> means averaging over all atoms and initial configurations (or time series synthesis). The n represents the number of diffusing particles [13]. In order to make a quantitative comparison of the molecular diffusive motility, the diffusion coefficients (D) were further calculated, which are expressed by Equation (2) [34,35].

$$D = \frac{1}{6n} \lim_{t\to\infty} \frac{d}{dt} \sum_{i=1}^{n} \left| \vec{r_i}(t) - \vec{r_i}(0) \right|^2 \quad (2)$$

when t is large enough, D can be naturally obtained through the slope of the MSD [36], which is expressed by Equation (3),

$$D = m/6 \quad (3)$$

The m in the equation is the MSD curve slope against time obtained after fitting. After analysis, the data for each model are presented in Table 2.

Table 2. Diffusion Coefficients of Water Molecules.

Oxygen Concentration	m	D	R-Square
0%	0.4314	0.0719	0.9980
2%	0.5539	0.0923	0.9987
4%	0.5755	0.0959	0.9989
6%	0.6484	0.1081	0.9978
8%	0.6772	0.1129	0.9995
10%	0.7637	0.1273	0.9993

Table 2 demonstrates that the goodness of fit of the water molecule MSD curves are all greater than 0.9, indicating that the results are robust. Figure 3 shows that the D of water molecules gradually rose from 0.0718 to 0.1273, which means that the water molecules in the model possess more flexibility and their movement in the cellulose chain is enhanced. This may be due to the fact that the water molecules possess more kinetic energy as the oxygen concentration increases during the heat treatment.

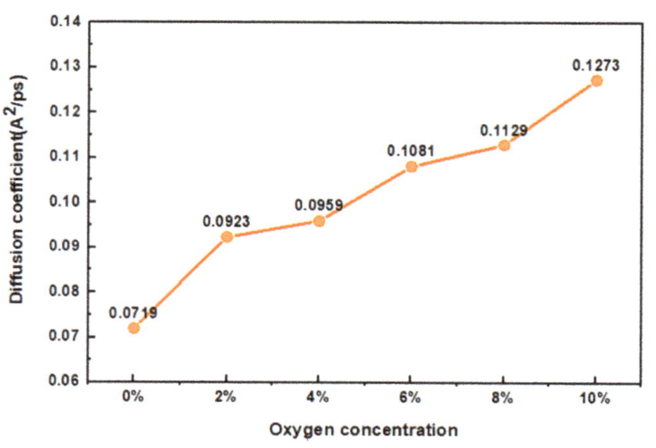

Figure 3. Diffusion coefficient of water molecules.

3.3. MSD of Wood Cellulose Chains

In the process of studying the effect of hydrogen sulfide on the performance of transformer cellulose insulating paper, Du et al. used the MSD of cellulose chains to represent its movement [37]. The main components of wood and transformer insulating paper are cellulose, so the MSD of wood cellulose can be used to represent the movement of cellulose chains. The degree of violent motion of the cellulose chains determines their stability; the more violent their motion, the less stable they are, thus reducing the overall structural stability of the wood.

Figure 4 reveals that the MSD of the cellulose chains decreased and then rose as the oxygen concentration increased. Compared to the non-oxygen model, the MSD of the cellulose molecular chains was smaller at low oxygen concentrations, indicating that their movement was less displaced and more thermally stable at this time. When the oxygen concentration was greater than 4%, the MSD of wood cellulose chains increased as the oxygen concentration increased. At this time, the motion of cellulose chains was more violent, and the stability and resistance to deformation decreased [38]. This is because the increase in the diffusion coefficient of water molecules accelerates the thermal movement of molecules, making it easier for water molecules to affect the internal structure of cellulose molecular chains, which in turn leads to deformation and cracking of wood [39].

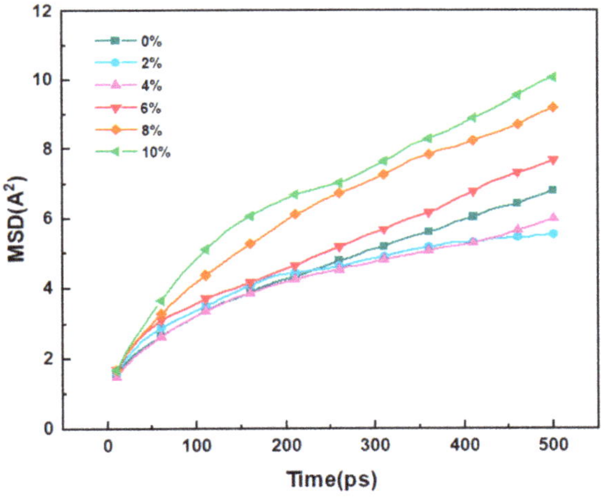

Figure 4. Mean square displacement curves of cellulose chains under different oxygen concentrations.

3.4. Lattice Parameters and Density

The cell formed by this system is a cube. Its size can be expressed by the lattice parameters. After molecular dynamics simulation, the cell parameters and volume variation of the model in different environments are illustrated in Table 3.

The cell size and density are closely related, and as the cell volume increases, its density decreases accordingly [39]. The density variation of the cellulose-water-oxygen system after water vapor heat treatment at different oxygen concentrations is shown in Table 4.

Where Final and Average represent the final and average values of the model after dynamic simulation, respectively. The final value represents the behavior situation at the last moment, and the average value shows the behavior variation situation throughout the process. The standard deviation (Std. Dev.) is a measure of the dispersion of the data distribution. If the standard deviation is lower, these values deviate less from the mean and are therefore more credible.

Table 3. Cell parameter and volume variation of the cellulose-water-oxygen system.

Oxygen Concentration	Cell Parameters			Volume (A^3)		
	the Length	the Width	the Height	Final	Average	Std.Dev.
0%	21.11	21.11	21.11	9404.397	9483.821	145.470
2%	21.16	21.16	21.16	9475.331	9535.602	121.343
4%	21.19	21.19	21.19	9520.700	9597.114	109.067
6%	21.24	21.24	21.24	9577.419	9633.354	125.889
8%	21.31	21.31	21.31	9671.309	9655.824	115.276
10%	21.37	21.37	21.37	9776.236	9756.931	162.700

Table 4. Density of the cellulose-water-oxygen system.

Oxygen Concentration	Density (g/cm^3)		
	Final	Average	Std.Dev.
0%	1.361	1.350	0.020
2%	1.353	1.345	0.017
4%	1.349	1.339	0.015
6%	1.344	1.336	0.017
8%	1.333	1.335	0.016
10%	1.321	1.324	0.021

From the values in Tables 3 and 4, it can be seen that the cell size gradually increased from 21.11 to 21.37 with increasing oxygen concentration, the average volume increased from 9483.821 to 9756.931, and correspondingly, its average density gradually decreased from 1.350 to 1.324. This may be the result of an increasing diffusion coefficient of water molecules. As the flexibility of the water molecules increases, the kinetic energy of the system increases, which in turn leads to an increase in the size of the model and a decrease in density.

3.5. Hydrogen Bonding

Hydrogen bonding refers to the powerful non-bonding interaction between a hydrogen atom bonded to an electronegative atom and a neighboring electronegative atom with a lone pair of electrons, which is essentially an electrostatic attraction between a hydrogen nucleus on a strong polar bond and an electronegative atom with a lone pair of electrons [40,41]. A hydrogen bond can be expressed as "X-H...". X is called the donor, Y is called the acceptor, and X-H is called the proton donor. Hydrogen bonds are saturated and directional, and generally, one H-atom may form one or two hydrogen bonds. The mechanical properties of cellulose are greatly influenced by the hydrogen bonding situation. There are numerous hydrogen bonds within and between cellulose molecules [42]. The change of various types of hydrogen bonds at different oxygen concentrations is shown in Table 5.

Table 5 shows that the hydrogen bonding number among cellulose chains first increases and then decreases, indicating that the low oxygen environment enhances the interactions between cellulose chains, which in turn enhances the structural stability. This may be due to the fact that the energy loss during the heat treatment was reduced under the low oxygen concentration, which helped to improve the processing efficiency of the wood heat treatment, resulting in an increase in the number of hydrogen bonds inside the cellulose chains. However, as the oxygen concentration increased, the cellulose molecular chains become more susceptible to the impact of water molecules, leading to a decrease in the number of hydrogen bonds within them. The structure's stability is not only strengthened by the formation of hydrogen bonds inside the cellulose chains, but it also alleviates the adverse effects of water molecules on the mechanical properties of cellulose to a certain extent. This also confirms the trend that the MSD of cellulose chains first decreases and then increases. At the same time, the hydrogen bonds between water and cellulose gradually decreased and the interaction between water molecules and cellulose was weakened [43], leading to a weaker adsorption of water molecules by cellulose, which also confirmed the rising tendency of the diffusion coefficient of water molecules.

Table 5. Alterations in the various hydrogen bonds.

Oxygen Concentration	Number of Hydrogen Bonds			
	between Cellulose Chains	between Water Molecular	between Water–Cellulose	Total
0%	73	30	91	194
2%	96	29	90	215
4%	91	25	81	197
6%	87	28	75	190
8%	90	21	70	181
10%	83	22	70	175

3.6. Mechanical Properties

Mechanical properties are one of the important properties of the material, directly affecting the processing and production of materials and their safe use. This is an important performance related to wood processing, production, and use. According to the principle of elastic mechanics [44], when the material is subjected to external forces, the most common connection of stress and strain satisfies the generalized Hooke's law. For a completely isotropic body, there are only two independent elastic coefficients, C_{11} and C_{12}. The reason is that the correlation of the elastic coefficients increases with the symmetry between the bodies, and their stress-strain behavior can be characterized through two individual constants, such that $C_{12} = \lambda$ and $C_{11} - C_{12} = \mu$. The rigidity matrix can be written as Equation (4),

$$[C_{ij}] = \begin{bmatrix} \lambda+2\mu & \lambda & \lambda & 0 & 0 & 0 \\ \lambda & \lambda+2\mu & \lambda & 0 & 0 & 0 \\ \lambda & \lambda & \lambda+2\mu & 0 & 0 & 0 \\ 0 & 0 & 0 & \mu & 0 & 0 \\ 0 & 0 & 0 & 0 & \mu & 0 \\ 0 & 0 & 0 & 0 & 0 & \mu \end{bmatrix} \quad (4)$$

where λ and μ are known as the Lamé coefficients, which serve to derive physical quantities, including Young's modulus (E), bulk modulus (K), shear modulus (G), and Poisson's ratio (γ).

$$E = \frac{\mu(3\lambda + 2\mu)}{\mu + \lambda} \quad (5)$$

$$G = \mu \quad (6)$$

$$K = \lambda + \frac{2}{3\mu} \quad (7)$$

$$\gamma = \frac{\lambda}{2(\lambda + \mu)} \quad (8)$$

where E is the ratio of tensile stress to tensile strain in the elastic deformation of the material and is used to measure the stiffness of the material; a larger value means greater resistance to deformation. Where γ is the proportion of lateral to lengthwise deformation during material stretching, which can reflect the plasticity. K/G denotes the proportion of the bulk modulus to the shear modulus that measures a material's resilience. The higher its value, the greater the material's resiliency [45]. Through MD calculations, the mechanical properties of each hybrid model at different oxygen concentrations were obtained, as presented in Table 6.

Table 6. Analysis of Mechanical Parameters.

Oxygen Concentration	λ	μ	G (GPa)	E (GPa)	γ (GPa)	K/G
0%	7.15	5.66	5.66	14.48	0.28	1.28
2%	8.24	8.28	8.28	20.70	0.25	1.00
4%	10.13	6.52	6.52	17.01	0.30	1.5
6%	5.41	6.52	6.52	15.99	0.23	0.85
8%	4.47	4.95	4.95	12.26	0.24	0.93
10%	3.70	4.03	4.03	9.99	0.24	0.96

Table 6 shows that the values of γ and K/G of cellulose chains did not change significantly as oxygen concentration increased. However, it can be seen from Figure 5 that E and G showed an obvious trend of first increasing and then decreasing, indicating that the deformation resistance and rigidity of cellulose chains first strengthened and then weakened. At the oxygen concentration of 2%, the shear modulus of cellulose chains was 8.28 and Young's modulus was 20.70, both of which reached maximum values. These results are consistent with the conclusion that water vapor heat treatment under the appropriate oxygen concentration can improve the flexural elastic modulus and stiffness of wood [12]. In addition, the trends of E and G are in good agreement with the trends of MSD and hydrogen bonding of cellulose chains, indicating that the internal structure of cellulose chains is relatively stable at low oxygen concentrations, but that its internal structure is damaged and becomes increasingly unstable at high oxygen concentrations.

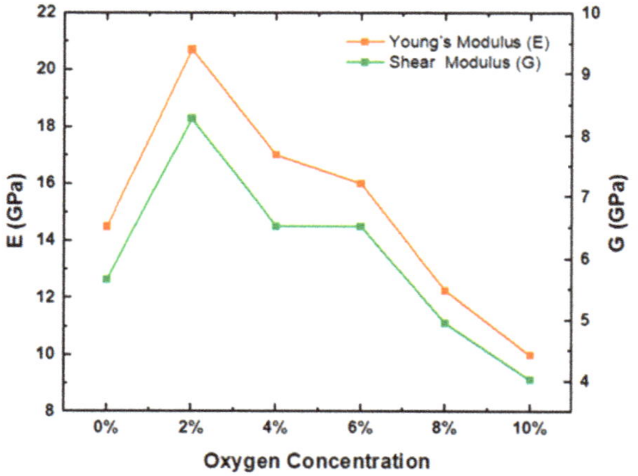

Figure 5. Young's modulus and shear modulus in the model.

There are many methods of modification of wood heat treatment, and previous studies have analyzed the effects of medium, temperature, and pressure on the mechanical properties of wood from macroscopic and microscopic perspectives. In this paper, the effect of oxygen concentration on the mechanical properties of wood during water vapor heat treatment was studied for the first time from the microscopic perspective of molecular dynamics simulation. The results demonstrate that an appropriate increase in oxygen concentration can significantly enhance the rigidity and deformation resistance of wood. This provides more theoretical support for wood modification at the microscopic level, which can better meet the requirements of wood rigidity in various industries.

4. Conclusions

This research examined how varied oxygen concentrations during water vapor heat treatment impacted the mechanical characteristics of wood cellulose. The diffusion coefficient of water molecules, cell volume and density, mean square displacement of cellulose chains, hydrogen bonds, and mechanical parameters were analyzed. The following conclusions are obtained:

1. Water molecules become more flexible in the presence of oxygen, which causes the diffusion coefficient of water molecules to gradually rise with oxygen concentration. This aids in minimizing energy loss during heat treatment, thus improving the processing efficiency. At the same time, the increase in water molecule diffusion coefficient causes a simultaneous rise in cell volume and a corresponding fall in density.
2. The MSD of cellulose chains decreases and then increases with the increase in oxygen concentration, which indicates that the thermal stability of cellulose chains is better at low oxygen concentrations, and oxygen concentrations that are too high will lead to the destruction of the internal structure of cellulose chains and greatly reduce the stability. This is related to the number of hydrogen bonds within the cellulose chain. The formation of intermolecular hydrogen bonds increases the molecular interactions within the cellulose chains and enhances the stability of the structure.
3. Young's modulus and shear modulus of cellulose chains first rise and then decline with oxygen concentration, indicating that the rigidity and distortion resistance of cellulose chains improve and then fall, peaking at 2% oxygen concentration. This change trend is compatible with the cellulose chains' hydrogen bonding and MSD change trends, which further indicate that the internal structure of cellulose chains is more stable at low oxygen concentrations. These results indicate that an appropriate increase in oxygen concentration can help to potentially improve the stiffness and resistance to deformation of wood, and also confirm the significance of this paper's research, which provides additional theoretical support for the development of wood heat treatment processes.

Author Contributions: Conceptualization, Y.G.; methodology, Y.G.; software, Y.G.; validation, Y.G., W.W. and X.J.; formal analysis, Y.G.; investigation, X.J.; resources, Y.G.; data curation, Y.G.; writing—original draft preparation, Y.G.; writing—review and editing, Y.G.; visualization, Y.G.; supervision, X.J.; project administration, Y.G.; funding acquisition, W.W. All authors have read and agreed to the published version of the manuscript.

Funding: This research was funded by the Fundamental Research Funds for the Central Universities, grant number 2572019BL04, and the Scientific Research Foundation for the Returned Overseas Chinese Scholars of Heilongjiang Province, grant number LC201407.

Institutional Review Board Statement: Not applicable.

Informed Consent Statement: Not applicable.

Data Availability Statement: Data are available upon request from the corresponding author.

Conflicts of Interest: The authors declare no conflict of interest.

References

1. Kamperidou, V.; Barboutis, I.; Vasileiou, V. Effect of Thermal Treatment on Colour and Hygroscopic Properties of Poplar Wood. In Proceedings of the 23rd International Scientific Conference: Wood is Good—With Knowledge and Technology to a Competitive Forestry and Wood Technology Sector, Zagreb, Croatia, 12 October 2012; pp. 59–67.
2. Mohebby, B.; Ilbeighi, F.; Kazemi-Najafi, S. Influence of hydrothermal modification of fibers on some physical and mechanical properties of medium density fiberboard (MDF). *Eur. J. Wood Wood Prod.* **2008**, *66*, 213–218. [CrossRef]

3. Zhou, F.; Fu, Z.; Gao, X.; Zhou, Y. Changes in the wood-water interactions of mahogany wood due to heat treatment. *Holzforschung* **2020**, *74*, 853–863. [CrossRef]
4. Fu, Z.; Zhou, Y.; Gao, X.; Liu, H.; Zhou, F. Changes of water related properties in radiata pine wood due to heat treatment. *Constr. Build. Mater.* **2019**, *227*, 116692. [CrossRef]
5. Burmester, A. Effect of heat-pressure-treatments of semi-dry wood on its dimensional stability. *Holz Als Roh- Werkst.* **1973**, *31*, 237–243. [CrossRef]
6. Cao, Y.; Lu, J.; Huang, R.; Jiang, J. Increased dimensional stability of Chinese fir through steam-heat treatment. *Eur. J. Wood Wood Prod.* **2012**, *70*, 441–444. [CrossRef]
7. Cao, Y.; Lu, J.; Huang, R.; Zhao, Y.; Wu, Y. Evaluation of decay resistance for steam-heat-treated wood. *Bioresources* **2011**, *6*, 4696–4704.
8. Giebeler, E. Dimensional stabilization of wood by moisture-heat-pressure-treatment. *Holz Als Roh- Werkst.* **1983**, *41*, 87–94. [CrossRef]
9. Bruno, E.; Helena, F.; Hélder, V.; José, F.; Idalina, D.; Luísa, C.; Dennis, J.; Lina, N. Termite Resistance, Chemical and Mechanical Characterization of *Paulownia tomentosa* Wood before and after Heat Treatment. *Forests* **2021**, *12*, 1114.
10. Lee, S.H.; Ashaari, Z.; Jamaludin, F.R.; Yee, C.N.; Ahamad, W.N. Physico-mechanical properties of particleboard made from heat-treated rubberwood particles. *Eur. J. Wood Wood Prod.* **2017**, *75*, 655–658. [CrossRef]
11. Lu, C.; Liu, Y.; Jiang, H.; Lu, Q. Impact of heat treatment on the surface color and glossiness of young eucalyptus wood. *Wood Res.* **2022**, *67*, 348–360. [CrossRef]
12. Jiang, J.; Lu, J.; Zhou, Y.; Huang, R.; Zhao, Y.; Jiang, J. Optimization of processing variables during heat treatment of oak (*Quercus mongolica*) wood. *Wood Sci. Technol.* **2014**, *48*, 253–267. [CrossRef]
13. Lee, S.H.; Rossky, P.J. A comparison of the structure and dynamics of liquid water at hydrophobic and hydrophilic surfaces—A molecular dynamics simulation study. *J. Chem. Phys.* **1994**, *100*, 3334–3345. [CrossRef]
14. Khazraji, A.C.; Robert, S. Interaction Effects between Cellulose and Water in Nanocrystalline and Amorphous Regions: A Novel Approach Using Molecular Modeling. *J. Nanomater.* **2013**, *2013*, 409676. [CrossRef]
15. Meier, R.J.; Maple, J.R.; Hwang, M.J.; Hagler, A.T. Molecular Modeling Urea- and Melamine-Formaldehyde Resins. 1. A Force Field for Urea and Melamine. *J. Phys. Chem.* **1995**, *99*, 5445–5456. [CrossRef]
16. Hou, T.; Zhang, W.; Xu, X. Binding Affinities for a Series of Selective Inhibitors of Gelatinase-A Using Molecular Dynamics with a Linear Interaction Energy Approach. *J. Phys. Chem. B* **2001**, *105*, 5304–5315. [CrossRef]
17. Fukuda, M.; Kuwajima, S. Molecular dynamics simulation of water diffusion in atactic and amorphous isotactic polypropylene. *J. Chem. Phys.* **1998**, *108*, 3001–3009. [CrossRef]
18. Tanaka, F.; Fukui, N. The behavior of cellulose molecules in aqueous environments. *Cellulose* **2004**, *11*, 33–38. [CrossRef]
19. Liao, R.-J.; Zhu, M.-Z.; Zhou, X.; Yang, L.-J.; Yan, J.-M.; Sun, C.-X. Molecular Dynamics Simulation of the Diffusion Behavior of Water Molecules in Oil and Cellulose Composite Media. *Acta Phys.-Chim. Sin.* **2011**, *27*, 815–824.
20. Theodorou, D.N.; Suter, U.W. Detailed molecular structure of a vinyl polymer glass. *Macromolecules* **1985**, *18*, 1467–1478. [CrossRef]
21. Mazeau, K.; Heux, L. Molecular dynamics simulations of bulk native crystalline and amorphous structures of cellulose. *J. Phys. Chem. B* **2008**, *107*, 2394–2403. [CrossRef]
22. Paajanen, A.; Vaari, J. High-temperature decomposition of the cellulose molecule: A stochastic molecular dynamics study. *Cellulose* **2017**, *24*, 2713–2725. [CrossRef]
23. Wang, X.; Tang, C.; Wang, Q.; Li, X.; Hao, J. Selection of Optimal Polymerization Degree and Force Field in the Molecular Dynamics Simulation of Insulating Paper Cellulose. *Energies* **2017**, *10*, 1377. [CrossRef]
24. Wang, W.; Ma, W.; Wu, M.; Sun, L. Effect of Water Molecules at Different Temperatures on Properties of Cellulose Based on Molecular Dynamics Simulation. *Bioresources* **2022**, *17*, 269–280. [CrossRef]
25. Onyon, P.F. Polymer Handbook. *Nature* **1972**, *238*, 56. [CrossRef]
26. Liu, W.K.; Karpov, E.G.; Park, H.S. *Nano Mechanics and Materials: Theory, Multiscale Methods and Applications*; John Wiley & Sons: Chichester, UK, 2006.
27. Maple, J.R.; Hwang, M.J.; Stockfisch, T.P.; Hagler, A.T. Derivation of Class II Force Fields. III. Characterization of a Quantum Force Field for Alkanes. *Isr. J. Chem.* **2013**, *34*, 195–231. [CrossRef]
28. Nosé, S. Constant Temperature Molecular Dynamics Methods. *Prog. Theor. Phys. Suppl.* **1991**, *103*, 1–46. [CrossRef]
29. Berendsen, H.J.C.P.; Postma, J.; Gunsteren, W.; Dinola, A.D.; Haak, J.R. Molecular-Dynamics with Coupling to An External Bath. *J. Chem. Phys.* **1984**, *81*, 3684–3690. [CrossRef]
30. Andersen, H.C. Molecular dynamics simulations at constant pressure and/or temperature. *J. Chem. Phys.* **1980**, *72*, 2384–2393. [CrossRef]
31. Ewald, P.P. Die Berechnung optischer und elektrostatischer Gitterpotentiale. *Ann. Phys.* **1921**, *369*, 253–287. [CrossRef]
32. Allen, M.; Tildesley, D.; Press, U. *Computer Simulation of Liquids*; Oxford University: Oxford, UK, 1989.
33. Sangiovanni, D.G.; Hellman, O.; Alling, B.; Abrikosov, I.A. Efficient and accurate determination of lattice-vacancy diffusion coefficients via non equilibrium *ab initio* molecular dynamics. *Phys. Rev. B.* **2016**, *93*, 094305. [CrossRef]
34. Brown, W.R.; Jenkins, R.B.; Park, G.S. The sorption and diffusion of small molecules in amorphous and crystalline polybutadienes. *J. Polym. Ence Polym. Symp.* **2010**, *41*, 45–67. [CrossRef]

35. Einstein, A.; Einstein, E. Zur Elektrodinamik bewegter KÖrper. *Ann. Phys.* **1905**, *17*, 891–921. [CrossRef]
36. Yang, L.; Qi, C.; Wu, G.; Liao, R.; Wang, Q.; Gong, C.; Gao, J. Molecular dynamics simulation of diffusion behaviour of gas molecules within oil–paper insulation system. *Mol. Simul.* **2013**, *39*, 988–999. [CrossRef]
37. Du, D.; Tang, C.; Zhang, J.; Hu, D. Effects of Hydrogen Sulfide on the Mechanical and Thermal Properties of Cellulose Insulation Paper: A Molecular Dynamics Simulation. *Mater. Chem. Phys.* **2019**, *240*, 122153. [CrossRef]
38. Xu, B.; Chen, Z.; Ma, Q. Effect of high-voltage electric field on formaldehyde diffusion within building materials. *Build. Environ.* **2016**, *95*, 214–218. [CrossRef]
39. Ouyang, F.; Wang, W. Effect of Thermo-Hydro-Mechanical Treatment on Mechanical Properties of Wood Cellulose: A Molecular Dynamics Simulation. *Forests* **2022**, *13*, 903. [CrossRef]
40. Li, X.; Tang, C.; Wang, J.; Tian, W.; Hu, D. Analysis and mechanism of adsorption of naphthenic mineral oil, water, formic acid, carbon dioxide, and methane on meta-aramid insulation paper. *J. Mater. Sci.* **2019**, *54*, 8556–8570. [CrossRef]
41. Li, X.; Tang, C.; Wang, Q.; Li, X.P.; Hao, J. Molecular simulation research on the micro effect mechanism of interfacial properties of nano SiO2/meta-aramid fiber. *Int. J. Heat Technol.* **2017**, *35*, 123–129. [CrossRef]
42. Hinterstoisser, B.; Akerholm, M.; Salmen, L. Load distribution in native cellulose. *Biomacromolecules* **2003**, *4*, 1232–1237. [CrossRef] [PubMed]
43. Zhang, N.; Shen, Z.; Chen, C.; He, G.; Hao, C. Effect of hydrogen bonding on self-diffusion in methanol/water liquid mixtures: A molecular dynamics simulation study. *J. Mol. Liq.* **2015**, *203*, 90–97. [CrossRef]
44. Fu, S.Y.; Lauke, B.; Li, R.K.Y.; Mai, Y.W. Effects of PA6,6/PP ratio on the mechanical properties of short glass fiber reinforced and rubber-toughened polyamide 6,6/polypropylene blends. *Compos. Part B-Eng.* **2006**, *37*, 182–190. [CrossRef]
45. Pugh, S.F. XCII. Relations between the elastic moduli and the plastic properties of polycrystalline pure metals. *Philos. Mag.* **2009**, *45*, 823–843.

forests MDPI

Article

Physical and Mechanical Properties of High-Density Fiberboard Bonded with Bio-Based Adhesives

Aneta Gumowska and Grzegorz Kowaluk *

Institute of Wood Sciences and Furniture, Warsaw University of Life Sciences—SGGW, 02-776 Warsaw, Poland
* Correspondence: grzegorz_kowaluk@sggw.edu.pl

Abstract: The high demand for wood-based composites generates a greater use of wood adhesives. The current industrial challenge is to develop modified synthetic adhesives to remove harmful formaldehyde, and to test natural adhesives. The scope of the current research included the manufacturing of high-density fiberboards (HDF) using natural binders such as polylactic acid (PLA), polycaprolactone (PCL), and thermoplastic starch (TPS) with different resination (12%, 15%, 20%). The HDF with biopolymers was compared to a reference HDF, manufactured following the example of industrial technology, with commonly used adhesives such as urea-formaldehyde (UF) resin. Different mechanical and physical properties were determined, namely modulus of rupture (MOR), modulus of elasticity (MOE), internal bonding strength (IB), thickness swelling (TS), water absorption (WA), surface water absorption (SWA), contact angle, as well as density profile; scanning electron microscope (SEM) analysis was also performed. The results showed that increasing the binder content significantly improved the mechanical properties of the panels in the case of starch binder (MOR from 31.35 N mm^{-2} to 40.10 N mm^{-2}, IB from 0.24 N mm^{-2} to 0.39 N mm^{-2} for dry starch), and reduces these in the case of PLA and PCL. The wet method of starch addition improved the mechanical properties of panels; however, it negatively influenced the reaction of the panels to water (WA 90.3% for dry starch and 105.9% for wet starch after 24 h soaking). Due to dynamically evaporating solvents from the PLA and PCL binding mixtures, a development of the fibers' resination (blending) techniques should be performed, to avoid the uneven spreading of the binder over the resinated material.

Keywords: high density fiberboard; wood; mechanical properties; physical properties; polylactide; polycaprolactone; thermoplastic starch

1. Introduction

Progressive economic development and the growing population significantly affect both the natural environment and the global biomass resources, including wood, forestry, and agricultural by-products, based on which wood-based panels are produced. Wood-based panels can be classified into three basic groups depending on the form of their structural elements. The structure and technological process determine the mechanical and physical properties. The first group consists of fiber composites, including boards manufactured using the dry forming method, where the fiber-carrying medium is air; this includes high-density fiberboards (HDF), medium-density fiberboards (MDF); and low-density fiberboards (LDF). The wet forming method, where the medium is water, is used to produce softboards (SB) and hardboards (HB). Fiber composites also include wood-plastic composites (WPC). The second group consists of particleboard composites, such as particleboards (PB) and oriented strand boards (OSB). The third group is plywood, and laminated veneer lumber (LVL), representatives of layered composites. The production of wood-based panels increases from year to year, reaching approx. 167 million m^3 for plywood, approx. 110 million m^3 for fiberboards, and approx. 100 million m^3 for particleboards [1,2]. The high demand for wood-based panels generates a greater use of wood adhesives, which automatically translates into the use of synthetic adhesives. Formaldehyde-based adhesives account for approx. 90%–95% of all wood adhesives used in the industry, and urea formaldehyde (UF) is the leader in terms of consumption, accounting for approx. 11 million tons of resin dry weight per year [3,4]. The global wood adhesives and binders market size reached 15.8 billion US dollars (USD) in 2020, and it is projected to grow by approx. 4% to reach 21.9 billion USD by 2028 [5]. It is worth adding that the above data include UF, phenol-formaldehyde (PF) resins, and soy-based binders, the latter being a good indicator for new green adhesives.

Until the end of the 1950s, adhesives of natural origin were available, which had defects, e.g., low water resistance and durability, resulting in the development of petrochemical adhesives. Synthetic formaldehyde-based adhesives have gained popularity and lack competition from natural adhesives, due to their low production cost while maintaining perfect adhesive properties such as excellent stability, low curing temperature, and fast curing, colorless and resistance to microorganisms [6,7]. The biggest challenge for formaldehyde-based resin technology is to reduce the emission of free formaldehyde, which is harmful to health and the environment during production and in final products during their entire life cycle. While levels are generally low, increasingly stringent environmental regulations will require more environmentally friendly adhesives to produce bio-panels [8]. The turn of the 20th and 21st centuries is a period of return to the development of natural adhesives, the desire to look for an alternative to conventional petroleum-based adhesives by reducing or eliminating formaldehyde, and the sustainable development of raw materials and final products [9].

MDF is a wood-based board produced by pressing lignocellulose fibers together with synthetic resins under conditions of high temperature and pressure [10]. Wood fibers comprise the dispersed phase, which is responsible for the strength and stiffness of the board, while the matrix used is a construction binder, and can be a synthetic resin, an inorganic compound, or a biopolymer. [11]. Currently, commonly used resins in the production of MDF panels are UF, melamine urea formaldehyde (MUF), PF, and isocyanate compounds [12]. On a global production scale, 90% of MDF panels are made with UF resin [6]. Commercial MDF has a higher density and strength than particleboard or plywood. Unlike particleboard, which is also commonly and widely used in furniture and construction, the MDF board has a homogeneous structure over the entire cross-section; because of this, it has excellent mechanical workability in the milling process, a high precision finish which provides an excellent base for veneers, and high dimensional stability, making it ideal for use in building materials or furniture boards [13–15].

With the predominance of petrochemical-based products raising concerns about limited fossil resources and harm to health and the environment, it is becoming prudent to obtain materials that will have similar properties to the properties of commercial wood-based panels with one difference—they will come from renewable raw materials.

A future-proof alternative, and at the same time a response to the restrictions and needs of the market, is the production of bio-composites, which means that one of at least two component materials will be of natural origin. Recyclable composites are being developed more widely due to the possible use of renewable raw materials: by-products from the agricultural industry, such as sugarcane bagasse [16], kenaf stalks [17], rice husk [18], corn [19], tomato stalks [20], sugar beet pulp [21], walnut shell [22], and forest by-products such as cones [23,24], tree bark [25,26], and branches [27]. The current state-of-the-art of particleboard manufacturing which uses environmentally-friendly agricultural biomass was widely reviewed and described by researchers [28,29]. Another important aspect of bio-composites is natural and renewably sourced adhesives. Synthetic resin can be replaced with natural binders, i.a. starch and its modifications [30–34] lignin [35], cellulose [36], soybean and soybean protein [37–40], chitosan [41], tannins [42,43], vegetable oils [44], natural rubber [45,46], and citric acid [47,48]. The use of biopolymers as a binder in wood-based material technologies has also been extensively studied. Biopolymers such as polylactic acid (PLA), polyhydroxybutyrate (PHB), and polycaprolactone (PCL) are also used as alternatives to synthetic adhesives. PLA is the most popular and the only biopolymer currently produced on an industrial scale. The production scale was 150,000 metric tons in 2013, with an increase of 10% in 2020. Currently, the production process of PLA generates 60% fewer greenhouse gases and uses 50% less non-renewable energy than the production of traditional polymers such as polyethylene terephthalate (PET) or polystyrene (PS) [49].

Song et al. [50] manufactured board in a wet method process using soybean straw fiber, with varying amounts of PLA as a binder (0, 10, 30, and 50%). They proved that the addition of PLA weakened the mechanical properties of tested boards. However, when it came to hygroscopic properties, they showed that increasing the amount of PLA improved the water resistance. Ultimately, they confirmed the validity of using PLA as a binder for fiberboards; however, a coupling agent was needed to improve the mechanical properties. Ye et al. [51] produced fiberboards from softwood fiber, wheat, and soybean straws. They proved that fibers from wheat and soybean straws can be used in a 50/50% agricultural fiber/wood fiber ratio, because they could provide comparable mechanical properties and water resistance to fiberboards made of 100% wood fiber. Many literature reports on PLA, PCL, and PHB concerning fiber composites refer to wood-plastic composites (WPC), produced by methods such as direct injection molding, extrusion, compression molding, film and sheet formation [52–56].

The current state of the art is limited on the subject of using biopolymers, e.g., PLA, PCL, PHB, as a binder in wood-based panel technology, particularly with medium-density fiberboards. This investigation aimed to assess the impact of natural biopolymer binders—thermoplastic starch (TPS), PLA, and PCL—on selected physical and mechanical properties of HDF manufactured with different resination.

2. Materials and Methods

2.1. Materials and Their Characterization

In the present study, high-density fiberboards (HDF) were produced under laboratory conditions from pine (*Pinus sylvestris* L.) and spruce (*Picea abies* L.) industrial fibers (IKEA Industry Poland sp. z o. o. brand Orla, Szczecin, Poland). The fibers were dried to a moisture content (MC) of about 4%.

Composites were manufactured with four different binders: pure, laboratory-purpose polylactide (PLA; Sigma-Aldrich, Saint Louis, MO, USA, product no. 38534), polycaprolactone (PCL; Sigma-Aldrich, product no. 704105) in drops with a diameter of 3 mm, thermoplastic starch (TPS; Grupa Azoty S.A., Tarnów, Poland) in drops and powder form, and urea-formaldehyde resin (UF; Silekol S-123, Silekol Sp. z o.o., Kędzierzyn-Koźle, Poland), the latter of which is commonly used in industry.

2.2. Fiberboard Manufacturing

The HDF were manufactured with dimensions 320 mm × 320 mm, a target density of 900 kg m^{-3}, and an average thickness of 3 mm. The fiberboards were produced using different binders as well as different resination values. All composites with biopolymers as a binder were manufactured with 12%, 15%, and 20% resination. The reference variant was produced with 12% resination, as in industrial conditions. The resination values used in this study were previously used for particleboard with biopolymers as an adhesive [57]. Increasing the resination from 12% to 15% significantly improved the physical and mechanical properties of the produced particleboards. However, increasing from 15% to 20% seemed unjustified. Therefore, in this study, the same resination values were used, but not exceeding 20%. No paraffin emulsion or wax was added during the manufacturing of the high-density fiberboards. The binder liquid solutions were mixed in a laboratory mixer and sprayed by air gun onto the lignocellulosic fibers. The TPS powder was manually applied into fibers, and then water was sprayed using a spray gun, with the amount of water depending on the percentage of resination. After applying biopolymers and blending, the resinated fibers were stored in the laboratory fume hood for 3 days to evaporate the solvents. Then, the mats were formed and pre-densified manually. To improve the heat transfer into the core of HDF, 65 g m^{-2} of water was sprayed on the top and bottom sides before hot pressing. The HDF was pressed in two stages: cold pressed (unit pressure 0.8 MPa, 30 s) to thicken the mat, and then hot pressed for 10 min at 180 °C under 2.5 MPa unit pressure (AKE, Mariannelund, Sweden). For the reference panels, the hot pressing time was reduced to 60 s, while other pressing parameters remained unchanged. For the biopolymers, an extended pressing time was used to allow the thermoplastic binders to properly melt. In the process of manufacturing wood-based panels, the pressing temperature depends on the binder used. DSC and TGA analysis for PLA, PCL [58], and starch [59] allowed us to determine the "safe" temperature of 180 °C, which does not cause degradation of the biopolymer and is sufficient to melt.

As a result, thirteen types of fiberboards were produced with different resination (hereby referred to by the shortcuts listed in Table 1), and two panels of each binder type. The manufactured HDF were conditioned in ambient conditions (20 °C; 65% R.H.) until they reached a constant weight, before being cut according to the research schedule.

Table 1. Shortcuts for every elaborated HDF.

Variants of Binders and Resination	Shortcut of HDF
UF 12%	UF12
Powder TPS 12%—dry starch	DS12
Powder TPS 15%—dry starch	DS15
Powder TPS 20%—dry starch	DS20
Drops of TPS 12%—wet starch	WS12
Drops of TPS 15%—wet starch	WS15
Drops of TPS 20%—wet starch	WS20
PLA12%	PLA12
PLA15%	PLA15
PLA20%	PLA20
PCL12%	PCL12
PCL15%	PCL15
PCL20%	PCL20

2.3. Preparation of the Adhesives

The adhesive mass for PLA, PCL, and TPS (drops) was produced by dissolving the dry mass of polymers in solvent to obtain the consistency of a thick liquid. The following solvents have been used to achieve liquid-state binders: methylene chloride (CH_2Cl_2) for PLA (about 23% dry matter content of resulting mixture), toluene (C_7H_8) for PCL [57] (about 27% dry matter content of resulting mixture), and hot water for TPS (drops). The TPS drops with hot water were mixed using a low-speed mixer, until a homogeneous suspension was achieved (about 25% dry matter content). All solution concentrations were tuned to obtain the viscosity of solutions similar to the reference binder (in the range of 420–450 mPa s). The reference HDF was produced with UF industrial resin (65% dry matter content), where the hardener was a 10% water solution of ammonium sulfate ($(NH_4)_2SO_4$) in a weight ratio of 50:5:1.5, respectively: resin: water: hardener. The curing time of the reference bonding mixture at 100 °C was about 82 s.

2.4. Physical and Mechanical Properties

The modulus of rupture (MOR) and modulus of elasticity (MOE) were characterized according to EN 310 [60], and the internal bond (IB) was determined according to EN 319 [61]. The mechanical properties were analyzed with an INSTRON 3369 (Instron, Norwood, MA, USA) standard laboratory testing machine, and as many as 10 test specimens for each fiberboard type were analyzed for the tests mentioned. Thickness swelling (TS) and water absorption (WA) at two-time intervals, i.e., after 2 h and 24 h of immersion in water, were measured according to EN 317 [62], and 12 samples of HDF panels of each binder variant were used. Surface water absorption was conducted according to EN 382–2 [63], in two repetitions for each variant of the panels. The density profile (DP) of samples was analyzed using a DA-X measuring instrument (GreCon, Alfeld, Germany). The measurement based on direct scanning X-ray densitometry was carried out at a speed of 0.05 mm s^{-1} across the panel thickness with a sampling step of 0.02 mm. The density profile was performed using three samples of each variant for a type of binder.

Contact angle measurements were made using the contact angle analyzer PHOENIX 300 (SEO—Surface & Electro Optics Co., Gyeonggi-do, Ltd., Suwon City, Korea) equipment, using the method of distilled water sessile drop. A Quanta 200 (FEI, Hillsboro, OR, USA) scanning electron microscope was used to define the surface morphology of the produced fiberboards.

2.5. Statistical Analysis

Analysis of variance (ANOVA) and t-test calculations were used to test for significant differences ($\alpha = 0.05$) between factors and levels where appropriate, using the IBM SPSS statistic base (IBM, SPSS 20, Armonk, NY, USA). A comparison of the means was performed when the ANOVA indicated a significant difference, by employing the Duncan test. The detailed *p*-values have been attached as Supplementary Material.

3. Results and Discussion

The density profile was characterized, and the results are summarized in Figure 1. The average densities of all samples are ~900 kg m^{-3}. The density profiles for individual samples of HDF were symmetrical to the middle of the thickness of the board; therefore, the graph presents the density profiles to their axis of symmetry to facilitate the analysis. Regardless of the biopolymer type or the resination used, the shape of all profiles did not differ significantly. Each of the profiles presented in the graph shows a characteristic density profile of HDF, where the core layer has a slightly lower density than the surface layers [64]. The boards produced were 3 mm thick; therefore, the difference between the surface layer density (SLD) and core density (CD) is smaller than in the case of 10 mm thick MDF boards, where the density distribution shows clear differences between the layers [65]. The effect of fiber compression is more even over the entire cross-section of the thinner board, as the set temperature of the press shelves reaches the core layer faster, causing hardening or binder liquefaction depending on the binder used [66]. The average density of the PCL12 samples on the surface achieved a value of 1090 kg m^{-3} and recorded a decrease in density in the core layer by 24%. The most homogeneous density profile was recorded for the PLA20 samples, where density decreases in the core layer by about 6%. The HDF was initially cold-pressed to densify the mat and then hot-pressed to achieve a more even density distribution. However, it should be pointed out that the mats were sprayed with water on the surfaces to improve the heat transfer during hot-pressing. This promotes a higher densification of the face layers of the produced panels with biopolymers. Wong et al. [66] manufactured 12 mm MDF which was subjected to a two-step hot pressing process. They showed that the final profile is significantly dependent on the thickness achieved in the first stage of pressing. Obtaining a smaller thickness in the first stage resulted in a greater difference between maximum surface layer density (SLD) and core density (CD). There is no noticeable difference between the profiles for thermoplastic polymers and the UF resin used in the reference panel. Figure 2 shows the changes in the appearance of the surface of the produced HDF, depending on the binder used and the resination. HDF with PLA and PCL as a binder has visible adhesive stains, which were caused by the too fast evaporation of the solvent during the covering of the fibers. Stains indicated that the biopolymer did not evenly cover the fibers in the entire board. In the boards there are places with a large amount of biopolymer, and there are also places where this biopolymer has been minimally distributed on the fibers. The graph shows that for PLA and PCL, the density profile is more homogeneous with increasing of resination.

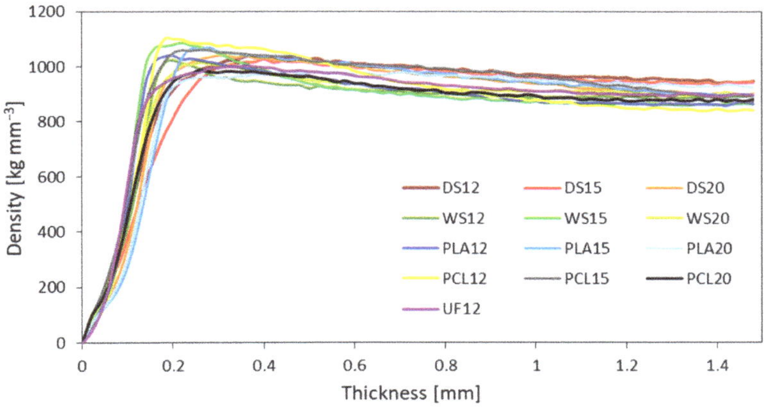

Figure 1. Density profiles of fiberboards.

Figure 2. Changes in the appearance of the surface of the produced HDF, depending on the binder used and the resination.

The average values of modulus of rupture (MOR) and modulus of elasticity (MOE) under the three-point bending stress of the tested composites are presented in Figure 3. The error bars in the graphs represent the standard deviation. The highest value of MOR (59.18 N mm^{-2}) was recorded for the reference HDF, while the lowest (23.64 N mm^{-2}) was for PLA20. In the case of MOE, the highest average value was obtained for PS20 (4225 N mm^{-2}), and the lowest was for PCL12 (2596 N mm^{-2}), as well as for MOR. Analyzing the obtained results for DS, it can be observed that with the increase in resin content from 12, 15 to 20%, the average value of both MOR and MOE increases. For the WS samples, an increase in the MOR and MOE values can be seen with an increase in the resin content from 12 to 15%. The addition of 20% starch (WS) caused a decrease in strength and MOE. In the case of PLA and PCL, we can observe the opposite situation as in the case of DS12-DS20 boards; the average values of MOR and MOR decrease with the increase in resination. It can also be seen that between 12–15% differences in the average MOR and MOE values for PLA and PCL are smaller than at 15–20%. Statistically significant differences among all variants of panels with PLA and PCL can be noted between variants PCL15 and PCL20. The explanation for the decrease in the MOR and MOE values for PLA and PCL with increasing resin content can be found in the structure of the produced HDF. As already shown above in Figure 2, there was a problem with the uniformity of the distribution of biopolymers on the fibers. The WS20 samples also show a decrease in the MOR and MOE values, although they increased from WS12 to WS15. The photo shows surface defects; thus, we can predict what the structure looks like in a cross-section of the samples. The more biopolymer, the more stain and weak areas in the board. Xiaowen et al. [50] confirmed that the addition of PLA as a binder weakened the mechanical properties (bending strength and tensile strength) of fiberboard produced using a wet method, using soybean straw fiber. They used different pressure values, which affected the final results of the tests. Of the two forms (powder, drops) of applying TPS to the fibers, the starch drops in the form of a suspension show better strength properties. The MOR value for WS12 relative to DS12 and WS15 to DS15 is higher by 32% and 30%, respectively. There are no statistically significant differences between DS20 and WS20. According to statistical analysis, there are statistically significant differences between the average values of the MOR results for UF12 and the rest of the HDF variants, also between the DS, WS, PLA, and PCL20 samples. Taking into account statistical analysis, there are no statistically significant differences between the average values of the MOE results from UF12, WS15, and WS20; DS12-DS20, WS20, PLA12, PLA15, and PCL12.

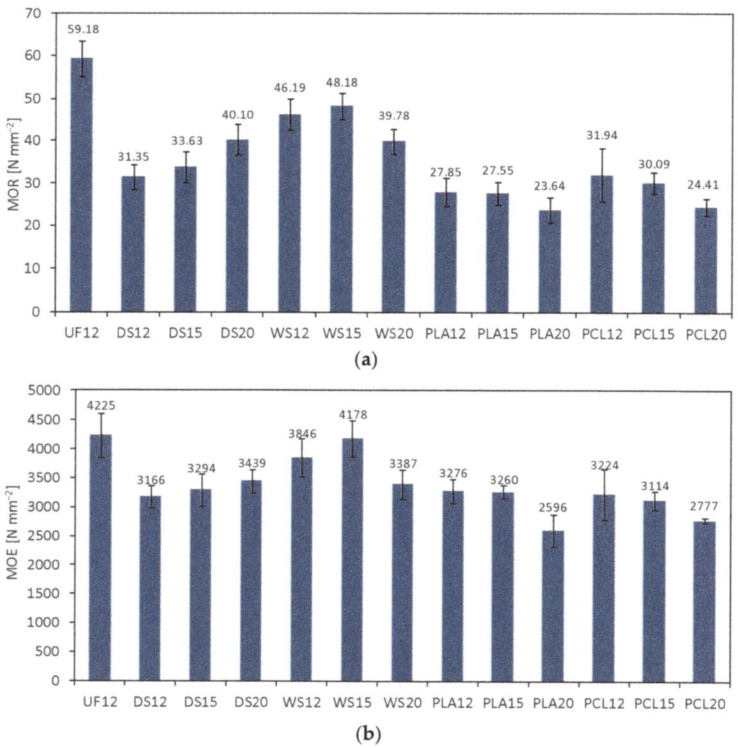

Figure 3. (**a**) Modulus of rupture (MOR) and (**b**) modulus of elasticity (MOE) of tested HDF.

The obtained results of internal bonding (IB) tests are presented in Figure 4. The error bars in the graph represent the standard deviation. The outcomes show that the highest average value of IB was that of the reference HDF (1.73 N mm^{-2}), and the lowest value was for DS12 (0.24 N mm^{-2}). Analyzing the results, it can be seen that the lowest IB values were recorded for DS and PLA samples, regardless of the level of resination. The same relation refers to IB as to MOR and MOE. For DS, an increase of resination from 15% to 20%, and for WS12 to WS15, causes a slight increase in average IB values. The PLA and PCL board variants show a decrease in the IB with an increase in resination. Better results in this test were also obtained by starch in the form of drops, which were applied to the fibers in the form of a thick liquid using a pneumatic gun. This method of the application allowed us to obtain better coverage of the fibers. The starch powder probably did not fully mix with the introduced water, leaving zones where the binder and fibers did not mix well. No significant influence of the density profile on the IB has been found. According to statistical analysis, statistically significant differences exist between the average values of IB results for the reference HDF and the rest of the panels produced; between DS12-DS20, PLA12-PLA20, and variants WS12-WS20, PCL12-PCL20. The representative forms of damage of the samples resulting after the IB test are compiled in Figure 5. The first group of destruction took place in the middle of the composite thickness. The first form of destruction was obtained by UF12, WS12, WS15, WS20, PCL12, PCL15, and PCL20. The second group includes samples in which the destruction took place in the near-surface zone; it includes samples with an average IB below 0.48 N mm^{-2}. The HDF with damage in the near-surface layers had the weakest strength of all tested samples, which is caused by more effective bonding in the core layer. Ji et al. [41] developed a method for manufacturing medium-density fiberboard (MDF) adhesives using chitosan as the main component and glutaraldehyde as a crosslinking agent. They prove that an increasing amount of glutaraldehyde could be contributing toward the deterioration of the mechanical properties of MDF. The optimal IB value was 1.22 N mm^{-2}, with the board sanded to reduce the low-density surface area which results in an improved surface quality and finish.

Baskaran et al. (2012) [67] conducted tests with the addition of freeze-dried and pure polyhydroxyalkanoates (PHA) in fine particles of oil palm trunk at different amounts. They proved that MOR and MOE increased as the amount of both forms of PHA increased.

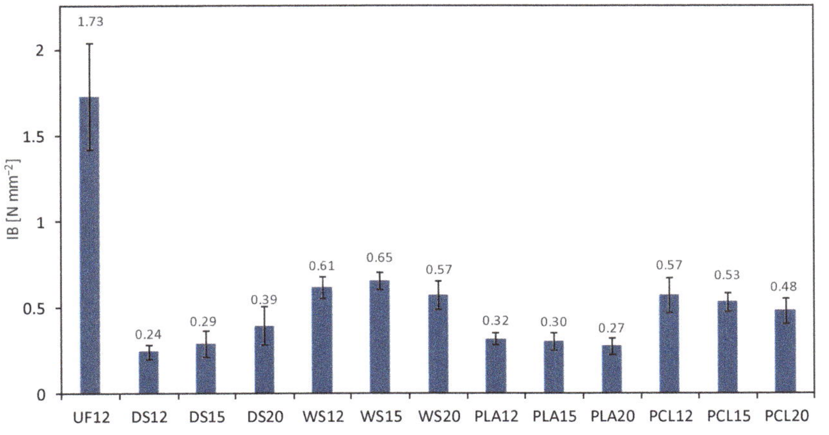

Figure 4. Internal bond (IB) of tested HDF.

(a) (b)

Figure 5. Two representative forms of damage after the IB test (**a**) in the core layer; (**b**) near-surface zone.

Scanning electron microscopy (SEM) was used to analyze the bonding quality of HDF with biopolymers and UF resin directly at the microstructure scale. The photos in Figure 6 are presented for each of the produced variants of panels, and were marked with arrows to exemplify the characteristic areas that clearly show the wood fiber covered by the binder used, and in the case of DS samples, how powder grains are visible. There is a visible increase in the number of biopolymers covering the fibers with increasing resination. The SEM pictures show the difference between the bonding of PLA and PCL with fibers, and the reference board based on UF resin. The binder was not distributed uniformly, and places where the fibers are fully covered by the biopolymer were observed in HDF with PLA and PCL, which potentially affected the mechanical and physical properties. SEM photos of PLA-based composites with the addition of agricultural by-product fibers show poor fiber-to-matrix adhesion, voids, and fiber breakage [68]. Song et al. [50], performed an SEM analysis, proving that the addition of 10% PLA in a fiberboard covers insignificant part fibers, which causes gaps between the fibers. The addition of 30–50% PLA causes the penetration of PLA into the gaps between fibers, which has a positive effect on water resistance and dimensional stability. Sivakumar et al. [69] observed an even distribution of banana leaf fiber with thermoplastic cassava starch as a strong adhesive matrix, which improved the mechanical properties of the bio-composites. The composite–water interaction (like TS, WA, and SWA) can be dependent on the tested panels' microstructure and binder distribution.

The average values of thickness swelling (TS) and water absorption (WA) on HDF after 2 h and 24 h are presented in Figure 7a–b. The error bars in the graphs represent the standard deviation. The average TS of the specimens after 2 h of immersion ranged from 10.4% for UF12 to 37.6% for DS12. After 24 h of immersion, the results were between 21.7% and 80.4% for UF12 and WS12, respectively.

Under industrial conditions, for the production of MDF, about 1% of paraffin emulsion is added to ensure water resistance; in contrast, no hydrophobic agents were used in these studies. The reference composites use UF resin, which is intended to be used in dry conditions. For each tested group of biopolymers, it can be observed that TS decreases with increasing resination. Biopolymers act as a binder but also fill the voids between the fibers, acting in favor of water resistance. Increasing the resination was the least effective for HDF with PLA as a binder, both after 2 h and 24 h of soaking in water. The SEM analysis confirmed that the increase in resination visibly shows a greater fiber surface coverage by the biopolymer, which translates into a higher water resistance. The results of water absorption (WA) of the tested HDF after 2 h and 24 h of immersion in water show a trend consistent with the results of TS. The lowest value of WA either after 2 h and 24 h was achieved for the reference HDF, while the highest obtained after 2 h was DS12 and after 24 h was WS12.

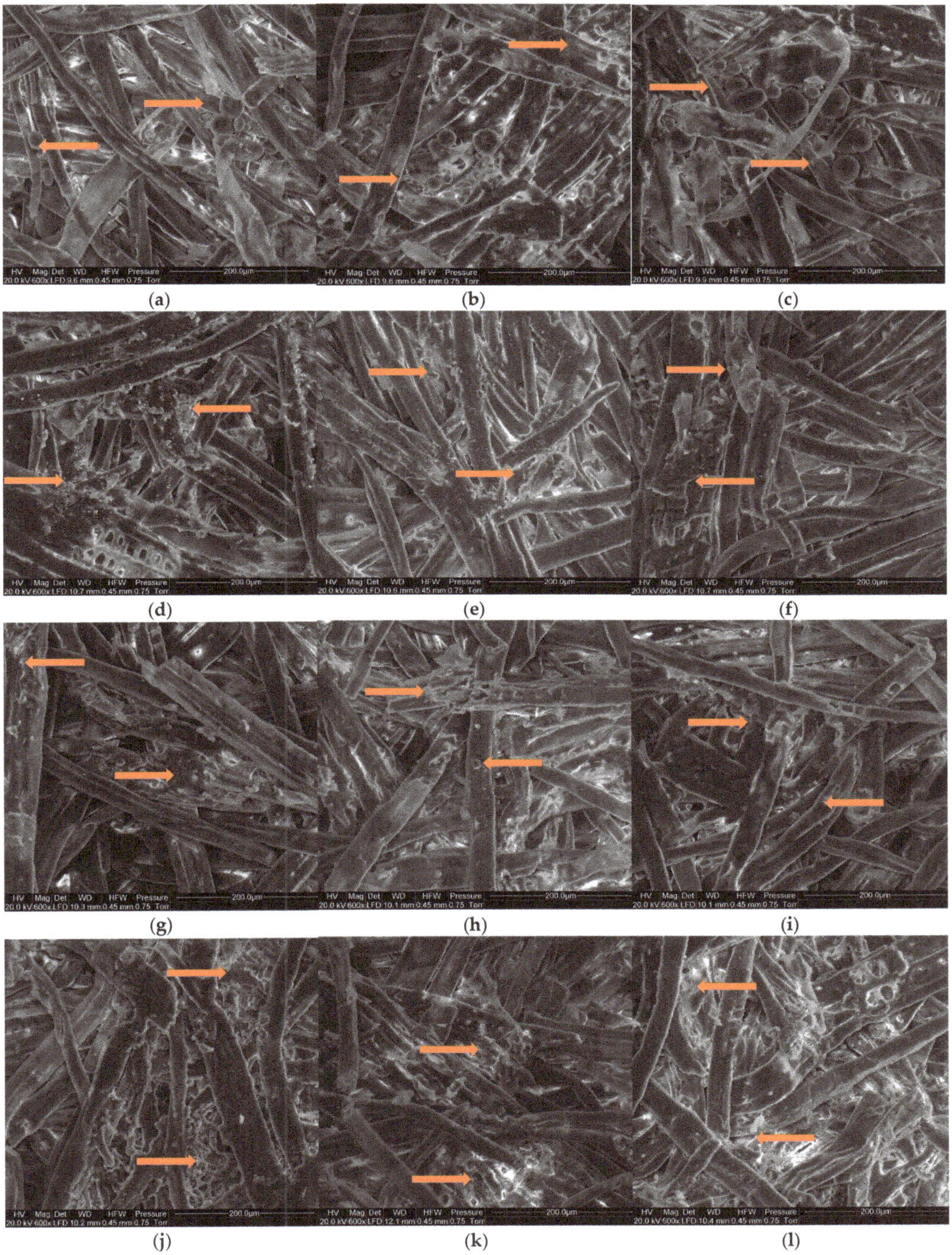

Figure 6. *Cont.*

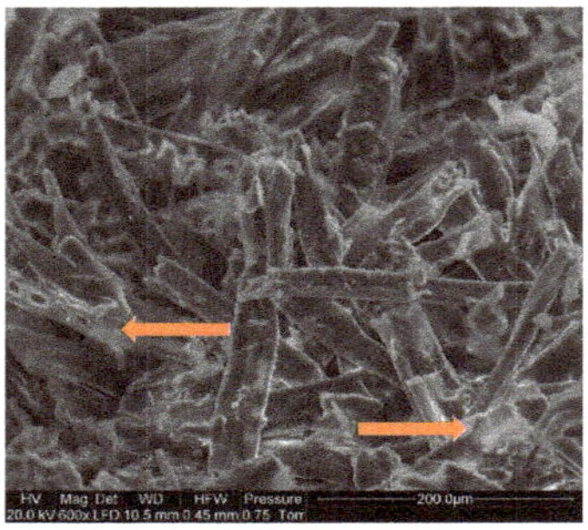

(m)

Figure 6. Scanning electron microscope images of HDF with, (**a**) DS12, (**b**) DS15, (**c**) DS20, (**d**) WS12, (**e**) WS15, (**f**) WS20, (**g**) PLA12, (**h**) PLA15, (**i**) PLA20, (**j**) PCL12, (**k**) PCL15, (**l**) PCL20, (**m**) UF12 as a binder (the arrows indicate the binder presence).

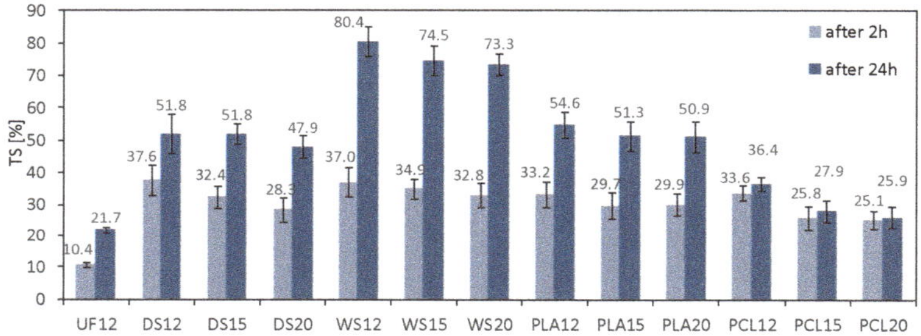

(**a**)

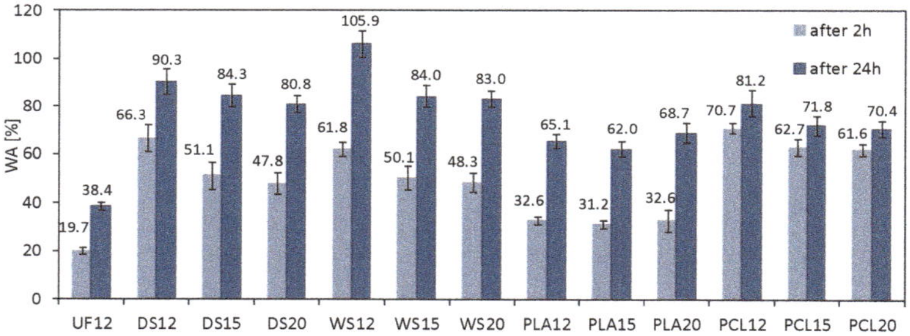

(**b**)

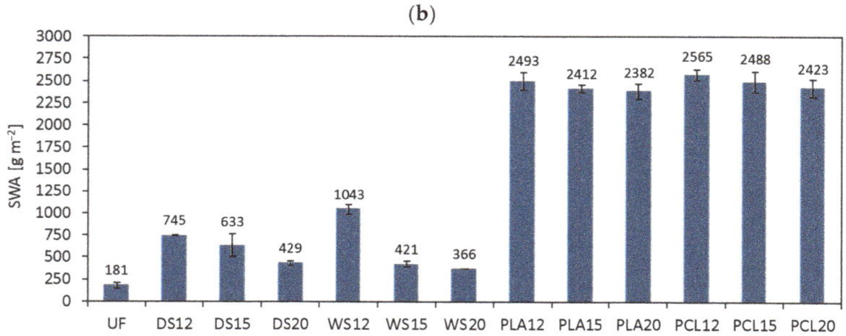

(**c**)

Figure 7. (**a**) Thickness swelling (TS), (**b**) water absorption (WA), and (**c**) surface water absorption (SWA) of tested HDF.

The results achieved for TS and WA are further proved in surface water absorption tests (SWA) (Figure 7c). The error bars in the graph represent the standard deviation. According to the results, the SWA decreases with the rise of the binder content. The lowest SWA average value was noted for UF12 (181 g m^{-2}) and the highest for PCL12 (2565 g m^{-2}) and PLA12 (2493 g m^{-2}). Significantly higher average values of SWA were found for the PLA and PCL samples. This may be the effect of the uneven spread of the binder over the fibers during blending, which causes the occurrence of local zones of binder-less fibers (Figure 6g–m). Despite the action of water on only the top surface for 2 h, PLA and PCL were impregnated with water over the entire thickness of the sample. Increasing the resination caused a slight decrease in the average SWA values. Reducing the resin content for PLA in extreme values from 12 to 20% resulted in a decrease in SWA by 4.6%, and for PCL by 5.5%. The statistical analysis shows that there are no statistically significant differences between the average values of the SWA results and the resination for the type of PLA and PCL binder used. The decreasing reaction to water of the panels made with the increasing amount of starch has been confirmed by [70].

Figure 8 presents the results of the contact angle measurements for all produced fiberboards. The error bars in the graph represent the standard deviation. The water sessile drop test conducted on DS and WS samples shows the highest hydrophobic behavior, possibly due to decreased porosity of the HDF with the binder with 20% resination. Therefore, as the resination increased for both 1 s and 60 s, the average values of the contact angle increased. For PLA and PCL, an increase in the contact angle in the range of resination from 12 to 15% was observed, both for 1 s and 60 s. The changes were highest for PCL20 (19.2% of reduction), PCL15, DS12, DS15, DS20—10.5%, 17.0%, 13.3%, 10.4%, respectively. The smallest changes in contact angle after 60 s were achieved for TPS: 1.9% for WS12, 1.8% for WS15, and 2.7% for WS20. A resination increase of 20% in HDF causes a decrease in the angle value for 1 s and 60 s, which means better surface wettability and higher surface energy for PLA20 and PCL20 [71,72]. The decrease in the wetting angle value was small; the increase in wettability from 15 to 20% for PLA resulted in a decrease in the contact angle by 1% after 1 s, while after 60 s the average value of the angle remained at the same level as for the 15% resin, at 102°. The PCL biopolymer caused a decrease after 1 s by 10%, and after 60 s by 7%. The increase in surface wettability with the increase in resination was probably caused by the uneven biopolymer coverage of the fibers (Figure 2). This is also explained by the analysis of SEM images, where voids and places where there is an excess of biopolymer are visible (Figure 6). Gumowska et al. [57] confirmed that with the increase in resination (12, 15, 20%) in particleboard bonded by PLA and PCL, the average values of the contact angle increased both after 1 s and after 60 s. The highest average contact angle after 1 s was recorded for DS12, DS15, DS20, reaching values of 112°, 113°, 115°, and respectively; the lowest after 1 s was recorded for PCL20, at 95°. The highest average contact angle after 60 s was recorded for UF12, WS12, WS15, and WS20, reaching values 105°, 105°, 107°, and 107°, respectively; the lowest after 60 s was recorded for PCL12, at 84°. For each tested HDF sample, after the water droplets remained on the surface for 60 s, the contact angle decreased.

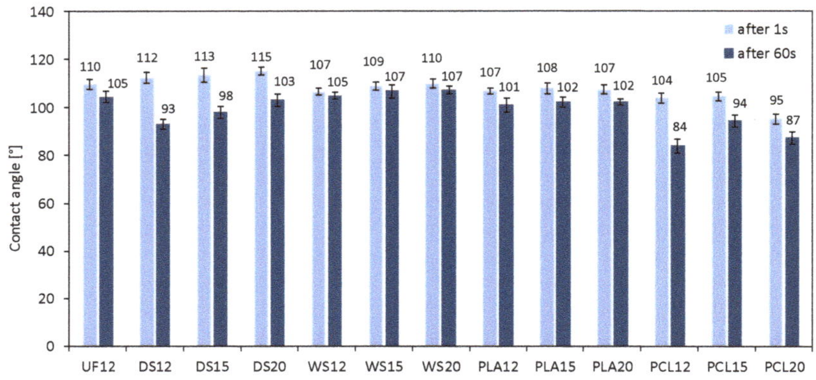

Figure 8. The contact angle of tested HDF.

4. Conclusions

On the basis of the completed research and analysis of the results, the following conclusions can be drawn:

A slightly higher densification of the face layer was found for panels where the water spray was applied over the mat surfaces before hot-pressing.

The bending features (MOR and MOE) depended on the binder type and resination level; there was increasing MOR and MOE with a rise in the DS resination; the maximum MOR and MOE values for WS were found for 15% resination; the bending properties of panels made with PLA and PCL decreased with the resination increase; however, the radical drop in MOR and MOE occurred when resination grew from 15% to 20%.

A similar relation between the binder type and resination level was found for IB; however, a significantly higher IB among the panels with biopolymers was registered for WS and PCL; the IB was not influenced by the density profile.

A slight, statistically insignificant reduction in TS, WA, and SWA was found for panels with increasing biopolymer binder content; an intensive increase in TS and WA was noted for WS after 24 h of soaking where the starch as a binder in wet form was mixed with fibers; the PLA panels had the lowest WA, and the lowest SWA among the biopolymer-bonded samples was for WS.

With the increase in resination for TPS from 12 to 20%, the value of the contact angle increased both after 1 s and 60 s. The increase in resination for PLA from 12 to 15% caused a statistically insignificant increase in the contact angle and a statistically insignificant decrease when resination grew from 15% to 20% for both 1 s and 60 s. The average value of the contact angle for PCL increased with the increase in resin content, while for PCL20 an increase in surface wettability was observed concerning PCL15.

In the case of dynamically evaporating solvents from binding mixtures, a development of fibers' resination (blending) techniques should be carried out, to avoid the uneven spreading of the binder over the resinated material. Further research on improving the strength of wood-based panels bonded with biopolymers should focus on improving their adhesion to the surface of wood/wood particles or fibers.

Supplementary Materials: The following are available online at https://www.mdpi.com/article/10.3390/f14010084/s1, Supplementary Material: Significant differences ($\alpha = 0.05$) between factors and levels.

Author Contributions: Conceptualization, A.G. and G.K.; methodology, A.G. and G.K.; formal analysis, A.G.; investigation, A.G.; resources, G.K.; writing—original draft preparation, A.G. and G.K.; writing—review and editing, A.G. and G.K.; visualization, A.G. All authors have read and agreed to the published version of the manuscript.

Funding: This research received no external funding.

Data Availability Statement: Not applicable.

Conflicts of Interest: The authors declare no conflict of interest.

References

1. World Health Organization. *Safety Evaluation of Certain Food Additives: Prepared by the Eighty-Sixth Meeting of the Joint FAO/WHO Expert Committee on Food Additives (JECFA)*; World Health Organization: Geneva, Switzerland, 2020.
2. UNECE/FAO. *Forest Products Annual Market Review 2019–2020*; United Nations Publications: Geneva, Switzerland, 2020.
3. Kristak, L.; Antov, P.; Bekhta, P.; Lubis, M.A.R.; Iswanto, A.H.; Reh, R.; Sedliacik, J.; Savov, V.; Taghiyari, H.R.; Papadopoulos, A.N.; et al. Recent progress in ultra-low formaldehyde emitting adhesive systems and formaldehyde scavengers in wood-based panels: A review. *Wood Mater. Sci. Eng.* **2022**, 1–20. [CrossRef]
4. Hussin, M.H.; Abd Latif, N.H.; Hamidon, T.S.; Idris, N.N.; Hashim, R.; Appaturi, J.N.; Brosse, N.; Ziegler-Devin, I.; Chrusiel, L.; Fatriasari, W.; et al. Latest advancements in high-performance bio-based wood adhesives: A critical review. *J. Mater. Res. Technol.* **2022**, *21*, 3909–3946. [CrossRef]
5. Global Wood Adhesives and Binders Market Size by Product (Urea-Formaldehyde, Phenol-Formaldehyde, Soy-Based), by Application (Flooring and Plywood, Furniture and Subcomponents, Windows and Doors), by Geographic Scope and Forecast. Available online: https://www.verifiedmarketresearch.com/product/wood-adhesives-and-binders-market/ (accessed on 4 December 2022).
6. Boran, S.; Usta, M.; Gümükaya, E. Decreasing formaldehyde emission from medium density fiberboard panels produced by adding different amine compounds to urea formaldehyde resin. *Int. J. Adhes. Adhes.* **2011**, *31*, 674–678. [CrossRef]
7. Grigsby, W.J.; Thumm, A.; Carpenter, J.E.P.; Hati, N. Investigating the extent of urea formaldehyde resin cure in medium density fibreboard: Characterisation of extractable resin components. *Int. J. Adhes. Adhes.* **2014**, *50*, 50–56. [CrossRef]
8. Hemmilä, V.; Adamopoulos, S.; Karlsson, O.; Kumar, A. Development of sustainable bio-adhesives for engineered wood panels-A Review. *RSC Adv.* **2017**, *7*, 38604–38630. [CrossRef]
9. Yu, C.W.F.; Crump, D.R. Review: Testing for Formaldehyde Emission from Wood-Based Products—A Review. *Indoor Built Env.* **1999**, *8*, 280–286. [CrossRef]
10. *EN 316*; Wood Fibre Boards—Definition, Classification and Symbols. European Committee for Standardization: Brussels, Belgium, 2009.
11. Dai, D.; Fan, M. Wood fibres as reinforcements in natural fibre composites: Structure, properties, processing and applications. *Nat. Fibre Compos. Mater. Process. Appl.* **2014**, *2014*, 3–65. [CrossRef]
12. Halvarsson, S.; Edlund, H.; Norgren, M. Manufacture of non-resin wheat straw fibreboards. *Ind. Crops Prod.* **2009**, *29*, 437–445. [CrossRef]
13. Lee, T.C.; Mohd Pu'ad, N.A.S.; Selimin, M.A.; Manap, N.; Abdullah, H.Z.; Idris, M.I. An overview on development of environmental friendly medium density fibreboard. *Mater. Today Proc.* **2019**, *29*, 52–57. [CrossRef]
14. Sedlecký, M.; Gašparík, M. Power consumption during edge milling of medium-density fiberboard and edge-Glued panel. *BioResources* **2017**, *12*, 7413–7426. [CrossRef]
15. Prakash, S.; Mercy, J.L.; Goswami, K. A systemic approach for evaluating surface roughness parameters during drilling of medium density fiberboard using Taguchi method. *Indian J. Sci. Technol.* **2014**, *7*, 1888–1894. [CrossRef]
16. Iswanto, A.H.; Aritonang, W.; Azhar, I.; Supriyanto, S.; Fatriasari, W. The physical, mechanical and durability properties of sorghum bagasse particleboard by layering surface treatment. *J. Indian Acad. Wood Sci.* **2017**, *14*, 1–8. [CrossRef]
17. Juliana, A.H.; Paridah, M.T.; Rahim, S.; Nor Azowa, I.; Anwar, U.M.K. Properties of particleboard made from kenaf (*Hibiscus cannabinus* L.) as function of particle geometry. *Mater. Des.* **2012**, *34*, 406–411. [CrossRef]
18. Nicolao, E.S.; Leiva, P.; Chalapud, M.C.; Ruseckaite, R.A.; Ciannamea, E.M.; Stefani, P.M. Flexural and tensile properties of biobased rice husk-jute-soybean protein particleboards. *J. Build. Eng.* **2020**, *30*, 101261. [CrossRef]
19. Ramos, A.; Briga-Sá, A.; Pereira, S.; Correia, M.; Pinto, J.; Bentes, I.; Teixeira, C.A. Thermal performance and life cycle assessment of corn cob particleboards. *J. Build. Eng.* **2021**, *44*, 102998. [CrossRef]
20. Taha, I.; Elkafafy, M.S.; El Mously, H. Potential of utilizing tomato stalk as raw material for particleboards. *Ain Shams Eng. J.* **2018**, *9*, 1457–1464. [CrossRef]
21. Borysiuk, P.; Jenczyk-Tolloczko, I.; Auriga, R.; Kordzikowski, M. Sugar beet pulp as raw material for particleboard production. *Ind. Crops Prod.* **2019**, *141*, 111829. [CrossRef]
22. Khanjanzadeh, H.; Pirayesh, H.; Sepahvand, S. Influence of walnut shell as filler on mechanical and physical properties of MDF improved by nano-SiO2. *J. Indian Acad. Wood Sci.* **2014**, *11*, 15–20. [CrossRef]
23. Ayrilmis, N.; Buyuksari, U.; Avci, E.; Koc, E. Utilization of pine (*Pinus pinea* L.) cone in manufacture of wood based composite. *For. Ecol. Manag.* **2009**, *259*, 65–70. [CrossRef]
24. Santosa, J.; Pereira, J.; Ferreira, N.; Paiva, N.; Ferra, J.; Magalhães, F.D.; Martins, J.M.; Dulyanska, Y.; Carvalho, L.H. Valorisation of non-timber by-products from maritime pine (*Pinus pinaster*, Ait) for particleboard production. *Ind. Crops Prod.* **2021**, *168*, 113581. [CrossRef]

25. Efe, F.T. Investigation of some physical and thermal insulation properties of honeycomb-designed panels produced from Calabrian pine bark and cones. *Eur. J. Wood Wood Prod.* **2022**, *80*, 705–718. [CrossRef]
26. Casas-Ledón, Y.; Daza Salgado, K.; Cea, J.; Arteaga-Pérez, L.E.; Fuentealba, C.A. Life cycle assessment of innovative insulation panels based on eucalyptus bark fibers. *J. Clean. Prod.* **2020**, *249*, 119356. [CrossRef]
27. Kowaluk, G.; Szymanowski, K.; Kozlowski, P.; Kukula, W.; Sala, C.; Robles, E.; Czarniak, P. Functional Assessment of Particleboards Made of Apple and Plum Orchard Pruning. *Waste Biomass Valorization* **2020**, *11*, 2877–2886. [CrossRef]
28. Lee, S.H.; Lum, W.C.; Boon, J.G.; Kristak, L.; Antov, P.; Pędzik, M.; Rogoziński, T.; Taghiyari, H.R.; Lubis, M.A.R.; Fatriasari, W.; et al. Particleboard from agricultural biomass and recycled wood waste: A review. *J. Mater. Res. Technol.* **2022**, *20*, 4630–4658. [CrossRef]
29. Pędzik, M.; Janiszewska, D.; Rogoziński, T. Alternative lignocellulosic raw materials in particleboard production: A review. *Ind. Crops Prod.* **2021**, *174*, 114162. [CrossRef]
30. Sulaiman, N.S.; Hashim, R.; Sulaiman, O.; Nasir, M.; Amini, M.H.M.; Hiziroglu, S. Partial replacement of urea-formaldehyde with modified oil palm starch based adhesive to fabricate particleboard. *Int. J. Adhes. Adhes.* **2018**, *84*, 1–8. [CrossRef]
31. Lomelí-Ramírez, M.G.; Barrios-Guzmán, A.J.; García-Enriquez, S.; Rivera-Prado, J.D.J.; Manríquez-González, R. Chemical and Mechanical Evaluation of Bio-composites Based on Thermoplastic Starch and Wood Particles Prepared by Thermal Compression. *BioResources* **2014**, *9*, 2960–2974. [CrossRef]
32. Hikmah, N.; Susanto, D.; Suganda, E. Physical properties of medium density fiberboard from pineapple leaf fiber (PALF) with cassava peel starch and citric acid. *AIP Conf. Proc.* **2021**, *2376*, 040012. [CrossRef]
33. Lubis, M.A.R.; Park, B.D.; Hong, M.K. Tailoring of oxidized starch's adhesion using crosslinker and adhesion promotor for the recycling of fiberboards. *J. Appl. Polym. Sci.* **2019**, *136*, 47966. [CrossRef]
34. Lubis, M.A.R.; Park, B.D.; Hong, M.K. Tuning of adhesion and disintegration of oxidized starch adhesives for the recycling of medium density fiberboard. *BioResources* **2020**, *15*, 5156–5178. [CrossRef]
35. Domínguez-Robles, J.; Tarrés, Q.; Delgado-Aguilar, M.; Rodríguez, A.; Espinach, F.X.; Mutjé, P. Approaching a new generation of fiberboards taking advantage of self lignin as green adhesive. *Int. J. Biol. Macromol.* **2018**, *108*, 927–935. [CrossRef] [PubMed]
36. Khanjanzadeh, H.; Behrooz, R.; Bahramifar, N.; Pinkl, S.; Gindl-Altmutter, W. Application of surface chemical functionalized cellulose nanocrystals to improve the performance of UF adhesives used in wood based composites—MDF type. *Carbohydr. Polym.* **2019**, *206*, 11–20. [CrossRef]
37. Li, X.; Wang, D.; Ratto, J.A.; Sun, X.S. Production and characterization of high strength, thin-layered, pulp fiberboard using soy protein adhesives. *J. Adhes. Sci. Technol.* **2013**, *27*, 2065–2074. [CrossRef]
38. Li, X.; Li, Y.; Zhong, Z.; Wang, D.; Ratto, J.A.; Sheng, K.; Sun, X.S. Mechanical and water soaking properties of medium density fiberboard with wood fiber and soybean protein adhesive. *Bioresour. Technol.* **2009**, *100*, 3556–3562. [CrossRef] [PubMed]
39. Zhong, Z.; Susan Sun, X.; Wang, D.; Ratto, J.A. Wet Strength and Water Resistance of Modified Soy Protein Adhesives and Effects of Drying Treatment. *J. Polym. Environ.* **2003**, *11*, 137–144. [CrossRef]
40. Eslah, F.; Jonoobi, M.; Faezipour, M.; Afsharpour, M.; Enayati, A.A. Preparation and development of a chemically modified bio-adhesive derived from soybean flour protein. *Int. J. Adhes. Adhes.* **2016**, *71*, 48–54. [CrossRef]
41. Ji, X.; Li, B.; Yuan, B.; Guo, M. Preparation and characterizations of a chitosan-based medium-density fiberboard adhesive with high bonding strength and water resistance. *Carbohydr. Polym.* **2017**, *176*, 273–280. [CrossRef] [PubMed]
42. Wu, H.; Liao, D.; Chen, X.; Du, G.; Li, T.; Essawy, H.; Pizzi, A.; Zhou, X. Functionalized Natural Tannins for Preparation of a novel non-isocyanate polyurea-based adhesive. *Polym. Test.* **2023**, *117*, 107853. [CrossRef]
43. Barbosa, A.P.; Mano, E.B.; Andrade, C.T. Tannin-based resins modified to reduce wood adhesive brittleness. *For. Prod. J.* **2000**, *50*, 89–92.
44. Fernandes, F.C.; Kirwan, K.; Wilson, P.R.; Coles, S.R. Sustainable Alternative Composites Using Waste Vegetable Oil Based Resins. *J. Polym. Environ.* **2019**, *27*, 2464–2477. [CrossRef]
45. Akbari, S.; Gupta, A.; Khan, T.A.; Jamari, S.S.; Ani, N.B.C.; Poddar, P. Synthesis and characterization of medium density fiber board by using mixture of natural rubber latex and starch as an adhesive. *J. Indian Acad. Wood Sci.* **2014**, *11*, 109–115. [CrossRef]
46. Juliet, O.I.; Dilim, I.C.; Okpara, O.C.; Ejike, O.M. Natural Rubber Based Adhesive Modified with Starch and Reinforcer. *AASCIT J. Chem.* **2018**, *4*, 1–6.
47. Indrayani, Y.; Setyawati, D.; Munawar, S.S.; Umemura, K.; Yoshimura, T. Evaluation of Termite Resistance of Medium Density Fiberboard (MDF) Manufacture from Agricultural Fiber Bonded with Citric Acid. *Procedia Environ. Sci.* **2015**, *28*, 778–782. [CrossRef]
48. Widyorini, R.; Umemura, K.; Kusumaningtyas, A.R.; Prayitno, T.A. Effect of starch addition on properties of citric acid-bonded particleboard made from bamboo. *BioResources* **2017**, *12*, 8068–8077. [CrossRef]
49. NatureWorks. Available online: https://www.natureworksllc.com/News-and-Events/Press-Releases/2022/2022-11-09-NatureWorks-Selects-General-Contractor-for-Ingeo-Manufacturing-Facility (accessed on 4 December 2022).
50. Song, X.; Guo, Z.; Wu, J.; Zhang, Y.; Cai, Z.; Wang, X. Fabrication and characterization of soybean straw and polylactide acid-based hybrid bio-board. *J. Adhes. Sci. Technol.* **2022**, 1–18. [CrossRef]
51. Ye, X.; Julson, J.; Kuo, M.; Myers, D. Biocomposite hardboard from renewable biomass bonded with soybean-based adhesive. *Trans. Am. Soc. Agric. Eng.* **2005**, *48*, 1629–1635. [CrossRef]

52. Rajeshkumar, G.; Seshadri, S.A.; Devnani, G.L.; Sanjay, M.R.; Siengchin, S.; Maran, J.P.; Al-Dhabi, N.A.; Karuppiah, P.; Mariadhas, V.A.; Sivarajasekar, N.; et al. Environment friendly, renewable and sustainable poly lactic acid (PLA) based natural fiber reinforced composites—A comprehensive review. *J. Clean. Prod.* **2021**, *310*, 127483. [CrossRef]
53. Wang, B.; Qi, Z.; Chen, X.; Sun, C.; Yao, W.; Zheng, H.; Liu, M.; Li, W.; Qin, A.; Tan, H.; et al. Preparation and mechanism of lightweight wood fiber/poly(lactic acid) composites. *Int. J. Biol. Macromol.* **2022**, *217*, 792–802. [CrossRef]
54. Peltola, H.; Pääkkönen, E.; Jetsu, P.; Heinemann, S. Wood based PLA and PP composites: Effect of fibre type and matrix polymer on fibre morphology, dispersion and composite properties. *Compos. Part A Appl. Sci. Manuf.* **2014**, *61*, 13–22. [CrossRef]
55. Faludi, G.; Dora, G.; Renner, K.; Móczó, J.; Pukánszky, B. Improving interfacial adhesion in pla/wood biocomposites. *Compos. Sci. Technol.* **2013**, *89*, 77–82. [CrossRef]
56. Csizmadia, R.; Faludi, G.; Renner, K.; Móczó, J.; Pukánszky, B. PLA/wood biocomposites: Improving composite strength by chemical treatment of the fibers. *Compos. Part A Appl. Sci. Manuf.* **2013**, *53*, 46–53. [CrossRef]
57. Gumowska, A.; Robles, E.; Kowaluk, G. Evaluation of functional features of lignocellulosic particle composites containing biopolymer binders. *Materials* **2021**, *14*, 7718. [CrossRef] [PubMed]
58. Gumowska, A.; Robles, E.; Bikoro, A.; Wronka, A.; Kowaluk, G. Selected Properties of Bio-Based Layered Hybrid Composites with Biopolymer Blends for Structural Applications. *Polymers* **2022**, *14*, 4393. [CrossRef] [PubMed]
59. Méndez, P.A.; Méndez, Á.M.; Martínez, L.N.; Vargas, B.; López, B.L. Cassava and banana starch modified with maleic anhydride-poly (ethylene glycol) methyl ether (Ma-mPEG): A comparative study of their physicochemical properties as coatings. *Int. J. Biol. Macromol.* **2022**, *205*, 1–14. [CrossRef] [PubMed]
60. *EN 310*; Wood-Based Panels. Determination of Modulus of Elasticity in Bending and of Bending Strength. European Committee for Standardization: Brussels, Belgium, 1993.
61. *EN 319*; Particleboards and Fibreboards. Determination of Tensile Strength Perpendicular to the Plane of the Board. European Committee for Standardization: Brussels, Belgium, 1993.
62. *EN 317*; Particleboards and Fibreboards. Determination of Swelling in Thickness after Immersion in Water. European Committee for Standardization: Brussels, Belgium, 1993.
63. *EN 382-2*; Fibreboards—Determination of Surface Absorption—Part 2: Test Method for Hardboards. European Committee for Standardization: Brussels, Belgium, 1993.
64. Garcia, R.A.; Cloutier, A.; Riedl, B. Dimensional stability of MDF panels produced from fibres treated with maleated polypropylene wax. *Wood Sci. Technol.* **2005**, *39*, 630–650. [CrossRef]
65. Segovia, F.; Blanchet, P.; Essoua, G.G.E. Potential of the crude glycerol and citric acid mixture as a binder in medium-density fiberboard manufacturing. *Eur. J. Wood Wood Prod.* **2021**, *79*, 1141–1151. [CrossRef]
66. Wong, E.D.; Zhang, M.; Wang, Q.; Han, G.; Kawai, S. Formation of the density profile and its effects on the properties of fiberboard. *J. Wood Sci.* **2000**, *46*, 202–209. [CrossRef]
67. Baskaran, M.; Hashim, R.; Said, N.; Raffi, S.M.; Kunasundari, B.; Sudesh, K.; Sulaiman, O.; Arai, T.; Kosugi, A.; Mori, Y.; et al. Properties of binderless particleboard from oil palm trunk with addition of polyhydroxyalkanoates. *Compos. Part B Eng.* **2012**, *43*, 1109–1116. [CrossRef]
68. Nyambo, C.; Mohanty, A.K.; Misra, M. Polylactide-based renewable green composites from agricultural residues and their hybrids. *Biomacromolecules* **2010**, *11*, 1654–1660. [CrossRef]
69. Sivakumar, A.A.; Canales, C.; Roco-Videla, Á.; Chávez, M. Development of Thermoplastic Cassava Starch Composites with Banana Leaf Fibre. *Sustainability* **2022**, *14*, 12732. [CrossRef]
70. Wronka, A.; Rdest, A.; Kowaluk, G. Influence of starch content on selected properties of hardboard. *Ann. Wars. Univ. Life Sci.-SGGW For. Wood Technol.* **2020**, *109*, 48–52. [CrossRef]
71. Jirkovec, R.; Erben, J.; Sajdl, P.; Chaloupek, J.; Chvojka, J. The effect of material and process parameters on the surface energy of polycaprolactone fibre layers. *Mater. Des.* **2021**, *205*, 109748. [CrossRef]
72. Jothi Prakash, C.G.; Prasanth, R. Approaches to design a surface with tunable wettability: A review on surface properties. *J. Mater. Sci.* **2021**, *56*, 108–135. [CrossRef]

forests

MDPI

Article

Dimensional Stability of Treated Sengon Wood by Nano-Silica of Betung Bamboo Leaves

Istie Rahayu [1,*], Fitria Cita Dirna [1], Akhiruddin Maddu [2], Wayan Darmawan [1], Dodi Nandika [1] and Esti Prihatini [1]

[1] Department of Forest Products, Faculty of Forestry and Environment, IPB University, Bogor 16680, Indonesia; fitriacita@gmail.com (F.C.D.); wayandar@indo.net.id (W.D.); nandikadodi@gmail.com (D.N.); esti@apps.ipb.ac.id (E.P.)

[2] Department of Physics, Faculty of Mathematics and Natural Sciences, IPB University, Bogor 16680, Indonesia; maddu3@gmail.com

* Correspondence: istiesr@apps.ipb.ac.id

Abstract: Sengon (*Falcataria moluccana* Miq.) is one of the fastest growing wood that is broadly planted in Indonesia. Sengon wood has inferior wood properties, such as a low density and dimensional stability. Therefore, sengon wood requires a method to improve its wood quality through wood modification. One type of wood modification is wood impregnation. On the other hand, Betung Bamboo leaves are considered as waste. Betung Bamboo leaves contain silica. Based on several researches, nano-SiO_2 could improve fast-growing wood qualities. According to its perfect solubility in water, monoethylene glycol (MEG) is used in the study. The objectives are to evaluate the impregnation treatment (MEG and nano-silica originated from betung bamboo leaves) in regard to the dimensional stability and density of 5-year-old sengon wood and to characterize the treated sengon wood. MEG, MNano-Silica 0.5%, MNano-Silica 0.75%, and MNano-Silica 1% were used as impregnation solutions. The impregnation method was started with 0.5 bar of vacuum for 60 min, followed by 2.5 bar of pressure for 120 min. The dimensional stability, density, and characterization of the samples were studied through the use of scanning electron microscopy (SEM), energy dispersive X-ray spectroscopy, X-ray diffraction (XRD), and Fourier transform infrared spectroscopy (FTIR). The results show that the treatment had a significant effect on the dimensional stability and density of sengon wood. Alterations in the morphology of treated sengon wood were observed through the full coverage of the pits on the vessel walls (SEM analysis results) and the detection of ethylene (FTIR analysis results) and silica (XRD and FTIR analysis results). Overall, the 0.75% MNano-Silica treatment was the most optimal treatment for increasing the dimensional stability and density of 5-year-old sengon wood.

Keywords: betung bamboo; dimensional stability; impregnation; MEG; nano-silica; sengon

1. Introduction

Degradation and deforestation decrease wood production from our natural forests. Therefore, we need alternative wood resources to overcome the wood supply shortage through the utilization of fast-growing species [1] or a wood biomass [2–4]. Sengon (*Falcataria moluccana* Miq.) is a type of fast-growing tree that is widely planted in community forests and community plantation forests in Indonesia. Sengon has a short harvest time. Owing to the rapid growth of the tree, the wood has a low density, strength, and durability, as well as a high portion of it being juvenile wood [5]. Based on the research results of [6], 5- to 7-year-old jabon and sengon trees contain up to100% juvenile wood. The density and hardness of sengon wood are 0.3–0.5 g/cm^3 and 112–122 kg/cm^2, respectively. Sengon wood is classified into a durability class of IV–V and strength class of IV–V [7]. Nowadays, sengon wood in Indonesia is used for pulpwood, light construction, furniture, and wood composite. Sengon wood production in Indonesia in 2019 was 5.47 million m^3 [8]. Therefore, the quality of fast-growing wood needs to be improved through technology. Wood modification technology has been discussed in many journals in recent decades; however, only a few technologies have led to industrial application. One of them is thermal modification. It is a potential technology to be developed in industries. It improves the wood's hygroscopicity, dimensional stability, and durability without harming the environment [9–11]. However, heat treatment degrades mechanical properties [12,13], decreases wood durability against termites [14–16], and causes it to not be resistant to weathering [17]. In line with thermal modification, chemical modification is a promising alternative for improving wood qualities. There are two chemical modifications with the potential to be developed on an industrial scale: acetylation and furfurylation [18]. The impregnation of molecular compounds into the wood structure is also another option being able to increase the density and dimensional stability of wood. Research on impregnation technology for fast-growing wood has been carried out in Indonesia and other countries, using methyl methacrylate [19–21], formaldehyde-based chemical compounds such as phenol formaldehyde [22,23], rosin [24], furfuryl alcohol [25], paraffin [26], and aqueous solutions [27]. The utilization of chemical substances and formaldehyde material for wood modification could have a hazardous consequence to our health and environment. Therefore, we need to find other material alternatives that are more environmentally friendly.

Silica is a chemical that has wide applications in various fields, including being used as a polymer in wood impregnation. Silica can be obtained from the commercial markets or extracted from natural materials, such as bamboo leaf ash, as shown in research [28]. Bamboo leaves are considered waste [29]. Researchers analyzed bamboo leaf ash and found a silica content of 79.93%, which was the second largest content after rice husk ash (93.2%) [30]. Based on the findings of [31], nano-SiO_2 could effectively improve the dimensional stability and density of jabon wood. Similar outcomes were reported by [32] using sengon wood and nano-SiO_2 and [33] using poplar wood (Populus spp.) treated with furfuryl alcohol and nano-SiO_2. Therefore, in this study, we produced nano-silica created from betung bamboo leaves.

Monoethylene Glycol (MEG) is perfectly soluble in water, colorless, liquid, odorless, has a low volatility and has a 62.07 g/mol molecule weight [34,35]. According to these characteristics, MEG was used. In the current study, we aim to evaluate the impregnation treatment (MEG and nano-silica derived from betung bamboo leaves) with regards to the weight percent gain (WPG), anti-swelling efficiency (ASE), bulking effect (BE), water uptake (WU), and density of 5-year-old sengon wood, and to characterize the treated sengon wood.

2. Materials and Methods

This research was divided into several stages, namely, the preparation of wood samples; preparation of impregnation solutions; the impregnation process; the calculations of the dimension stability and density. In this study, the materials used were bamboo betung leaves; 5-year-old sengon wood from community forests in Sukabumi, West Java, Indonesia; monoethylene glycol (MEG); demineralized water. The analytical techniques included scanning electron microscopy (SEM), energy dispersive X-ray spectroscopy (EDX), X-ray diffraction (XRD), and Fourier transform infrared spectroscopy (FTIR). Bamboo betung leaves used in the form of nano-silica in this study [28] had a particle size of about 234.49–851.36 nm, where the average size was 472.67 nm with a PDI value (Particle Dispersion Index) of 0.0670.

2.1. Preparation of Wood Samples

The sample size of the sengon wood for testing was 2 cm × 2 cm × 2 cm [36]. Testing encompassed WPG, ASE, WU, BE, and oven-dried density. Each treatment had a total of 5 replicated samples.

2.2. Preparation of Impregnation Solutions

Impregnation solutions were prepared by mixing MEG and nano-silica by using an ultrasonic processor (amplitude of 40% for 60 min). In Table 1, it can be seen that there were several treatments for the composition of the MEG and MEG solutions and nano-silica (MNano-Silica).

Table 1. The volume of MEG and nano-silica as impregnation solutions.

Treatment	MEG 50% (mL)	Nano-Silica (g)
MEG	1000	0
MNano-Silica 0.5%	1000	5
MNano-Silica 0.75%	1000	7.5
MNano-Silica 1%	1000	10

2.3. The Impregnation Process

The impregnation process with MEG and the MNano-Silica solutions was carried out by adapting a previously reported method [32], in which sengon wood was impregnated using MEG and nano-SiO_2. Dimensional measurements and weighing were carried out to test the WPG, BE, and WU [37], ASE [38], and the oven-dried density.

2.4. Calculation of Dimensional Stability and Oven-Dried Density

Calculations for WPG, ASE, BE, WU, and oven-dried density were conducted using the following formulas:

$$\text{WPG}\ (\%) = [(\text{W1} - \text{W0})/\text{W0}] \times 100\ \text{ASE}\ (\%) = [(\text{Su} - \text{St})/\text{Su}] \times 100$$

$$\text{WU}\ (\%) = [(\text{W2} - \text{W1})/\text{W1}] \times 100\ \text{BE}\ (\%) = (\text{V1} - \text{V0})/\text{V0} \times 100$$

$$\rho\left(\text{g}/\text{cm}^3\right) = \frac{\text{B}}{\text{V}} \times 100$$

where:

W0 = the oven-dried weight of sample before treatment;
W1 = the oven-dried sample weight after treatment;
W2 = the weight of the sample after immersion in water for 24 h;
V0 = the oven-dried volume of sample before treatment;
V1 = the oven-dried sample volume after treatment;
Su = the volume shrinkage of untreated wood;
St = the volume shrinkage of treated wood;
B = the weight of the sample before or after treatment;
V = the volume of the sample before or after treatment.

ASE values were determined by cycled water soaking method [38]. It was repeated 3 times, while oven-dried density after treatment was evaluated by taking into account WPG and BE values and untreated oven-dried density as a baseline.

2.5. Data Analysis Procedure

The simple, completely randomized design with 1 factor was used in this study, the factor being, namely, the treatment variation factor at 4 levels, namely, MEG, MNano-Silica 0.5%, MNano-Silica 0.75%, and MNano-Silica 1%. Statistical analyses were carried out by using IBM SPSS Statistics software (version 25.0), and the Duncan test at $\alpha = 1\%$ was conducted if there was a significant difference.

3. Results

Dimensional stability testing on sengon wood included the calculation of the WPG, ASE, WU, BE, and oven-dried density. The results of the dimensional stability tests for sengon wood can be seen in Table 2.

Table 2. Testing the dimensional stability of sengon wood.

Treatment	WPG (%)	ASE (%)	WU (%)	BE (%)	Oven-dried Density (g/cm^3)
MEG	27.18 ^{a} (±3.42)	53.28 ^{a} (±6.73)	101.68 ^{d} (±1.91)	2.89 ^{a} (±0.84)	0.38 ^{a} (±0.02)
MNano-Silica 0.5%	44.61 ^{b} (±2.29)	72.98 ^{b} (±3.88)	76.16 ^{c} (±3.30)	4.85 ^{b} (±0.63)	0.42 ^{a} (±0.15)
MNano-Silica 0.75%	50.20 ^{c} (±1.43)	83.36 ^{c} (±4.19)	67.99 ^{b} (±2.81)	9.78 ^{c} (±0.94)	0.42 ^{a} (±0.03)
MNano-Silica 1%	58.03 ^{d} (±2.39)	87.69 ^{c} (±6.15)	56.34 ^{a} (±3.88)	10.85 ^{c} (±1.37)	0.46 ^{b} (±0.01)

Note: Values in parentheses indicate the standard deviations. $^{a–c}$ Values followed by the different letters show real difference according to the Duncan test.

The WPG value for sengon wood increased along with the nano-silica concentration, presumably because the MEG and nano-silica entered the sengon wood structure. The ASE value of sengon wood also increased with each treatment. These results show that the MEG and nano-silica solutions increased in concentration in each treatment. The WU value decreased as treatments used a greater concentration of nano-silica. This result was correlated with the high WPG and ASE value. The BE value of sengon wood also increased with each treatment using a high nano-silica concentration, indicating that the added MEG and nano-silica functioned as bulking agents. The oven-dried density value of sengon wood increased with each treatment using a higher level of nano-silica. The percentage of increased oven-dried density after being treated was 26% (MEG), 40% (MNano-Silica 0.5% and 0.75%), and 53% (MNano-Silica 1%). Overall, the analysis of variance showed that each MEG and nano-silica impregnation treatment had a significant effect on sengon wood. Results were influenced by the various concentrations (0.5%, 0.75%, and 1%) of nano-silica used.

3.1. Characteristics of Impregnated Sengon Wood

3.1.1. SEM-EDX Analysis

Figure 1a shows pits in the vessel wall of sengon wood that were not closed. In Figure 1b, through the addition of 0.5% MNano-Silica, the sengon wood underwent morphological changes. Figure 2a shows that, with the addition of 0.75% MNano-Silica, nano-silica could enter the pits in the sengon wood, adhere to the vessel walls, and almost entirely cover all of them. A similar result was found with the addition of MNano-Silica 1% (Figure 2b), with nano-silica covering the entire surface. Additional information obtained from the analysis by SEM included EDX data, which are described in Table 3.

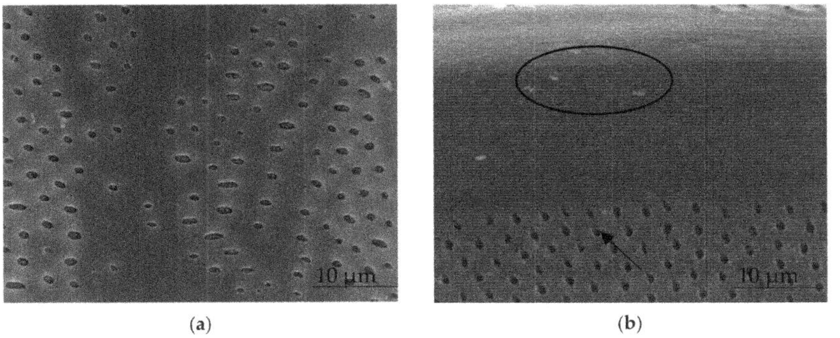

(**a**) (**b**)

Figure 1. Morphology of wood (**a**) sengon MEG and (**b**) sengon MNano-Silica 0.5%.

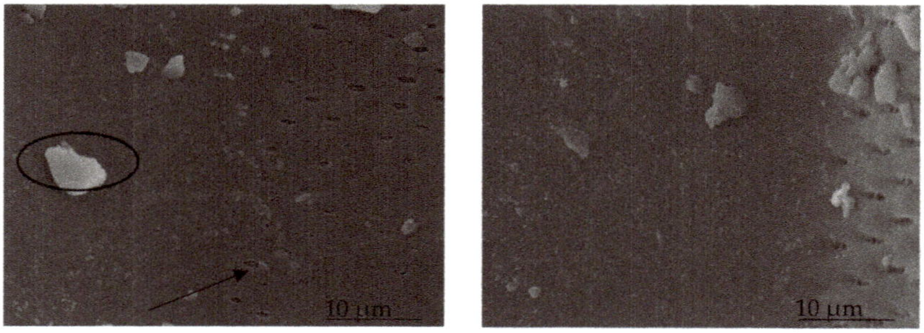

(a) (b)

Figure 2. Morphology of wood (**a**) sengon MNano-Silica 0.75% and (**b**) sengon MNano-Silica 1%. Magnification 1000×.

Table 3. Chemical content of treated sengon wood.

Treatment	Silicon (wt. %)
MEG	0
MNano-Silica 0.5%	0.35
MNano-Silica 0.75%	0.59
MNano-Silica 1%	0.62

3.1.2. XRD Analysis

In sengon wood (Figure 3), MEG treatment was detected as high-intensity cellulose peaks with an angle of 2θ = 22.47°. The silica peaks were detected with a high-intensity treatment of MNano-Silica 0.5%, MNano-Silica 0.75%, and MNano-Silica 1% at angles of 2θ = 20.59°, 20.51°, and 21.06°, respectively, which were close to the silica peak in the JCPDS 44-1394 database, namely, angles of 2θ = 20.54° and 21.13°. In the MNano-Silica 0.5%, MNano-Silica 0.75%, and MNano-Silica 1% treatments, cellulose peaks were detected at angles of 2θ = 22.61°, 22.67°, and 22.19°, respectively, which corresponded to the cellulose peaks in the JCPDS 03-0226 database, namely an angle of 2θ = 22.26°. Furthermore, cellulose peaks were also detected in the MEG treatment for an angle of 2θ = 17.04°. This result was close to the peak of cellulose in the JCPDS 03-0226 database, namely, an angle of 2θ = 17.13°.

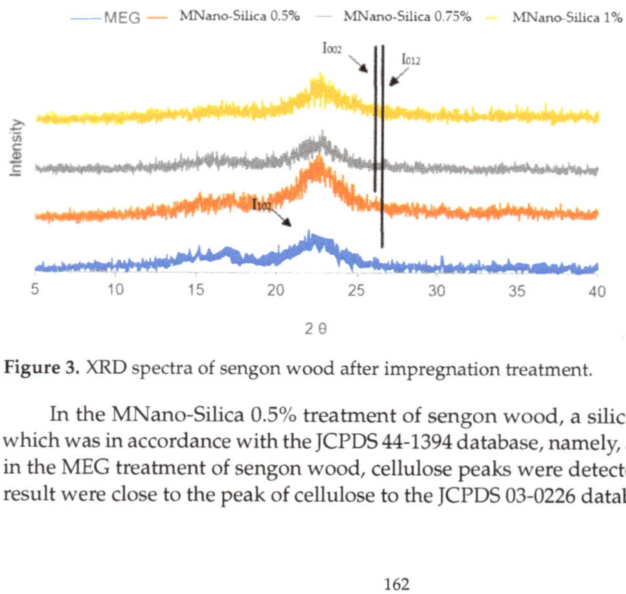

Figure 3. XRD spectra of sengon wood after impregnation treatment.

In the MNano-Silica 0.5% treatment of sengon wood, a silica peak of 2θ = 18.61° was detected, which was in accordance with the JCPDS 44-1394 database, namely, an angle of 2θ = 18.54°. Furthermore, in the MEG treatment of sengon wood, cellulose peaks were detected for an angle of 2θ = 17.04°. These result were close to the peak of cellulose to the JCPDS 03-0226 database, namely, an angle of 2θ = 17.13°.

This peak disappeared with the addition of nano-silica used in the treatment. Additional information obtained during XRD testing included data on the degree of crystallinity, which can be seen in Table 4. The results showed that the crystallinity degree of the sengon wood, augmented with the MEG treatment compared to the MNano-Silica 0.75% treatment, indicated that the addition of nano-silica could increase crystallinity.

Table 4. Crystallinity degree of treated sengon wood.

Treatment	Crystallinity Degree (%)
MEG	24.00
MNano-Silica 0.5%	27.94
MNano-Silica 0.75%	30.79
MNano-Silica 1%	29.22

The crystallinity degree in sengon wood decreased with the addition of 1% (29.22%) nano-silica, presumably because the resulting silica had an amorphous structure and the resulting nano-silica had a semi-crystalline structure. The nano-silica produced was semi-crystalline, which was likely influenced by the sonication treatment used.

3.1.3. FTIR Analysis

Figure 4 shows the FTIR spectrum of five sengon wood samples with a wavenumber ranging from 500 to 4000 cm^{-1}. The peak of the Si–O–Si vibrational silica was detected at the wavenumber 1060.45 cm^{-1} with the addition of MNano-Silica 0.5% and strengthened by the appearance of wave numbers 1056.39 cm^{-1} and 1112.19 cm^{-1} with the addition of MNano-Silica 0.75%. Very similar results were also shown for the addition of MNano-Silica 1%, with wavenumbers 1058.11 cm^{-1} and 1113.15 cm^{-1} (Figure 4).

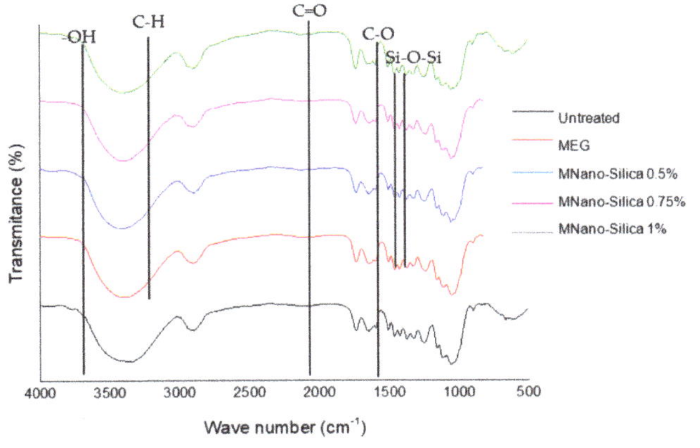

Figure 4. FTIR analysis of sengon wood after impregnation treatment.

4. Discussion

The impregnation using a polymer caused it to penetrate the wood, which could be followed by bonding with the constituent components of cell walls as described in [37]. Due to the increased ASE value, the wood's ability to adsorb water was reduced to vapor from the surroundings [28]. This finding indicated that the wood's ability to absorb water became lower with each treatment. The higher the BE value, the more polymer filling the cell walls in the wood, which could increase the dimensional stability of the wood. This outcome was in line with the increase in the WPG value from one treatment to the next. The increase in the WPG value of sengon wood was directly comparable to the BE value. The higher the WPG values, the higher the oven-dried density of sengon wood.

This result corresponded with research reported by [32], who showed that the replenishment of nano-silica in the impregnation of sengon wood had a significant effect on each treatment. For sengon wood treated by nano-silica derived from the leaves of bamboo betung, the 0.75% MNano-Silica treatment was optimal for increasing the dimensional stability and the oven-dried density of fast-growing sengon wood.

4.1. Characteristics of Impregnated Sengon Wood

4.1.1. SEM-EDX Analysis

MNano-Silica was deposited into vessel walls and covered the cell wall of sengon wood. The image generated from the SEM-EDX test showed that the MNano-Silica 0.5% had a significant effect on the morphology of sengon wood. The MNano-Silica 0.75% resulted in better morphological changes compared with a concentration of 0.5%. Overall, these results indicate that the replenishment of nano-silica at various concentrations could affect the morphology of sengon wood.

The SEM analysis above explains the cause of this treated sengon wood that underlies the decrease in the WU value and the increase in the value of the dimensional stability parameters and oven-dried density. The nano-silica addition could improve the distribution of the MNano-Silica solution. Based on the above results, the MNano-Silica 0.75% treatment showed optimal results compared with other treatments.

4.1.2. XRD Analysis

Similar results were shown by [32], who reported that the degree of crystallinity of sengon wood as a result of MEG and nano-SiO_2 impregnation increased from the MEG treatment up to $MEGSiO_2$ 1%. The XRD analysis results showed the effect of adding nano-silica to values for dimensional stability parameters and the resulting oven-dried density. This effect was assisted by the appearance of silica peaks detected in the diffraction plane based on the JCPDS database, and it was concluded that the added nano-silica increased the crystallinity degree.

4.1.3. FTIR Analysis

In the wood sample, the functional group detected at wave number 3400 cm^{-1} was a functional group -OH stretching; at wave number 1730 cm^{-1}, it was a functional group C=O stretching from the acetyl group; then, at wave number 1261 cm^{-1}, it was a functional group CO of guaiasil ring stretching [39]. In sengon wood, a C–H stretching peak of the -CH_2 group was detected, which appeared at wave number 2905.34 cm^{-1}, which corresponded to the ethylene glycol peak [40]. The presence of the Si–O–Si functional group indicated the presence of silica in jabon and sengon wood. This was in line with [41], who found that the range of wavenumbers was 1050–1115 cm^{-1}. This range of wavenumbers is called the Si–O–Si vibration.

5. Conclusions

The impregnation of an MEG and nano-silica mixture had a significant effect on the resulting WPG, WU, ASE, BE, and oven-dried density of sengon wood. Changes in the morphology of the treated sengon wood were indicated by the closure of the pits on the vessel walls (SEM analysis results) and the detection of ethylene glycol (FTIR analysis results) and silica (XRD and FTIR analysis results) in the wood samples. Overall, the MNano-Silica 0.75% treatment was the optimal method to increase the dimensional stability of 5-year-old sengon wood.

Author Contributions: Conceptualization, I.R.; methodology, I.R. and F.C.D., A.M. and E.P.; software, F.C.D. and E.P.; validation, I.R., A.M., W.D. and D.N.; formal analysis, F.C.D.; investigation, I.R., F.C.D. and E.P.; resources, I.R.; data curation, F.C.D. and E.P.; writing—original draft preparation, F.C.D.; writing—review and editing, I.R., W.D. and D.N. All authors have read and agreed to the published version of the manuscript.

Funding: This research was funded by the Ministry of Education, Culture and Research and Technology of Indonesia (grant no. 1/E1/KP.PTNBH/2021 and grant no. 8/E1/KPT/2021).

Institutional Review Board Statement: Not applicable.

Informed Consent Statement: Not applicable.

Data Availability Statement: Not applicable.

Acknowledgments: The authors are grateful for the support of the Ministry of Education, Culture and Research and Technology of Indonesia (grant no. 1/E1/KP.PTNBH/2021 and grant no. 8/E1/KPT/2021).

Conflicts of Interest: The authors declare no conflict of interest.

References

1. Adi, D.S.; Risanto, L.; Damayanti, R.; Rullyati, S.; Dewi, L.M.; Susanti, R.; Dwianto, W.; Hermiati, E.; Watanabe, T. Exploration of Unutilized Fast Growing Wood Species from Secondary Forest in Central Kalimantan: Study on the Fiber Characteristic and Wood Density. *Procedia Environ. Sci.* **2014**, *20*, 321–327. [CrossRef]
2. Szostak, A.; Bidzińska, G.; Ratajczak, E.; Herbeć, M. Biomasa drzewna z upraw drzew szybkorosna{ogonek}cych jako alternatywne źródło surowca drzewnego w Polsce. *Drewno* **2013**, *190*, 85–113. [CrossRef]
3. Bredemeier, M.; Busch, G.; Hartmann, L.; Jansen, M.; Richter, F.; Lamersdorf, N.P. Fast growing plantations for wood production—Integration of ecological effects and economic perspectives. *Front. Bioeng. Biotechnol.* **2015**, *3*, 1–14. [CrossRef] [PubMed]
4. Gałazka, A.; Szadkowski, J. Enzymatic hydrolysis of fast-growing poplar wood after pretreatment by steam explosion. *Cellul. Chem. Technol.* **2021**, *55*, 637–647. [CrossRef]
5. Kojima, M.; Yamamoto, H.; Okumura, K.; Ojio, Y.; Yoshida, M.; Okuyama, T.; Ona, T.; Matsune, K.; Nakamura, K.; Ide, Y.; et al. Effect of the lateral growth rate on wood properties in fast-growing hardwood species. *J. Wood Sci.* **2009**, *55*, 417–424. [CrossRef]
6. Rahayu, I.; Darmawan, W.; Nugroho, N.; Nandika, D.; Marchal, R. Demarcation point between juvenile and mature wood in sengon (Falcataria moluccana) and jabon (Antocephalus cadamba). *J. Trop. For. Sci.* **2014**, *26*, 331–339.
7. Martawijaya, A.; Kartasujana, I.; Kadir, K.; Prawira Among, S. *Indonesian Wood Atlas Volume 1*; Forest Research and Development Agency, Ministry of Forestry Republik of Indonesia: Jakarta, Indonesia, 2005; pp. 1–171.
8. [BPS] Central Bureau of Statistics. *Statistic of Forestry Production 2019*; Central Bureau of Statistics: Jakarta, Indonesia, 2019.
9. Popper, R.; Niemz, P.; Eberle, G. Investigations on the sorption and swelling properties of thermally treated wood. *Holz. Roh. Wer.* **2005**, *63*, 135–148. [CrossRef]
10. Kocaefe, D.; Younsi, R.; Chaudry, B.; Kocaefe, Y. Modeling of heat and mass transfer during high temperature treatment of aspen. *Wood Sci. Technol.* **2006**, *40*, 371–391. [CrossRef]
11. Pratiwi, L.A.; Darmawan, W.; Priadi, T.; George, B.; Merlin, A.; Gérardin, C.; Dumarçay, S.; Gérardin, P. Characterization of thermally modified short and long rotation teaks and the effects on coatings performance. *Maderas Cienc. Y Tecnol.* **2019**, *21*, 209–222. [CrossRef]
12. Bekhta, P.; Niemz, P. Effect of high temperature on the change in color, dimensional stability and mechanical properties of spruce wood. *Holzforschung* **2003**, *57*, 539–546. [CrossRef]
13. Gökhan, G.; Deniz, A. The influence of mass loss on the mechanical properties of heat-treated black pine wood. *Wood Res. Slovak.* **2007**, *54*, 33–42.
14. Shi, J.L.; Kocaefe, D.; Amburgey, T.; Zhang, J. A comparative study on brown-rot fungus decay and subterranean termite resistance of thermally-modified and ACQ-C-treated wood. *Holz. Als. Rohund Werkst.* **2007**, *65*, 353–358. [CrossRef]
15. Sivrikaya, H.; Can, A.; de Troya, T.; Conde, M. Comparative biological resistance of differently thermal modified wood species against decay fungi, Reticulitermes grassei and Hylotrupes bajulus. *Maderas Cienc. Y Tecnol.* **2015**, *17*, 559–570. [CrossRef]
16. Salman, S.; Thévenon, M.F.; Pétrissans, A.; Dumarçay, S.; Candelier, K.; Gérardin, P. Improvement of the durability of heat-treated wood against termites. *Maderas Cienc. Y Tecnol.* **2017**, *19*, 317–328. [CrossRef]
17. Yildiz, S.; Tomak, E.D.; Yildiz, U.C.; Ustaomer, D. Effect of artificial weathering on the properties of heat treated wood. *Polym. Degrad. Stab.* **2013**, *98*, 1419–1427. [CrossRef]
18. Gérardin, P. New alternatives for wood preservation based on thermal and chemical modification of wood—A review. *Ann. For. Sci.* **2016**, *73*, 559–570. [CrossRef]
19. Wardani, L.; Risnasari, I.; Yasni, H.Y.S. Resistance of jabon timber modified with styrene and methyl methacrylate against drywood termites and subterranean termites. In Proceedings of the 9th Pacific Rim Termite Research Group Conference, Hanoi, Vietnam, 27–28 February 2012; Science and Technics Publishing House: Hanoi, Vietnam, 2012; pp. 73–78.
20. Hadi, Y.S.; Rahayu, I.S.; Danu, S. Physical and mechanical properties of methyl methacrylate impregnated jabon wood. *J. Indian Acad. Wood Sci.* **2013**, *10*, 77–80. [CrossRef]
21. Hadi, Y.S.; Rahayu, I.S.; Danu, S. Termite resistance of jabon wood impregnated with methyl methacrylate. *J. Trop. For. Sci.* **2015**, *27*, 25–29.
22. Gabrielli, C.P.; Kamke, F.A. Phenol-formaldehyde impregnation of densified wood for improved dimensional stability. *Wood Sci. Technol.* **2010**, *44*, 95–104. [CrossRef]
23. Fukuta, S.; Watanabe, A.; Akahori, Y.; Makita, A.; Imamura, Y.; Sasaki, Y. Bending properties of compressed wood impregnated with phenolic resin through drilled holes. *Eur. J. Wood Wood Prod.* **2011**, *69*, 633–639. [CrossRef]
24. Dong, Y.; Yan, Y.; Wang, K.; Li, J.; Zhang, S.; Xia, C.; Shi, S.Q.; Cai, L. Improvement of water resistance, dimensional stability, and mechanical properties of poplar wood by rosin impregnation. *Eur. J. Wood Wood Prod.* **2016**, *74*, 177–184. [CrossRef]
25. Pfriem, A.; Dietrich, T.; Buchelt, B. Furfuryl alcohol impregnation for improved plasticization and fixation during the densification of wood. *Holzforschung* **2012**, *66*, 215–218. [CrossRef]

26. Esteves, B.; Nunes, L.; Domingos, I.; Pereira, H. Improvement of termite resistance, dimensional stability and mechanical properties of pine wood by paraffin impregnation. *Eur. J. Wood Wood Prod.* **2014**, *72*, 609–615. [CrossRef]
27. Tsioptsias, C.; Panayiotou, C. Thermal stability and hydrophobicity enhancement of wood through impregnation with aqueous solutions and supercritical carbon dioxide. *J. Mater. Sci.* **2011**, *46*, 5406–5411. [CrossRef]
28. Dirna, F.C.; Rahayu, I.; Maddu, A.; Darmawan, W.; Nandika, D.; Prihatini, E. Nanosilica synthesis from betung bamboo sticks and leaves by ultrasonication. *Nanotechnol. Sci. Appl.* **2020**, *13*, 131–136. [CrossRef] [PubMed]
29. Priyanto, A. *Synthesis and Application of Silica Derived from Leaves ash of Betung Bamboo (Dendrocalamus Asper (Schult F.) Backer Ex Heyne) to Reduce Ammonium and Nitrat Content on Liquid Tofu Waste [Undergraduate].*; Walisongo Islamic State University: Semarang, Indonesia, 2015.
30. Dwivedi, V.N.; Singh, N.P.; Das, S.S.; Singh, N.B. A new pozzolanic material for cement industry: Bamboo leaf ash. *Int. J. Phys. Sci.* **2006**, *1*, 106–111. [CrossRef]
31. Dirna, F.C.; Rahayu, I.; Zaini, L.H.; Darmawan, W.; Prihatini, E. Improvement of fast-growing wood species characteristics by MEG and nano SiO_2 impregnation. *J. Korean Wood Sci. Technol.* **2020**, *48*, 41–49. [CrossRef]
32. Rahayu, I.; Darmawan, W.; Zaini, L.H.; Prihatini, E. Characteristics of fast-growing wood impregnated with nanoparticles. *J. For. Res.* **2020**, *31*, 677–685. [CrossRef]
33. Dong, Y.; Yan, Y.; Zhang, S.; Li, J. Wood/polymer nanocomposites prepared by impregnation with furfuryl alcohol and Nano-SiO_2. *BioResources* **2014**, *9*, 6028–6040. [CrossRef]
34. Eisenreich, S.J.; Looney, B.B.; Thornton, J.D. Airborne organic contaminants in the Great Lakes ecosystem. *Environ. Sci. Technol.* **1981**, *15*, 30–38. [CrossRef]
35. [ATSDR] Agency for Toxic Substances and Disease. *Agency Toxicological Profile for Ethylene Glycol and Propylene Glycol*; [ATSDR] Agency for Toxic Substances and Disease: Atlanta, GA, USA, 1997; p. 249.
36. [BS] British Standard. Methods of Testing Small Clear Specimen of Timber. BS 373:1957. In *Annual Book of BS Standard*; British Standard Institution: London, UK, 1957.
37. Hill, C.A.S. *Wood Modification: Chemical, Thermal and Other Processes*; Wiley Series in Renewable Resources; Wiley: Chichester, UK, 2006; ISBN 9780470021729.
38. Rowell, R.M.; Ellis, W.D. Determination of dimensional stabilization of wood using the water-soak method. *Wood Fiber* **1978**, *10*, 104–111.
39. Pandey, K.K. A Study of Chemical Structure of Soft and Hardwood and Wood Polymers by FTIR Spectroscopy. *J. Appl. Polym. Sci.* **1999**, *71*, 1969–1975. [CrossRef]
40. Sukirno, E.; Shofiyani, A. Nurlina. Fabrication of si/pva/peg composit membrane derived from silica of singkup stone for reducing solution phosphat ion contentration. *J. Equatoria.l Chem.* **2017**, *6*, 1–9.
41. Patil, R.C.; Dongre, R.; Meshram, J. Preparation of silica powder from rice husk. *Agric. Eng. Int. CIGR J.* **2017**, *19*, 158–161.

forests MDPI

Article

The Influence of Processing Conditions on the Quality of Bent Solid Wood from European Oak

Aleš Straže [1], Jure Žigon [1], Stjepan Pervan [2], Mislav Mikšik [2] and Silvana Prekrat [2,*]

[1] Biotechnical Faculty, University of Ljubljana, Jamnikarjeva 101, 1000 Ljubljana, Slovenia; ales.straze@bf.uni-lj.si (A.S.); jure.zigon@bf.uni-lj.si (J.Ž.)

[2] Faculty of Forestry, University of Zagreb, Svetošimunska cesta 23, 10000 Zagreb, Croatia; spervan@sumfak.unizg.hr (S.P.); mmiksik@sumfak.unizg.hr (M.M.)

* Correspondence: sprekrat@sumfak.unizg.hr; Tel.: +385-1-235-2408

Abstract: Bending of solid wood from European oak is one of the most demanding technological processes due to its specific structural and physical properties and variability. We investigated the influence of wood moisture content (MC) and stiffness, determined by NDT, as well as previous drying methods on the bending ability of the wood. The best quality was obtained with bending specimens bent at a moisture content of at least 16% and quarter- or semi-quartersawn. The number of rejected specimens increased slightly when HF bending was used. Single-stage predrying of oak to a final MC of 8% resulted in a high rejection rate (>70%) regardless of drying technique. The acceptance rate was higher for less stiff specimens where the ratio of ultrasonic velocity in the straight (v_S) and bent region (v_B) was less than 0.5 (v_B/v_S).

Keywords: wood; solid wood bending; quality; nondestructive testing

1. Introduction

European oak (*Quercus robur* L., *Quercus petraea* (Matt.) Liebl.) is an important commercial tree species and a widely used industrial wood for a variety of products such as veneers, furniture, interior and exterior structures, and many other items [1]. In modern production of solid oak furniture, the need for complex spatial shapes of furniture elements is common. In these cases, the reduced mechanical properties of the wood in the transverse direction can be a limiting technological and applicable factor. Difficulties in the production of complex 3D solid wood shapes also arise from the low material yield during sawing and from the demanding technological processing [2,3]. One solution that overcomes these limitations is solid oak bending.

Solid wood bending has been practiced for centuries, and the quality of the bending is judged by the proper deformation achieved without apparent failure of the wood [4]. Of particular interest in solid wood bending are the effects of creep and, to a lesser extent, relaxation. Creep is greatly enhanced by the absolute value of moisture content (MC) and by the changes in MC of the wood under bending load due to the mechanosorptive effect [5–7]. The rate and extent of mechanosorptive deformation correspond to the extent of MC changes and are usually independent of the time during which the MC changes occurred [8,9].

In the practice of commercial wood bending, moistened solid wood is usually bent at higher temperatures, which leads to a corresponding plasticity of the material [10]. It is assumed that under such conditions the glass transition temperature (T_g), at which the modulus of elasticity (MOE) decreases significantly, is exceeded for some or all of the basic polymeric constituents of the wood. The highest T_g in the dry state, above 200 °C, was found for cellulose, slightly lower, between 150 °C and 180 °C, for hemicelluloses, and below 150 °C for lignin [11,12]. However, several studies show that the transition from elastic to plastic mechanical behavior of wood occurs at temperatures well below 100 °C, especially at higher moisture contents [13,14]. Most explanations for this phenomenon state that the low-molecular-weight water molecules in wood act as diluents and plasticizers [7].

However, the correct MC of specimens before bending is controversial. Conditioning the material in the range of 12 to 25% MC covers most bending applications and methods [15]. Specific bending radii and severity of deformation may require different MCs, achieved by different drying and presteaming methods [16,17]. Heat and moisture plasticizing by presteaming increases the compressibility of the wood by up to 40% of the compressive strain, but has virtually no effect (up to 2%) on the tensile ductility of the wood [10,15,18]. The softening treatment shifts the neutral axis of the bent parts towards the convex side, which is axially stressed during bending. This shift significantly improves the bending deformation of the wood. In addition, a metal band with end stop is used, which is wrapped around the convex side of the sample, a method discovered by Thonet in 1856 [18].

Due to the wide variability of chemical, structural, and mechanical properties of wood species, the bending quality of wood varies, so selection is important. The highest bending quality is obtained with straight-grained pieces, free of crossgrain, of generally fast-growing and less dense species [18,19]. Some studies also report that ring-porous woods generally give better results than diffuse-porous woods. In practice, quartersawn lumber is preferred for bending, but some sources also indicate that flatsawn lumber bends better in severe bends [10,16,18].

In this study, we aimed to investigate the possibility of bending of European oak wood, which is less commonly used in industrial practice because of the good bending properties of other European hardwoods. We were particularly interested in the influence of wood moisture content and previous drying methods on the bending ability of the wood. We also wanted to see if it was possible to determine the quality of the bending using non-destructive techniques.

2. Materials and Methods

2.1. Sampling

The European oak (*Quercus robur* L.) wood used in the study was obtained by sawing 1st-grade oak logs. The test pieces were 1300 mm long, 60 mm wide, and 38 mm thick and were straight-grained and sawn from heartwood. A total of 120 elements were taken from the mill's own lumberyard (Spin Valis d.d., Požega, Croatia; N 45.338396°, E 17.690455°, 311 m a.s.l.), with 40% of the elements oriented radially (R; quartersawn) and semi-radially (RT) and 20% of the elements oriented tangentially (T; flatsawn) (Figure 1). The initial moisture content of the samples was 56.3% (St.dev. = 10.2%; St.dev.—Standard deviation).

Figure 1. Sampling and sorting of green oak wood in the woodyard (**a**), wax protection of the front and back ends before drying (**b**), and orientation of the wood grain of test pieces (**c**).

2.2. Material Processing

2.2.1. Wood Drying

To optimize the organization of the industrial production process, 3 different sawn-wood drying techniques were included in the study: air drying (AD), convection kiln drying (KD), and vacuum drying (VD). To achieve sufficiently short processing times, we have combined these available industrial drying techniques. We have tested the possibility of single-phase processing of oak elements to a final moisture content (MC) of 8% before bending (G_1, G_2) and two-phase processing (G_3, G_4). In the latter case, phase 1 is performed up to a wood MC of 16%, at which the wood was bent. For high wood MCs, only air drying and convection kiln drying were used, while for drying at low MCs (<16%), vacuum drying was additionally tested. By combining different drying techniques (AD, KD, VD) and the final MC achieved before the bending process (16% and 8%), 4 test groups ($G_1 \ldots G_4$) were formed (Table 1).

Table 1. Classification into process groups ($G_1 \ldots G_4$) and treatment sequence of test oak samples, with target moisture content, before and/or after solid wood bending.

Test Group	Processing Procedure	No. of Samples
G_1	Air drying (20...25%) → Kiln drying (8%) → **Bending**	30
G_2	Kiln drying (20...25%) → Vacuum drying (8%) → **Bending**	30
G_3	Kiln drying (16%) → **HF Bending** → Kiln drying (8%)	30
G_4	Kiln drying (16%) → **Bending** → Kiln drying (8%)	30

Air drying of G_1 test samples took place in the first half of 2022 in the sawmill warehouse at the company's site. The average drying temperature increased from 3 °C in January to 18 °C at the end of May 2022. Subsequently, the G_1 samples were conventionally dried in the kiln dryer at normal temperature (<45 °C) and moderate drying gradient (<2.5) until a target value of 8% MC was reached. The solid wood was then steamed and bent.

For the G_2 samples, standard convection kiln drying was performed at low temperature from the initial green state to 22% MC. These samples were then placed in a vacuum drying chamber where heat transfer was performed using the hot plate method. Vacuum drying was performed at a temperature as high as 70 °C and an absolute vacuum of nearly 100 mbar to achieve a final value of 8% MC. This was followed by steaming and bending of dried raw oak wood pieces.

The samples in groups G_3 and G_4 were also kiln dried using the same drying equipment as for samples G_1 and G_2 (Figure 2a). However, the samples of groups G_3 and G_4 were dried in 2 stages. First, they were convection kiln-dried as raw elements at low temperature (following the same drying schedule as group G_2) until they had a final MC of 16%. The solid wood was then steamed and bent. The bent solid wood elements were then dried in the same kiln dryer at normal temperature to a final MC of 8%, as in G_1.

Figure 2. Normal temperature convection kiln drying (KD) of test oak samples (**a**) and vacuum drying (VD), using hot plate method for heat transfer (**b**).

2.2.2. Solid Wood Steaming

Before bending, the dried specimens were steamed. The specimens were placed in boilers heated with steam at 130 ± 3 °C and a pressure of 1.8 bar. The process took an average of 3.5 h (±0.5 h). The temperature at the end of steaming, just before the specimens were placed in the bending press, was checked with a non-contact thermal imaging camera (FLIR i60; FLIR Systems AB, Täby, Sweden). The surface temperature of the specimens was above 70 °C, while the temperature inside the specimens was even higher (Figure 3).

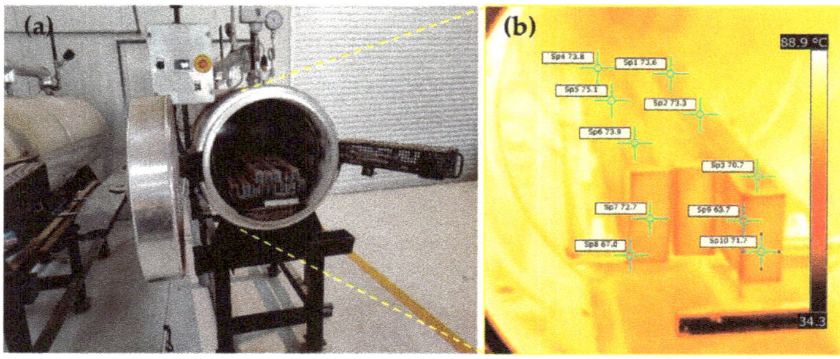

Figure 3. Steaming of solid oak specimens in boiler (**a**) and achieved surface temperature (FLIR i60) before the bending (**b**).

2.2.3. Solid Wood Bending

The bending of the oak specimens of groups G_1, G_2, and G_4 was carried out classically in the bending press using a metal belt with end stop on the convex side of the specimens. A mold with a center radius of 150 mm was used for bending. Three specimens each were placed in the press-bending mold. The angle between the two straight parts of the bent specimen was 105°, which corresponded to the desired geometry of the piece of furniture—the armchair (Figure 4). The pressing time was 30 min. Subsequently, the bent specimens were placed in stacks, either for further mechanical processing (G_1 and G_2) or for additional drying (G_4).

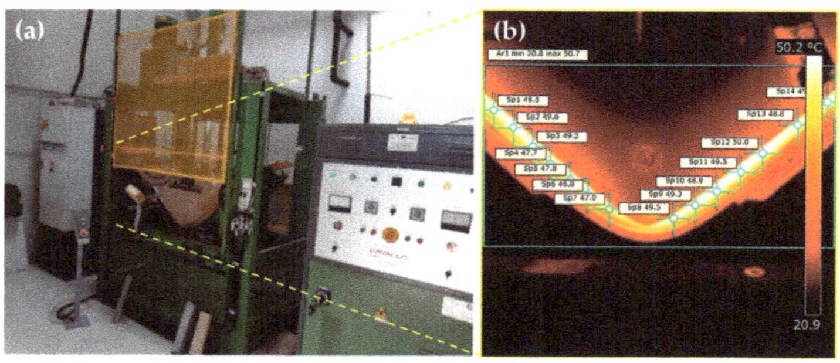

Figure 4. Press with bending mold (**a**) and achieved temperature (FLIR i60) in bent oak samples during the bending process of (**b**).

To increase productivity, G_3 samples were bent at high frequency (HF). Research on microwave drying of wood has been conducted since the early 1960s and is well-documented [20]. The same press and mold was used for HF bending as for conventional bending. With a 20 kW HF generator, the wood was bent at 10 MHz, heated, and finally dried to 8% MC. The pressing time was on average 30 min.

2.3. Quality Assessment of Bent Oak Elements

The quality of the bent oak was determined primarily visually. In the bending area, the tested specimens were carefully inspected in all planes. Bending was confirmed to be successful when the specimens showed no tensile cracks and failures on the convex side and a homogeneous structure on the concave side. Specimens with local bending, buckling, and kneading of the fibers on the concave side of the specimens were also rejected.

The velocity of ultrasound (v) has been added to the quality assessment of bent oak samples with twice-per-piece (v_S—straight part, v_B—bent region) measurement of the time of flight in the longitudinal direction by Proceq Pundit PL-200PE (Proceq Inc., Scharzenbach, Switzerland) pulse ultrasonic device. The exponential transducers with working 54 kHz frequency were used (Figure 5). The velocity of ultrasound in the straight part of the specimens (v_S) and wood density (ρ) were used to determine the modulus of elasticity (MOE) of bent oak specimens (Equation (1)).

$$MOE = \rho \cdot v_S^2 \tag{1}$$

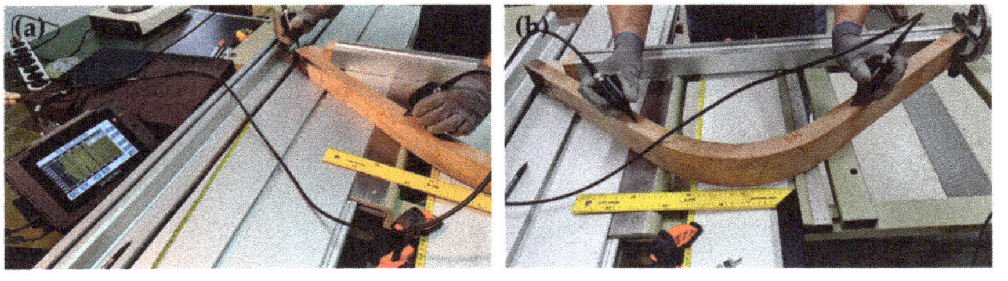

Figure 5. Determination of the velocity of ultrasound (Proceq PL-200PE) in the straight part (v_S) (**a**) and in the bent region (v_B) of solid oak specimens (**b**).

3. Results and Discussion

3.1. Process Kinetics

Air drying of G_1 specimens was the slowest and the longest process, depending on local climatic conditions. Air drying lasted 84 days, during which the wood reached an average MC of 21.9%. We then dried the wood in a convection kiln dryer for an additional 28 days to an average final MC of 8.7%. The drying of group G_2 was only slightly faster than that of group G_1. In the first part, drying took 56 days to an average MC of 20.8%, followed by 11 days of vacuum drying to a final MC of 8.6%. In both groups, specimen bending was performed at low MC (<10%) (Figure 6).

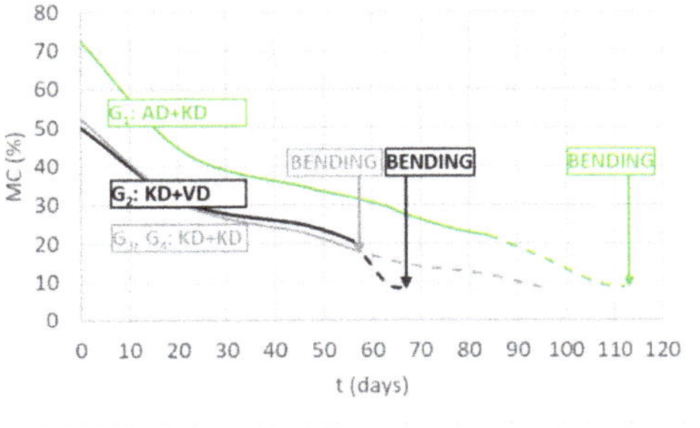

Figure 6. Kinetics of drying processes with bending of four tested groups of oak specimens, used for solid wood bending: G_1 sequence: 1. air drying (AD) → 2. kiln drying (KD) → 3. bending, G_2 sequence: 1. kiln drying (KD) → 2. vacuum drying (VD) → 3. bending, G_3, G_4 sequence: 1. kiln drying (KD) → 2. bending → 3. kiln drying (KD) (— 1st drying technique, - - - 2nd drying technique).

The drying of the G_3 and G_4 samples was carried out in the same way as that of the G_2 samples in the first phase. An average MC of 16.8% was obtained, and the wood was bent in this condition, using the HF field in group G_3 and the conventional method in group G_4. The drying of the bent wood of groups G_3 and G_4 in the second phase in a kiln dryer to a final MC of 8.6% took another 39 days.

3.2. Bending Success Rate of Oak Wood

Bending success rates were highest for groups G_3 (60.9%) and G_4 (42.9%) when specimens were bent at higher MCs (>16%). In these two groups, G_3 specimens appear to have lost some success at the expense of HF bending (Figure 7). As other studies have shown, excessive energy input of HF into the bent specimens can lead to an increase in temperature and, due to trapped moisture in the specimens, an increase in vapor pressure that can break down the cellular wood structure [17]. Especially in impermeable wood species such as European oak [21], this can lead to high internal stresses and possibly failure.

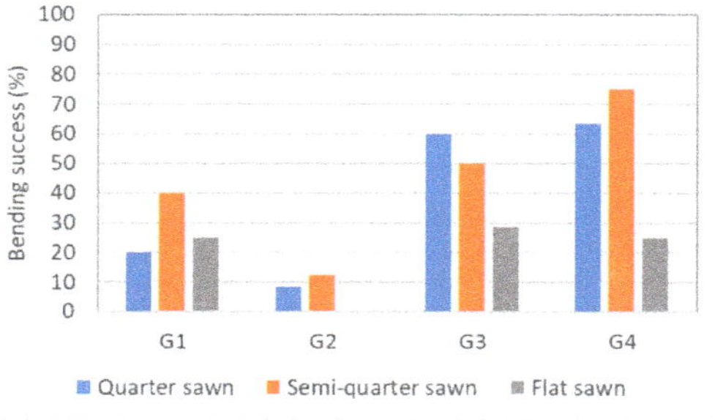

Figure 7. Bending success rate of oak wood in each test group ($G_1 \ldots G_4$) with respect to orientation of specimens (quartersawn, half-quartersawn, flatsawn).

However, we did not succeed in bending the G_1 (28.6%) and G_2 (8.3%) samples. It appears that the low moisture content of the bends (MC_{G1} = 8.7%; MC_{G2} = 8.6%) does not allow the necessary plastic deformability that can be achieved when the material exceeds the glass transition temperature [14,15]. At the same time, it is shown that accelerated vacuum drying in stage 2, from 20.8 to 8.6% MC (G_2 group), can cause high internal stresses due to the poor permeability of the oak wood tissue [22]. These can lead to additional microstructural defects, which then cause the material to fail during bending. This is reflected in the higher rejection rate of the G_2 group compared to G_1.

3.3. Appearance and Visual Assessment of Defects of Bent Oak Wood

Visual inspection of the rejected bent oak specimens revealed approximately three typical defects, with material failing on the concave side (I), on the convex side (II), or on both sides (III) (Figure 8). On the concave side, local wrinkling, buckling, and kneading of the fibers were observed, most pronounced at the bottom of the curve (Figure 8b, I—failure type). On the opposite, convex side, several of the rejected bends exhibited tensile failure, with both brittle and fiber breaks (Figure 8c, II—failure type). The worst case, where bending was unsuccessful, was represented by the test specimens where local fractures occurred on both the concave and convex sides (Figure 8d, III—failure type).

The various failure modes occurred to about the same extent in all specimen groups examined ($G_1 \ldots G_4$). For the individual orientations (quartersawn, quarter- to flatsawn, flatsawn), we could not identify a characteristic failure mode. The main cause of structural failure appears to be poor plasticization of the specimens before or during the bending process. This is also a commonly cited reason for solid wood bending in past research [4,10,18].

Figure 8. Visual assessment of defects on bent oak: (**a**) no defects present, (**b**) wrinkling, buckling, and kneading of fibers on concave side (I—failure type), (**c**) tensile brittle fiber break (II—failure type), (**d**) buckling and kneading of fibers on concave side with additional tensile fiber break on convex side (III—failure type).

3.4. Physical and Acoustic Properties of Bent Oak Wood

The average density of oak wood at an average final MC of 8.4% was 634 kg/m^3 (CoV = 8.1%; CoV—Coefficient of Variation), which is lower compared to the results of other studies on this species [23–26]. Some studies suggest that the lignin content decreases with increasing density of oak wood, which could affect the bending capacity of the wood [27,28]. No significant differences were found between the mean values of the groups (ρ_{G1} = 640 kg/m^3, ρ_{G2} = 635 kg/m^3, ρ_{G3} = 624 kg/m^3, ρ_{G4} = 638 kg/m^3) (Figure 9), which does not directly indicate possible chemical differences between the studied specimens of each group. The latter would have to be confirmed by further investigations.

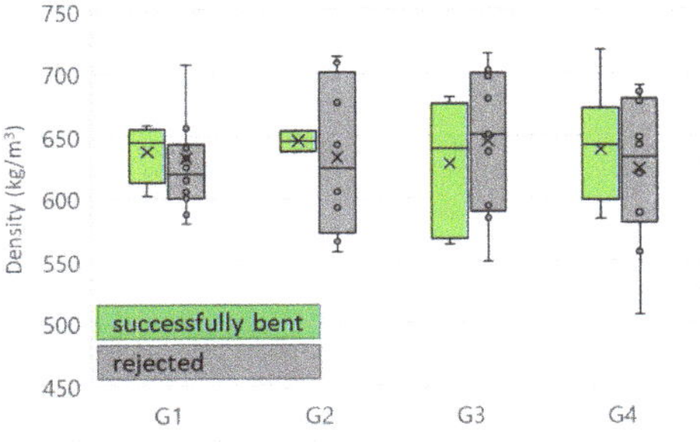

Figure 9. Density and its variation of oak wood specimens in tested groups $G_1 \ldots G_4$ (× mean, ○ outlier).

In the straight part of the successfully bent oak specimens, we determined a mean ultrasonic velocity of less than 4500 m/s (v_S) and a corresponding modulus of elasticity (MOE) of less than 13 GPa in all groups studied. There was a downward trend in the mean ultrasound velocity of specimens in groups G_1 to G_4 ($v_{S\text{-}G1}$ = 4470 m/s, $v_{S\text{-}G2}$ = 4450 m/s, $v_{S\text{-}G3}$ = 4440 m/s, $v_{S\text{-}G4}$ = 4220 m/s). We also confirmed a trend of decreasing stiffness from G_1 to G_4 (MOE_{G1} = 12.8 GPa, MOE_{G2} = 12.8 GPa, MOE_{G3} = 12.5 GPa, MOE_{G4} = 11.4 GPa). However, the differences between the means were not significant for both v_S and MOE, except for G_4, where the values were significantly lower than for the others (G_1, G_2, and

G_3; ANOVA, $p < 0.05$) (Figure 10). It seems that part of the best bending performance we have confirmed in group G_4 is also due to the slightly lower stiffness of the specimens.

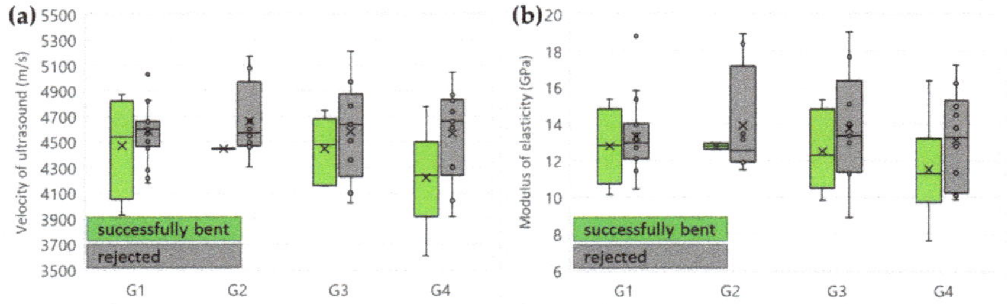

Figure 10. Ultrasound velocity (v_S) (**a**) and modulus of elasticity (MOE) (**b**) in tested groups $G_1 \ldots G_4$: (**a**) determined in the straight part of bent oak specimens (× mean, ○ outlier).

In a straight part of the rejected samples, we measured v_S above 4500 m/s and MOE above 13 GPa. We could not find any characteristic trend in the change of values between the groups of rejected specimens ($v_{S\text{-}G1}$ = 4550 m/s, $v_{S\text{-}G2}$ = 4620 m/s, $v_{S\text{-}G3}$ = 4650 m/s, $v_{S\text{-}G4}$ = 4540 m/s; MOE_{G1} = 13.4 GPa, MOE_{G2} = 13.7 GPa, MOE_{G3} = 13.6 GPa, MOE_{G4} = 13.3 GPa). However, when we compare the values of the rejected specimens with those of the successfully bent specimens, we find that v_S and MOE of these two categories are different in groups G_3 and G_4. Thus, it can be seen that the initial lower MOE of the specimens has a significant effect on the bending success. Only in groups G_3 and G_4 was the bending success significantly higher than in G_1 and G_2, which, in addition to the correspondingly high wood MC, could also be due to the lower MOE of the specimens in G_3 and G_4. Previous research has also shown that less stiff wood is easier to bend [4,15,18,19].

We measured a much lower average ultrasound velocity in the bent region of the specimens (v_B = 2270 m/s) than in their straight part (v_S = 4505 m/s) (Figure 11a). This is partly due to the curvature itself, as the sound waves can propagate along a shortcut between the two probes and thus across the wood grain where they would otherwise be slower [29–31]. The compression deformation of the wood tissue in the curved region of the specimens could also be a reason. Ultrasound velocities greater than 2300 m/s were measured in the G_1 and G_2 specimens for both the successfully bent and rejected specimens. No differences in mean values were observed between the two groups (G_1 and G_2) and the two categories (accepted and rejected). For specimens in groups G_3 and G_4, the v_B was lower and slightly below 2250 m/s. The differences between the two groups and between the two categories of specimens were also statistically insignificant (t-test; $p > 0.05$).

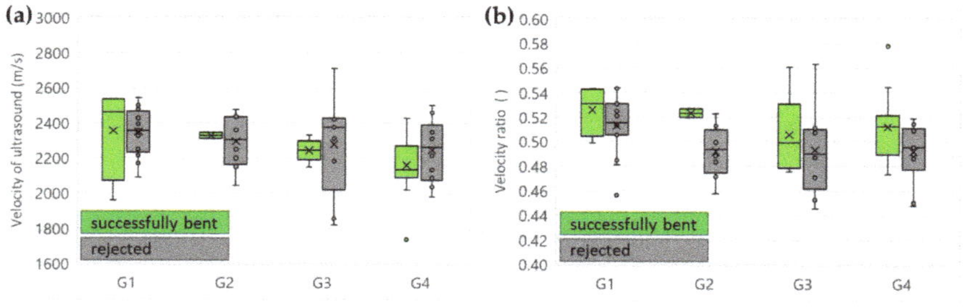

Figure 11. Ultrasound velocity in bent region (v_B) of oak specimens (**a**) and ratio of ultrasound velocities in straight and bent region (v_S/v_B) (**b**) of specimens in tested groups $G_1 \ldots G_4$ (× mean, ○ outlier).

It was found that the relative ultrasound velocity in the bent region of the specimens (v_B), represented as the ratio between the ultrasound velocity in the bent region and in the straight portion (v_B/v_S), was related to the bending success of the specimens. This ratio averaged above 0.51 for successfully bent specimens and below 0.50 for rejected specimens, regardless of test group ($G_1 \ldots G_4$). Due to considerable variability, the v_B/v_S ratio was significantly lower in the rejected compared to the accepted specimens only in groups G_2 and G_4. Shear ultrasound waves are commonly used to detect internal defects in some other materials [32], and we intend to include them in future studies.

4. Conclusions

Bending of European oak solid wood, by studying the influence of wood moisture content and previous drying methods on bending ability in industrial tests and also using non-destructive techniques, led to the following conclusions:

- In order to achieve low rejection rate in the bending process, the process parameters, i.e., the time and the final moisture content for the different processing stages, must be well-controlled, as the margin between time and moisture content for optimal bending is very narrow.
- A method in which the oak wood is predried in one step to a nominal final moisture content of 8% and the specimens are then bent is not practical because of the low bending deformability and low compressibility on the concave side and tensile ductility on the convex side.
- The study showed that the initial lower MOE in addition to the proper moisture content before bending ($MC \geq 16\%$) significantly affected the bendability and acceptance rate of the oak specimens.
- In addition to visual assessment, the acceptance rate of bent solid oak can be determined non-destructively from the ratio of the ultrasound velocity in the straight and bent region (v_B/v_S) of the specimens.

Author Contributions: Conceptualization and experiment design, A.S., M.M. and S.P. (Stjepan Pervan); M.M., A.S. and J.Ž. performed the experiments; A.S., M.M. and J.Ž. analyzed the data; validation and formal analysis, J.Ž., M.M. and S.P. (Silvana Prekrat); writing—original draft preparation, A.S. and S.P. (Silvana Prekrat); writing—review and editing, A.S., J.Ž., M.M., S.P. (Stjepan Pervan) and S.P. (Silvana Prekrat); project administration, S.P. (Stjepan Pervan) and S.P. (Silvana Prekrat); funding acquisition, S.P. (Stjepan Pervan), M.M. and S.P. (Silvana Prekrat). All authors have read and agreed to the published version of the manuscript.

Funding: The research is part of the project "Development of innovative products from modified Slavonian oak" (KK.01.2.1.02.0031) of Spin Valis d.d. and the partner University of Zagreb, Faculty of Forestry and Wood Technology. The total value of the project is HRK 55,064,343.84, while the amount co-financed by the EU is HRK 23,941,527.32. The project was co-financed by the European Union from the Operational Program Competitiveness and Cohesion 2014–2020, European Fund for Regional Development. This work was also supported by the Ministry of Higher Education, Science and Innovation of the Republic of Slovenia under the program P4-0430 (Forest timber chain and climate change: the transition to a circular bio-economy) and the research project CRP, V4-2016, funded by the Ministry of Agriculture, Forestry and Food of the Republic of Slovenia (MKGP) and the Slovenian Research Agency (ARRS).

Data Availability Statement: The data presented in this study are not publicly available due to a non-disclosure agreement.

Acknowledgments: Our special thanks to technical assistant Luka Krže for mechanical processing of the samples. We also thank Matija Straže for processing the samples and data collection.

Conflicts of Interest: The authors declare no conflict of interest. The funders had no influence on the design of the study, the collection, analysis, or interpretation of the data, the writing of the manuscript, or the decision to publish the results.

References

1. Krackler, V.; Keunecke, D.; Niemz, P.; Hurst, A. Possible fields of hardwood application. *Drewno* **2011**, *56*, 125–136.
2. Wieruszewski, M.; Turbanski, W.; Mydlarz, K.; Sydor, M. Economic Efficiency of Pine Wood Processing in Furniture Production. *Forests* **2023**, *14*, 688. [CrossRef]
3. Tomec, D.K.; Kariž, M. Use of Wood in Additive Manufacturing: Review and future prospects. *Polymers* **2022**, *14*, 1174. [CrossRef] [PubMed]
4. Taylor, Z. *Wood Bender's Handbook*; Sterling Publishing Inc.: New York, NY, USA, 2001.
5. Jin, F.; Jiang, Z.; Wu, Q. Creep behavior of wood plasticized by moisture and temperature. *BioResources* **2016**, *11*, 827–838. [CrossRef]
6. Ikuho, I.; Norimoto, M. Wood bending utilizing microwave heating–Bending creep in the direction perpendicular to grain. *Mokuzai Gakkaishi* **1984**, *9*, 1–12.
7. Gibson, E.J. Creep of wood: Role of water and effect of a changing moisture content. *Nature* **1965**, *206*, 213–215. [CrossRef]
8. Grossman, P.U.A. Requirements for a model that exhibits mechano-sorptive behaviour. *Wood Sci. Technol.* **1976**, *10*, 163–168. [CrossRef]
9. Hoffmeyer, P.; Davidson, R.W. Mechano-sorptive creep mechanism of wood in compression and bending. *Wood Sci. Technol.* **1989**, *23*, 215–227. [CrossRef]
10. Mikšik, M.; Pervan, S.; Klarić, M.; Čavlović, A.O.; Španić, N.; Prekrat, S. Factors Influencing Behaviour of Solid Wood Bending Process Čimbenici koji utječu na proces savijanja cjelovitog drva. *Drv. Ind.* **2023**, *74*, 105–114. [CrossRef]
11. Irvine, G.M. The glass transitions of lignin and hemicellulose and their measurement by differential thermal analysis. *Tappi J.* **1984**, *67*, 118–121.
12. Back, E.; Salmen, L. Glass transition of wood components hold implications for moulding and pulping processes. *Tappi J.* **1982**, *65*, 107–110.
13. Börcsök, Z.; Pásztory, Z. The role of lignin in wood working processes using elevated temperatures: An abbreviated literature survey. *Eur. J. Wood Wood Prod.* **2021**, *79*, 511–526. [CrossRef]
14. Belchschmidt, J.; Engert, P.; Stephan, M. The glass transition of wood from the viewpoint of mechanical pulping. *Wood Sci. Technol.* **1986**, *20*, 263–272. [CrossRef]
15. Rice, R.W.; Lucas, J. The effect of moisture content and bending rate on the work required to bend solid red oak. *For. Prod. J.* **2003**, *53*, 71–77.
16. Lemoine, T.J.; Koch, P. Steam-bending properties of southern Pine. *For. Prod. J.* **1971**, *21*, 34–42.
17. Sandberg, D.; Johansson, J. A new method for bending solid wood-high frequency heating of beech. In *Hardwood Research and Utilisation in Europe*; University of West Hungary Press: Sopron, Hungary, 2005; pp. 156–161.
18. Stevens, W.C.; Turner, N. *Wood Bending Handbook: Unlock the Secrets of Curving Wood*; Fox Chapel Publishing Inc.: Florin, CA, USA, 2007.
19. Peck, E.C. *Bending Solid Wood to Form*; U.S. Department of Agriculture: Madison, WI, USA, 1956.
20. Torgovnikov, G. *Dielectric Properties of Wood and Wood-Based Materials*; Springer: Haidelberg, Germany, 1993.
21. Emmerich, L.; Brischke, C.; Sievert, M.; Schulz, M.S.; Jaeger, A.-C.; Beulshausen, A.; Humar, M. Predicting the outdoor moisture performance of wood based on laboratory indicators. *Forests* **2020**, *11*, 1001. [CrossRef]
22. Espinoza, O.; Bond, B. Vacuum Drying of Wood—State of the Art. *Curr. For. Rep.* **2016**, *2*, 223–235. [CrossRef]
23. Rao, R.V.; Aebischer, D.P.; Denne, M.P. Latewood density in relation to wood fibre diameter, wall thickness, and fibre and vessel percentages in *Quercus robur* L. *IAWA J.* **1997**, *18*, 127–138. [CrossRef]
24. Zhang, S.-Y.; Owoundi, R.E.; Nepveu, G.; Mothe, F.; Dothe, J.-F. Modelling wood density in European oak (*Quercus petraea* and *Quercus robur*) in simulating the silvicultural influence. *Can. J. For. Res.* **1993**, *23*, 2587–2593. [CrossRef]
25. Kovryga, A.; Gamarra, J.O.C.; van de Kuilen, J.W.G. Strength and stiffness predictions with focus on different acoustic measurement methods. *Eur. J. Wood Wood Prod.* **2020**, *78*, 941–949. [CrossRef]
26. Jakubowski, M.; Dobroczyński, M. Allocation of wood density in european oak (*Quercus robur* L.) trees grown under a canopy of scots pine. *Forests* **2021**, *12*, 712. [CrossRef]
27. Kretschmann, D. Velcro mechanics in wood. *Nat. Mater.* **2003**, *2*, 775–776. [CrossRef] [PubMed]
28. Popescu, M.C.; Popescu, C.M.; Lisa, G.; Sakata, Y. Evaluation of morphological and chemical aspects of different wood species by spectroscopy and thermal methods. *J. Mol. Struct.* **2011**, *988*, 65–72. [CrossRef]
29. Straže, A.; Fajdiga, G.; Gospodarič, B. Nondestructive characterization of dry heat-treated fir (*Abies Alba* Mill.) timber in view of possible structural use. *Forests* **2018**, *9*, 776. [CrossRef]
30. Ozyhar, T.; Hering, S.; Sanabria, S.J.; Niemz, P. Determining moisture-dependent elastic characteristics of beech wood by means of ultrasonic waves. *Wood Sci. Technol.* **2013**, *47*, 329–341. [CrossRef]

31. Kurowska, A.; Kozakiewicz, P.; Gladzikowski, T. Ultrasonic waves propagation velocity and dynamic modulus of elasticity of European oak, European aspen, American cherry and wenge wood. *Ann. Warsaw Univ. Life Sci.* **2016**, *93*, 83–88.
32. Beall, F.C. Overview of the use of ultrasonic technologies in research on wood properties. *Wood Sci. Technol.* **2002**, *36*, 197–212. [CrossRef]

forests

MDPI

Article

A Study on the Susceptibility of PLA Biocomposites to Drilling

Piotr Borysiuk [1,*], Radosław Auriga [1], Jacek Wilkowski [1], Alicja Auriga [2], Adrian Trociński [3] and Lee Seng Hua [4]

1 Institute of Wood Sciences and Furniture, Warsaw University of Life Sciences (WULS-SGGW), 159C Nowoursynowska St., 02-776 Warszawa, Poland
2 Faculty of Biotechnology and Animal Husbandry, West Pomeranian University in Szczecin, Janickiego 33 St., 71-270 Szczecin, Poland
3 Department of Mechanical Wood Technology, Poznań University of Life Sciences, Wojska Polskiego 28 St., 60-637 Poznań, Poland
4 Laboratory of Biopolymer and Derivatives, Institute of Tropical Forestry and Forest Product, Universiti Putra Malaysia—UPM, Serdang 43400, Malaysia
* Correspondence: piotr_borysiuk@sggw.edu.pl; Tel.: +48-225938547

Abstract: Wood–plastic composites (WPCs) increase the range of applications of materials by creating new material solutions. As part of this research, PLA (polylactic acid)- and HDPE (high-density polyethylene)-based composites were manufactured. Softwood sawdust or conifer bark with different sizes (large and small) were used as filler. In selected cases, the addition of 3% additives, such as calcium oxide in the case of PLA or polyethylene-graft-maleic anhydride in the case of HDPE, were tested. The manufactured composites were examined for their density profile and their susceptibility to drilling, defined by the value of the axial force occurring during drilling. The obtained results revealed that the type of matrix had the greatest influence on the axial forces during drilling. Regardless of the composite formulation, composites based on PLA had 25% to 56% higher axial forces during drilling than those based on HDPE. Furthermore, increasing the proportion of lignocellulosic fillers resulted in a decrease in the value of axial forces during drilling, with PLA composites experiencing a greater decrease than HDPE composites. The type and size of the filler had a minor impact on the axial force values during drilling. The statistical analysis indicated that the additives had a greater influence on HDPE than on PLA.

Keywords: PLA; HDPE; lignocellulosic filler; biocomposites; drilling; machinability; density

1. Introduction

Owing to their properties, wood–plastic composites (WPCs) are widely used in a variety of industries. A significant portion of WPCs is made from polymers derived from fossil fuels. For example, polyethylene (PE) is used for decking, construction, and consumer goods; polypropene (PP) for automotive parts, construction, and consumer goods; and polyvinyl chloride (PVC) for decking and construction [1].

Biodegradable thermoplastics, such as poly (lactic acid) (PLA) [2,3], are gaining industry attention as an alternative to WPC production. PLA is easily compostable and has no negative effects on the natural environment [4]. The composite materials within PLA are also ecological due to the fact that the filling material is made of wood fibers. Furthermore, apart from solid wood, WPC fillers can be made of shredded postconsumer wood materials [5–7], recycled fibers [8,9], bark [3], bamboo fibers [10], or particles of annual plants [2].

WPCs are typically manufactured using extrusion or injection methods [2,11], resulting in the production of generally finished products that do not require any additional mechanical processing. However, as these materials become more widely used in a variety of products, the demand for machining grows. Drilling mounting holes, profile milling over an existing component edge, or surface grinding to smooth it are examples of such processes.

Little is known about the machinability properties of WPCs in terms of tool wear, cutting resistance, and machining quality. In general, WPC processing can be accomplished with standard tools and woodworking machines. Buehlmann et al. [12] emphasized, however, that increased pigment content in colored composites may result in faster tool wear. While investigating the parameters of the machining process of WPCs in the form of chipboards bonded with thermoplastics, Zbieć et al. [13,14] discovered that cutting parameters are comparable to those of chipboards of similar-density glued with UF resin.

At the same time, unlike chipboards glued with UF resin, thermoplastics in chip-polymer composites reduce tool wear. Wilkowski et al. [15] and Borysiuk et al. [16] both confirmed the good machinability of WPCs in the form of chipboards bonded with thermoplastic compared with traditional wood materials. Wilkowski et al. [15] investigated axial force and cutting torque while drilling three WPC variants (polyethylene (PE), polypropylene (PP), and polystyrene (PS)). The authors discovered a significant reduction in forces for all WPC variants compared with standard three-layer chipboards. The composites bonded with polyethylene had the greatest reduction in forces among the tested composites (PE). Furthermore, as forces decreased, the proportion of thermoplastic increased (30%, 50%, and 70%).

A study by Zajac et al. [17] on the quality assessment of turning machining based on a WPC's surface roughness parameters did not confirm the theoretical and practical regularity of machining quality reduction with increased feed speed [18]. This perplexing result was explained by the random distribution of the composite's individual component particles on the cross-section of the workpiece. Bajpai et al. [19] studied the drillability of laminates based on PLA and natural fiber (sisal and Grewia optiva fiber). The authors concluded that there was a significant impact on the quality and efficiency of machining drill types and cutting parameters. At the same time, the research did not indicate any influence of the type of natural fiber on the cutting forces.

To the best of our knowledge, research on the drilling characterization of PLA composites is still scarce and is thus worth investigating. Drilling is one of the most basic processes in processing wood-based materials that are intended for use in the furniture industry. We considered it necessary to study the susceptibility of drilling of WPCs with polylactic acid (PLA) and high-density polyethene (HDPE), as their use in the industry is constantly increasing. Based on the available literature, the hypothesis that the drilling process will be affected by the composition of WPCs was formed.

2. Materials and Methods

The polymer matrices used in the tested WPC panels were polylactic acid (PLA; Ingeo TM Biopolymer 2003D, NatureWorks LLC, Minnetonka, MN, USA) and high-density (HDPE; Hostalen GD 7255, Basell Orlen Polyolefins Sp. Zoo, Pock, Poland). Two types of lignocellulosic material were used as fillers: coniferous sawdust and bark. Calcium oxide (CaO; Avantor Performance Materials Poland SA, Gliwice, Poland) was applied in PLA composites and polyethylene-graft-maleic anhydride (MAHPE; SCONA TSPE 2102 GAHD, BYK-Chemie GmbH, Wesel, Germany) in HDPE. Tables 1 and 2 describe all 36 variants.

The sawmill lignocellulosic material was dried to 5% humidity before being mechanically comminuted and sorted into two size variants: >35 and 10–35 mesh.

The composites manufacture consisted of two stages:

(1) Production of WPC granules of an appropriate formulation (Tables 1 and 2) using an extruder (Leistritz Extrusionstechnik GmbH, Nürnberg, Germany), where temperatures in individual sections of the extruder were 170–180 °C, and an obtained continuous composite band was ground in a hammer mill;
(2) Manufacture of boards from the granules obtained, with nominal dimensions of 300 × 300 × 2.5 mm in the process of flat pressing in a mold. For this purpose, a one-shelf press (AB AK Eriksson, Mariannelund, Sweden) was used. The set-

tings of the process were as follows: 200 °C, a maximum unit pressing pressure p_{max} = 1.25 MPa, and pressing time 6 min. After hot pressing, the boards were cooled in the mold for 6 min in a cold press. Figure 1 shows examples of the manufactured panels. The boards were conditioned for 7 days under standard conditions (20 ± 2 °C; 65% ± 5% humidity).

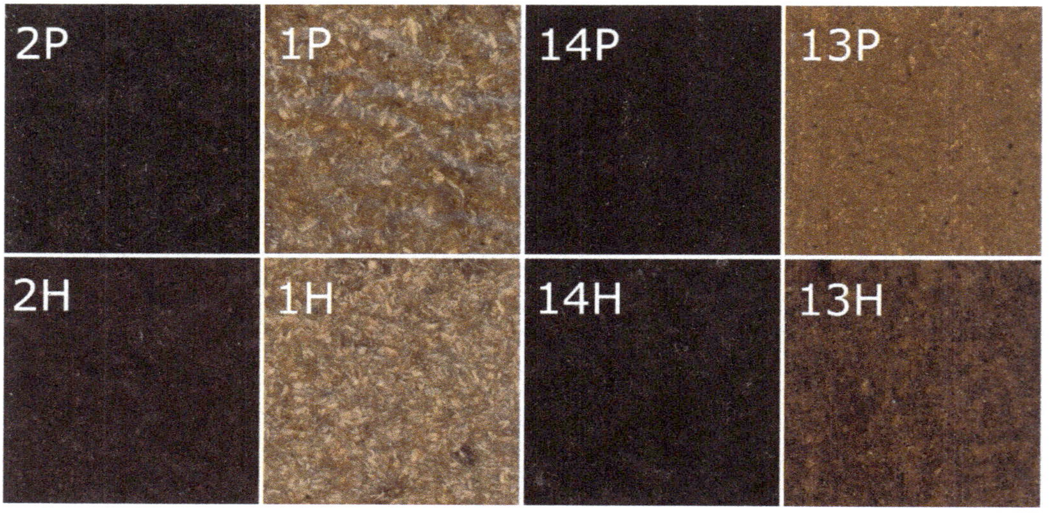

Figure 1. Examples of PLA- and HDPE-based composites.

Table 1. PLA-based composites characteristics.

Variant	Matrix	Share of Matrix	Share of Additives	Share of Filler	
				Small Particles (>35 Mesh)	Large Particles (10–35 Mesh)
1P	PLA	60	-	-	40 s
2P			-	-	40 b
3P			-	40 b	-
4P			-	40 s	-
5P		50	-	-	50 s
6P			-	-	50 b
7P			-	50 b	-
8P			-	50 s	-
9P		40	-	-	60 s
10P			-	-	60 b
11P			-	60 b	-
12P			-	60 s	-
13P		57	3	-	40 s
14P			3	40 b	-
15P		47	3	-	50 s
16P			3	50 b	-
17P		37	3	-	60 s
18P			3	60 b	-

s—sawdust, b—bark.

Table 2. HDPE-based composites characteristics.

Variant	Matrix	Share of a Matrix	Share of Additives	Share of a Filler	
				Small Particles (>35 Mesh)	Large Particles (10–35 Mesh)
1H	HDPE	60	-	-	40 s
2H			-	-	40 b
3H			-	40 b	-
4H			-	40 s	-
5H		50	-	-	50 s
6H			-	-	50 b
7H			-	50 b	-
8H			-	50 s	-
9H		40	-	-	60 s
10H			-	-	60 b
11H			-	60 b	-
12H			-	60 s	-
13H		57	3	-	40 s
14H			3	40 b	-
15H		47	3	-	50 s
16H			3	50 b	-
17H		37	3	-	60 s
18H			3	60 b	-

s—sawdust, b—bark.

The density according to EN 323: 1999 (EN 323, 1999) and density profile were determined for manufactured boards using a Laboratory Density Analyzer DAX GreCon (Fagus-Grecon Greten GmbH & Co. KG, Alfeld, Germany). A density measurement was recorded every 0.02 mm at a measurement speed of 0.05 mm/s.

Machinability tests were carried out on a CNC Busellato Jet 130 (Casadei-Busellato, Thiene, Italy) machining center. For through drilling (through the entire thickness of the board), a new 8 mm diameter polycrystalline diamond DPI single-point drill (Leitz) was used. The following cutting parameters were used: rotational speed of 6000 rpm, feed speed of 1.2 m/min, and 0.2 mm feed per revolution. During the test, through holes were drilled (through the entire thickness of 3 boards, in the package) in the center of the boards with dimensions of 50 × 50 mm. During drilling, the signals of the axial force Fz were recorded using a Kistler 9345A piezoelectric force sensor with a sampling frequency of 12 kHz. For each variant of the plate, 6 cuts were made. The effective value (RMS) of the axial force signals was analyzed.

Statistical Analysis

The results were statistically processed using Statistica 13.3 (TIBCO Software Inc., Palo Alto, CA, USA). The analysis of variance (MANOVA) was used for determining the significant differences ($\alpha = 0.05$) between factors, and Tukey's test was used for finding homogeneous groups at a significance level of 0.05.

3. Results and Discussion

The PLA boards had densities ranging from 1061 to 1232 kg/m^3, while the HDPE boards had densities ranging from 1002 to 1105 kg/m^3. Tables 3 and 4 show the values of the mean densities for individual variants. Individual variant density diversity within the same matrices did not exceed 17% for PLA and 11% for HDPE. There was a slight influence from the size of the filler particles, the type of filler, or the addition of additives in this regard. It should be noted, however, that PLA boards with the same filler share and type, as well as filler particle size, were 1 to 18% denser, depending on the variant.

Table 3. The mean density values for individual variants of panels without additives.

Matrix	Share of Filler	Bark Large Particles		Bark Small Particles		Sawdust Large Particles		Sawdust Small Particles	
		ρ	σ	ρ	σ	ρ	σ	ρ	σ
	(%)	(kg/m³)							
PLA	40	1167	47	1171	41	1092	65	1152	56
PLA	50	1159	65	1114	46	1118	44	1146	62
PLA	60	1182	58	1123	51	1061	49	1148	56
HDPE	40	1051	24	1053	26	1025	16	1025	25
HDPE	50	1078	25	1105	27	1035	22	1039	41
HDPE	60	1094	29	1077	38	1026	18	1076	25

ρ—density, σ—standard deviation.

Table 4. The mean density values for individual variants of panels with additives.

Matrix	Share of Filler	Bark Small Particles		Sawdust Large Particles	
		ρ	σ	ρ	σ
	(%)	(kg/m³)			
PLA	40	1180	34	1075	62
PLA	50	1232	62	1094	57
PLA	60	1123	54	1091	34
HDPE	40	1040	39	1002	26
HDPE	50	1043	27	1036	30
HDPE	60	1093	25	1081	22

ρ—density, σ—standard deviation.

This is because the PLA matrix has a higher density than the HDPE matrix. Andrzejewski et al. [20] discovered similar relationships while researching WPCs based on PLA and PP. The density distribution on the cross-section of all variants of the tested boards was generally uniform (Figure 2). The density differentiation in the thickness of the individual boards did not exceed 100 kg/m^3 (only in some variants did it reach 200 kg/m^3), demonstrating good homogenization of the composite components and uniform distribution of the filler particles in the polymer matrix. Borysiuk et al. [21] discovered that increasing the filler content in HDPE boards filled with sawdust reduces density in the central zone of the board.

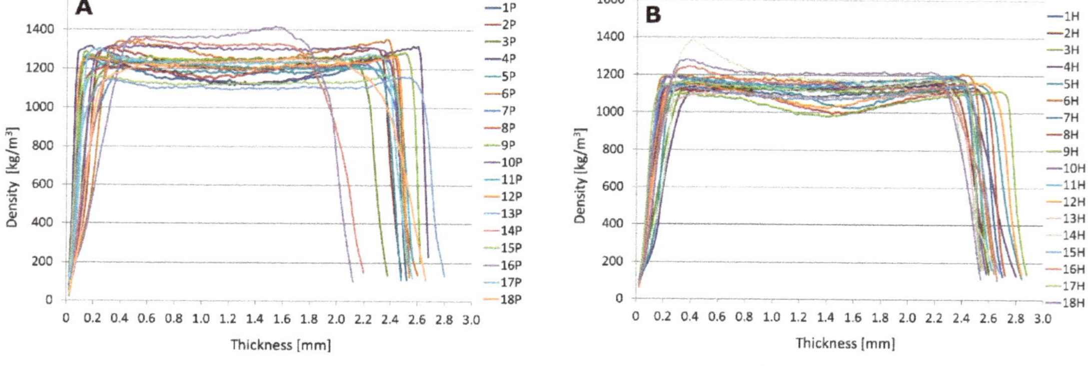

Figure 2. Density profiles of PLA-based composites (**A**) and HDPE-based composites (**B**).

The authors also concluded that the size of the filler particles had a slight influence on density or density profile. This was also confirmed in our research. Density variation may have an impact on board processing. According to Kowaluk [22], the majority of machining parameters (cutting force, feed force, and noise) decreases as the processed material density decreases. The recorded density changes in the tested WPCs had no significant effect on the values of axial forces during drilling (Figure 3).

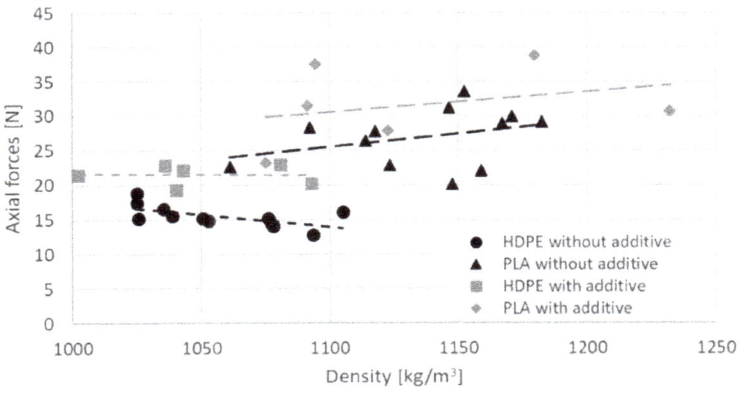

Figure 3. Relationship between axial forces during drilling and the density of WPCs.

Figure 4 shows the axial force values obtained while drilling individual composite variants. Regardless of variant content, composites made from PLA had from 25% to 56% higher values of axial forces during drilling. The observed differences were significant in each of the cases. PLA stiffness is associated with higher axial drilling forces for PLA-based composites compared with HDPE-based composites. PLA has a modulus of elasticity that is 2.5 to 5 times greater than that of HDPE [23,24].

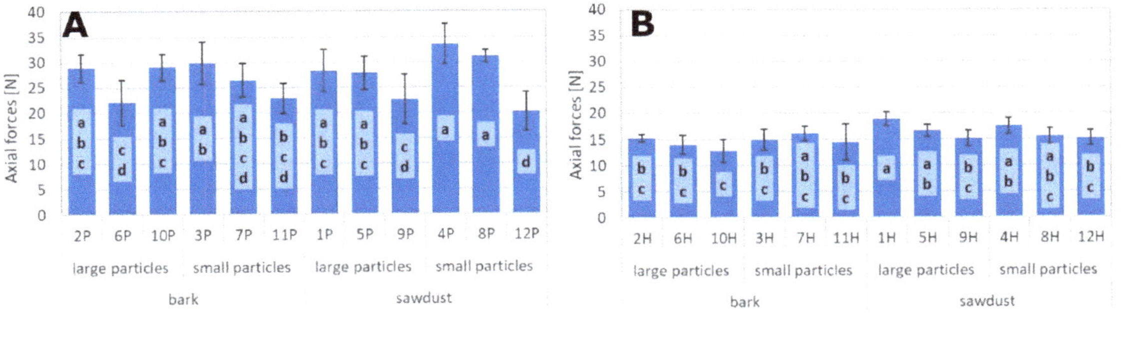

Figure 4. Values of axial forces during drilling of PLA-based composites (**A**) and HDPE-based composites (**B**) (a, b, c, d—homogeneous groups by Tukey's test).

It should also be noted that the type of matrix, compared with factors such as the type, proportion, and size of the filler, had a significant (67.05%) effect on the value of the axial force (Table 5). Wilkowski et al. [15] also noted the effect of matrix type (PE, PP, and PS) on cutting forces when drilling WPCs. The authors found that composites made of polyethylene baize had the lowest cutting forces. In our study, both PLA and HDPE composites had lower axial forces during drilling compared with the data available in the literature [25].

Table 5. Analysis of variance for selected factors and interactions between the factors influencing the axial force during drilling.

Factor	SS	Df	MS	F	p	X
Matrix (M)	4644.59	1	4644.59	561.268	0.000	67.05
Filler share (FSh)	458.87	2	229.44	27.726	0.000	6.62
Filler size (FSi)	9.23	1	9.23	1.116	0.293	0.13
Filler (F)	60.7	1	60.7	7.335	0.008	0.88
M × FSh	109.86	2	54.93	6.638	0.002	1.59
M × FSi	4.57	1	4.57	0.552	0.459	0.07
FSh × FSi	98.66	2	49.33	5.961	0.003	1.42
M × F	12.21	1	12.21	1.475	0.227	0.18
FSh × F	144.4	2	72.2	8.725	0.000	2.08
FSi × F	0.4	1	0.4	0.048	0.826	0.01
M × FSh × FSi	157.64	2	78.82	9.525	0.001	2.28
M × FSh × F	161.61	2	80.8	9.765	0.001	2.33
M × FSi × F	41.75	1	41.75	5.046	0.027	0.60
FSh × Fsi × F	23.66	2	11.83	1.43	0.243	0.34
M × FSh × Fsi × F	5.31	2	2.65	0.321	0.726	0.08
Error	993.02	120	8.28			14.34

SS—sum of the squares of deviations from the average value, Df—number of degrees of discretion, MS—average square of deviations (MS = SS/Df), F—test value, p—probability of error, X—percentage influence of factors on the examined property of particleboard.

When the type, share, and size of the filler were considered, only the size of the filler particles had a significant effect on the values of axial forces during drilling (Table 5). This is because uncontrolled mechanical division of larger filler particles into smaller elements and partial destruction of the cell wall structure may occur during the extrusion process [26–29]. The moisture content of the wood as well as the geometry and operating parameters of the extruder screw all play a role in this division. Furthermore, during extrusion, the lignocellulosic filler particles compacted to approximately 1300 kg/m^3 density [11].

Increased lignocellulosic filler content resulted in a decrease in the value of axial forces during drilling (the percentage effect was 6.62%), with a greater decrease observed in composites based on PLA. The percentage effect of filler type (0.88%) was relatively small compared with that of other factors significantly affecting the values of axial forces (type of matrix, filler share). It is also worth noting that the percentage effects of interactions between the studied factors on axial forces were relatively small, ranging from 0.60% to 2.33% in cases where they were significant (Table 5). Furthermore, the type, proportion, and size of the filler, as well as the interactions between the factors studied, had a smaller influence on the axial forces than the influence of factors not studied (error = 14.34%).

The addition of additives to the composites (CaO for PLA and MAHPE for HDPE) increased the value of axial forces during drilling in general (Figure 5). The percentage effect of CaO addition in PLA-based composites was 16.64% (Table 6), and, in most cases, this effect was insignificant (Figure 5A). This is because CaO is both a humectant and a biocide [30]. In the case of PLA-based composites, factors not included in the current study had a much greater percentage impact than the additive used (error = 70.84%).

In the case of HDPE-based composites, the percentage effect of MAHPE addition on the increase in the value of axial forces during drilling was 59.36% (Table 7). This effect was significant in the majority of cases (Figure 5B). This is because the addition of coupling agents (such as MAHPE) improves the mechanical properties of the WPCs produced by better bonding the matrix and filler particles [26,31–37]. This results in increased material resistance during the drilling process. The addition of MAHPE had a much larger percentage effect (59.36%) on the increase in the value of axial forces during drilling than the nonstudied factors (error = 34.67%) (Table 7).

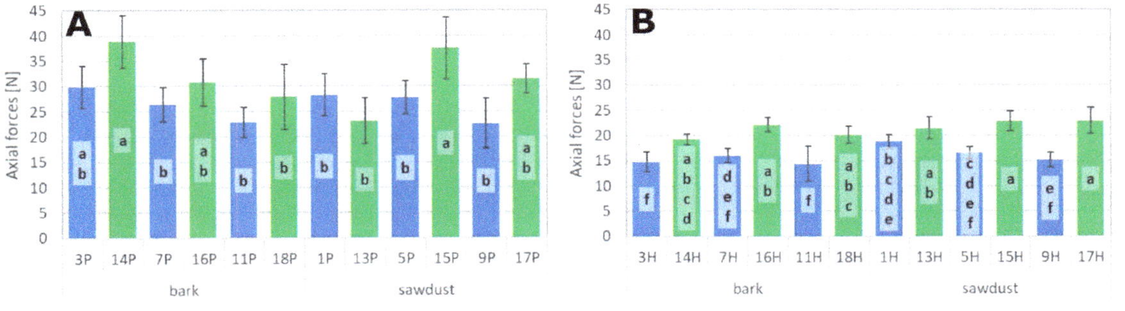

Figure 5. Values of axial forces during drilling: (**A**) PLA-based composites with (green) and without (blue) addition of CaO; (**B**) HDPE-based composites with (green) and without (blue) addition of MAHPE (a, b, c, d, e, f—homogeneous groups by Tukey's test).

Table 6. Analysis of variance for selected factors and interactions between the factors influencing the axial force during drilling PLA-based composite with addition of CaO.

Factors	SS	Df	MS	F	p	X
Filler Share (FSh)	277.92	2	138.96	4.24	0.018	9.10
Additive (A)	508.03	1	508.03	15.50	0.000	16.64
FSh × A	104.25	2	52.13	1.59	0.212	3.41
Error	2163.04	66	32.77			70.84

SS—sum of the squares of deviations from the average value, Df—number of degrees of discretion, MS—average square of deviations (MS = SS/Df), F—test value, p—probability of error, X—percentage influence of factors on the examined property of particleboard.

Table 7. Analysis of variance for selected factors and interactions between the factors influencing the axial force during drilling HDPE-based composite with addition of MAHPE.

	SS	Df	MS	F	p	X
Filler Share (FSh)	18.72	2	9.36	1.956	0.149	2.05
Additive (A)	540.72	1	540.72	112.993	0.000	59.36
FSh × A	35.68	2	17.84	3.728	0.029	3.92
Error	315.83	66	4.79			34.67

SS—sum of the squares of deviations from the average value, Df—number of degrees of discretion, MS—average square of deviations (MS = SS/Df), F—test value, p—probability of error, X—percentage influence of factors on the examined property of particleboard.

4. Conclusions

Drilling is one of the machining methods available for WPCs. The following conclusions could be drawn from the drilling of PLA and HDPE composites:

1. The type of matrix had the greatest impact on the axial force values during drilling. Regardless of the composite formulation, composites based on PLA had higher axial forces than composites based on HDPE.
2. The increase in the share of lignocellulosic fillers generally influenced the decrease in the value of axial forces during drilling, with PLA-based composites showing a more significant decrease than HDPE-based composites.
3. The type and size of the filler had a minor impact on the axial force values during drilling.
4. The increase in axial forces in the case of PLA was due to factors other than the presence of additives. In the case of HDPE, additives had a greater contribution to increasing the values of axial forces than other nonstudied factors.

Author Contributions: Conceptualization, P.B. and R.A.; methodology, P.B., R.A. and J.W.; validation, P.B., R.A. and J.W.; formal analysis, R.A., A.A. and J.W.; investigation, J.W. and A.T.; data curation, P.B. and R.A.; writing—original draft preparation, P.B., R.A. and A.T.; writing—review and editing, A.A., J.W. and L.S.H.; visualization, R.A. and A.A.; supervision, P.B., R.A. and L.S.H.; project administration, P.B. All authors have read and agreed to the published version of the manuscript.

Funding: The presented research was financed under the "Strategic research and development program: environment, agriculture, and forestry" (BIOSTRATEG, grant no. BIOSTRATEG3/344303/14/NCBR/2018). The funding institution was The National Centre for Research and Development, Poland.

Data Availability Statement: Not applicable.

Conflicts of Interest: The authors declare no conflict of interest.

References

1. Partanen, A.; Carus, M. Biocomposites, Find the Real Alternative to Plastic—An Examination of Biocomposites in the Market. *Reinf. Plast.* **2019**, *63*, 317–321. [CrossRef]
2. Kuciel, S.; Liber-Kneć, A.; Mikuła, J.; Kuźniar, P.; Korniejenko, K.; Żmudka, S.; Łagan, S.; Ryszkowska, J.; Gajewski, J.; Sałasiński, K.; et al. *Kompozyty Polimerowe Na Osnowie Recyklatów z Włóknami Naturalnymi: Praca Zbiorowa*; Kuciel, S., Ed.; Politechnika Krakowska: Kraków, Poland, 2010; ISBN 9788372425607.
3. Borysiuk, P.; Boruszewski, P.; Auriga, R.; Danecki, L.; Auriga, A.; Rybak, K.; Nowacka, M. Influence of a Bark-Filler on the Properties of PLA Biocomposites. *J. Mater. Sci.* **2021**, *56*, 9196–9208. [CrossRef]
4. Markarian, J. Biopolymers Present New Market Opportunities for Additives in Packaging. *Plast. Addit. Compd.* **2008**, *10*, 22–25. [CrossRef]
5. Gozdecki, C.; Kociszewski, M.; Zajchowski, S.; Patuszyński, K. Wood-Based Panels as Filler of Wood-Plastic Composites. *Ann. Warsaw Agric. Univ. For. Wood Technol.* **2005**, *56*, 255–258.
6. Chaharmahali, M.; Tajvidi, M.; Najafi, S.K. Mechanical Properties of Wood Plastic Composite Panels Made from Waste Fiberboard and Particleboard. *Polym. Compos.* **2008**, *29*, 606–610. [CrossRef]
7. Gozdecki, C.; Kociszewski, M.; Wilczyński, A.; Zajchowski, S. The Possibility of Using Wood Dust for Manufacturing Wood-Plastic Composites. In Proceedings of the XX International Symposium: "Adhesives in Woodworking Industry", Zvolen, Slovakia, 29 June–1 July 2011; pp. 86–91.
8. Myers, G.E.; Clemons, C.M. *Wastepaper Fiber in Plastic Composites Made by Melt Blending: Demonstration of Commercial Feasibility*; Final Report for Solid Waste Reduction and Recycling Demonstration Grant Program Project No 91-5; Wisconsin Department of Natural Resources, Forest Products Laboratory: Madison, WI, USA, 1993.
9. Ashori, A.; Nourbakhsh, A. Characteristics of Wood-Fiber Plastic Composites Made of Recycled Materials. *Waste Manag.* **2009**, *29*, 1291–1295. [CrossRef]
10. Nurul Fazita, M.R.; Jayaraman, K.; Bhattacharyya, D.; Mohamad Haafiz, M.K.; Saurabh, C.K.; Hazwan Hussin, M.; Abdul Khalil, H.P.S. Green Composites Made of Bamboo Fabric and Poly (Lactic) Acid for Packaging Applications—A Review. *Materials* **2016**, *9*, 435. [CrossRef]
11. Klysov, A.A. *Wood-Plastic Composites*; Wiley-Interscience: Hoboken, NJ, USA, 2007; ISBN 978-0-470-14891-4.
12. Buehlmann, U.; Saloni, D.; Lemaster, R.L. Wood Fiber-Plastic Composites: Machining and Surface Quality. In Proceedings of the 15th International Wood Machining Seminar, Anaheim, CA, USA, 30 July–1 August 2001; pp. 1–13.
13. Zbieć, M.; Borysiuk, P.; Mazurek, A. Polyethylene Bonded Composite Chipboard. In Proceedings of the 7th International Science Conference: "Chip and Chipless Woodworking Processes", Terchová, Slovakia, 9–11 September 2010; pp. 237–242.
14. Zbiec, M.; Borysiuk, P.; Mazurek, A. Thermoplastic Bonded Composite Chipboard Part 2—Machining Tests. In Proceedings of the Proceedings of the 8th International Science Conference: "Chip and Chipless Woodworking Processes", Zvolen, Slovakia, 6–8 September 2012; pp. 399–405.
15. Wilkowski, J.; Borysiuk, P.; Górski, J.; Czarniak, P. Analiza Wzgle{ogonek}dnych Wskaźników Skrawalności Płyt Wiórowych Spajanych Termoplastami Poużytkowymi. *Drewno* **2013**, *190*, 139–144. [CrossRef]
16. Borysiuk, P.; Wikowski, J.; Krajewski, K.; Auriga, R.; Skomorucha, A.; Auriga, A. Selected Properties of Flat-Pressed Wood-Polymer Composites for High Humidity Conditions. *BioResources* **2020**, *15*, 5141–5155. [CrossRef]
17. Zajac, J.; Mital', D.; Radchenko, S.; Kokul'a, P.; Olexa, L'.; Čep, R. Short-Term Testing of Cutting Materials Using the Method of Interrupted Cut. *Appl. Mech. Mater.* **2014**, *616*, 236–243.
18. Somsakova, Z.; Zajac, J.; Michalik, P.; Kasina, M. Machining of Wood Plastic Composite (Pilot Experiment). *Mater. Plast.* **2012**, *49*, 55–57.
19. Bajpai, P.K.; Debnath, K.; Singh, I. Hole Making in Natural Fiber-Reinforced Polylactic Acid Laminates: An Experimental Investigation. *J. Thermoplast. Compos. Mater.* **2017**, *30*, 30–46. [CrossRef]

20. Andrzejewski, J.; Szostak, M.; Barczewski, M.; Łuczak, P. Cork-Wood Hybrid Filler System for Polypropylene and Poly(Lactic Acid) Based Injection Molded Composites. Structure Evaluation and Mechanical Performance. *Compos. Part B Eng.* **2019**, *163*, 655–668. [CrossRef]
21. Borysiuk, P.; Auriga, R.; Kośka, P. Influence of the Filler on the Density Profile of Wood Polymer Composites. *Ann. WULS, For. Wood Technol.* **2019**, *106*, 31–37. [CrossRef]
22. Kowaluk, G. Machining Processes for Wood-Based Composite Materials. In *Machining Technology for Composite Materials*; Elsevier: Amsterdam, The Netherlands, 2012; pp. 412–425, ISBN 9780857090300.
23. Osswald, T.A.; Baur, E.; Brinkmann, S.; Oberbach, K.; Schmachtenbe, E. *International Plastics Handbook*; Carl Hanser Verlag GmbH & Co. KG: Munich, Germany, 2006; ISBN 9781569903995.
24. Farah, S.; Anderson, D.G.; Langer, R. Physical and Mechanical Properties of PLA, and Their Functions in Widespread Applications—A Comprehensive Review. *Adv. Drug Deliv. Rev.* **2016**, *107*, 367–392. [CrossRef]
25. Auriga, R.; Borysiuk, P.; Waracka, A.; Auriga, A. Susceptibility of Drilling Particleboard with Share of Hemp Shives. *Biul. Inf. OB-RPPD* **2021**, *3–4*. [CrossRef]
26. Bledzki, A.K.; Letman, M.; Viksne, A.; Rence, L. A Comparison of Compounding Processes and Wood Type for Wood Fibre—PP Composites. *Compos. Part A Appl. Sci. Manuf.* **2005**, *36*, 789–797. [CrossRef]
27. Bouafif, H.; Koubaa, A.; Perre, P.; Cloutier, A. Effects of Composite Processing Methods on Wood Particle Development and Length Distribution: Consequences on Mechanical Properties of Wood-Thermoplastic Composites. *Wood Fiber Sci.* **2010**, *42*, 62–70.
28. Gacitua, W.; Bahr, D.; Wolcott, M. Damage of the Cell Wall during Extrusion and Injection Molding of Wood Plastic Composites. *Compos. Part A Appl. Sci. Manuf.* **2010**, *41*, 1454–1460. [CrossRef]
29. Hietala, M.; Niinimäki, J.; Oksman, K. Processing of Wood Chip-Plastic Composites: Effect on Wood Particle Size, Microstructure and Mechanical Properties. *Plast. Rubber Compos.* **2011**, *40*, 49–56. [CrossRef]
30. Commission Delegated Regulation (EU) No 1062/2014. *Off. J. Eur. Union* **2014**, *294*, 20–30.
31. Li, T.Q.; Li, R.K.Y. A Fracture Mechanics Study of Polypropylene-Wood Flours Blends. *Polym. Plast. Technol. Eng.* **2001**, *40*, 1–21. [CrossRef]
32. Balasuriya, P.W.; Ye, L.; Mai, Y.W. Morphology and Mechanical Properties of Reconstituted Wood Board Waste-Polyethylene Composites. *Compos. Interfaces* **2003**, *10*, 319–341. [CrossRef]
33. Kuan, H.C.; Huang, J.M.; Ma, C.C.M.; Wang, F.Y. Processability, Morphology and Mechanical Properties of Wood Flour Reinforced High Density Polyethylene Composites. *Plast. Rubber Compos.* **2003**, *32*, 122–126. [CrossRef]
34. Lee, S.Y.; Yang, H.S.; Kim, H.J.; Jeong, C.S.; Lim, B.S.; Lee, J.N. Creep Behavior and Manufacturing Parameters of Wood Flour Filled Polypropylene Composites. *Compos. Struct.* **2004**, *65*, 459–469. [CrossRef]
35. Cai, X.; Riedl, B.; Bouaziz, M. Lignocellulosic Composites with Grafted Polystyrene Interfaces. *Compos. Interfaces* **2005**, *12*, 25–39. [CrossRef]
36. Cui, Y.; Lee, S.; Noruziaan, B.; Cheung, M.; Tao, J. Fabrication and Interfacial Modification of Wood/Recycled Plastic Composite Materials. *Compos. Part A Appl. Sci. Manuf.* **2008**, *39*, 655–661. [CrossRef]
37. San, H.P.; Nee, L.A.; Meng, H.C. Physical and Bending Properties of Injection Moulded Wood Plastic Composites Boards. *ARPN J. Eng. Appl. Sci.* **2008**, *3*, 13–19.

forests

Communication

The Effect of the Tree Dieback Process on the Mechanical Properties of Pine (*Pinus sylvestris* L.) Wood

Zbigniew Malinowski [1,*], Jakub Kawalerczyk [2], Joanna Walkiewicz [2], Dorota Dziurka [2] and Radosław Mirski [2,*]

[1] Regional Directorate of State Forests in Katowice, 40-453 Katowice, Poland
[2] Department of Mechanical Wood Technology, Faculty of Forestry and Wood Technology, Poznań University of Life Sciences, 60-637 Poznań, Poland
* Correspondence: zbyma@wp.pl (Z.M.); radoslaw.mirski@up.poznan.pl (R.M.)

Abstract: As a result of the progressing climate changes, there is an increase in the volume of pine deadwood harvested each year from Polish forests. Its presence is an important part of the forest ecosystem; however, there are some indications that the material obtained from dying trees can be characterized by lower quality and properties. Taking into account the growing issue of tree dieback, the volume of pine wood annually harvested in Poland, and the importance of wooden products from an economic standpoint, preliminary research aimed at recognizing the process and its effect on the mechanical properties was conducted. Model trees in Brzeg Forest District were selected based on the crown defoliation. The properties of wood obtained from trees representing three different categories of soundness were determined according to the relevant standards. Based on the results of density, modulus of elasticity, bending strength, and compressive strength, it was found that there were statistically significant differences in wood quality depending on the condition of the tree. The results were particularly interesting in the case of compressing strength, where a healthy tree of lower density showed a similar strength to a dying tree with a much higher density.

Keywords: pine; deadwood; tree dieback process; mechanical properties

1. Introduction

Scots pine (*Pinus sylvestris* L.) is the basic forest-forming species in Poland as it accounts for almost 60% of Polish forests. There was nearly 40 million m^3 of wood harvested in Poland in the recent period and pine wood formed the vast majority of it. Therefore, pine wood occupies a leading position in terms of suitability for industrial production. It is widely used in both a round state (construction stamps, mines, telecommunication poles, piles) and a sawn form [1–3].

In recent years, which were characterized by a significant decrease in precipitation, there was a rapid increase in the release of pine deadwood in the forests, which indicates the occurrence of a drought period. The appearance of drought as a phenomenon is not entirely new. The process of Scots pine dieback in a forest stands has been observed since the turn of the 20th and 21st centuries in many places, usually representing the area of its natural occurrence in the European climate zone. Pine dies especially at the border of its natural range, where the impact of environmental limiting factors may be stronger; however, there are also numerous locations distant from these borders, for example those in Poland. Moreover, the dieback process of pine is interpreted as the symptom of changes in the species structure in Europe [4–11]. The occurrence of drought periods was recorded in the years 1920–21, 1930–31, 1950–54, 1962–63, and 1989–92 [12]. A similar situation has been observed in recent years. The statistics obtained from the State Forests show an upward trend in the volume of harvested deadwood in the last five years. Until 2015, the amount of harvested dead pine wood oscillated around 2%. Then, in the following years, it gradually increased to reach the level of about 10% in 2020. However, the occurrence of deadwood does not apply to the entire region of Poland. The greatest weakening of stands, and hence the appearance of a large amount of deadwood, concerns mainly western and south-western Poland.

The presence of deadwood itself in forest stands is important for the entire ecosystem and it plays the crucial role in nutrient cycling, soil formation, and sustaining the biodiversity in general [13]. However, it has to be remembered that wood as a material has been indispensable in many applications for centuries; therefore, managing the deadwood, which represents a significant economic resource, is the issue that has to be thoroughly evaluated [14,15]. Literature reports indicate that deadwood may be characterized by a lower moisture content; lower density; and lower concentration of lignin, hemicelluloses, and extractives [16,17]. Thus, it may affect the properties of wood; cause technological difficulties in its processing; and, consequently, also limit the possibility of application. Moreover, the progressing dieback process can also affect the quality of chips for pulping, affect the quality of flakes for composites' production, and deteriorate both the physical and mechanical properties of manufactured particleboards [18].

The problem of the occurrence and role of deadwood in nature as well as, more broadly, understanding the issue of deadwood, recently became a scientific interest of many authors [19–24]. Unfortunately, there are no literature data describing the correlation between the reduction of the assimilation apparatus and the technical quality of wood. However, taking into account the complexity of the process, the effect of regional conditions, and the large number of influencing factors, more studies are needed to recognize the process and assess the possibility of slowing it down. Therefore, the presented study aimed to determine the mechanical properties of the wood originating from dying trees. Another aspect is the fact that, currently, there are no standards that define the condition of dying wood. Therefore, the conducted research is a part of an attempt to define the condition of trees during the process of dying. This is the introduction to extensive research and a further project that will allow for a multi-factor analysis of the impact of progressing dieback process on the properties of harvested wood.

2. Materials and Methods

In order to conduct the preliminary studies regarding the influence of the crown defoliation on changes in the properties of wood raw material, a stand located in the Brzeg Forest District (50°85′ N; 17°45′ E) was selected (Table 1).

Table 1. Characteristics of Brzeg Forest District.

Habitat	Share of Pine	Age	Abundance	Average Height	Average Diameter
Fresh, mixed coniferous forest	100%	45 years	334 m^3/ha	20 m	19 cm

On the selected area, 12 Scots pine trees in Kraft Class II and III were selected. The chosen individual trees represented classes I and II of thickness and differed in health condition. The condition of the trees was assessed based on the condition of the tree crown and the reduction of the assimilation apparatus. The crowns of the selected trees met the following assumptions (the appearance of the crowns is presented in Figure 1):

- Group A—trees with a green crown, no discoloration, healthy, without visible reduction in needles;
- Group B—trees weakened, partially discolored crowns with a notable reduction in needles up to 40%;
- Group C—trees weakened, partially discolored or completely discolored crowns with a notable reduction in needles from 30% to 50%.

Figure 1. Categories of pine based on the crown defoliations.

The classification into these groups was made on the basis of visual assessment.

Roundwood rollers 120 cm long with a diameter ranging from 30 to 50 cm were collected from the middle section of the log, which starts above 4 m from the lower (thicker) end (butt-end section) (Figure 2). The preliminary division of rollers was carried out with the use of a mechanical log splitter system in order to maintain the longitudinal course of the fibers for the obtained pieces. The obtained sawn timber was subjected to the chamber drying process until the equilibrium moisture content of approximately 12%–15% was reached. After cooling and seasoning, the obtained pieces were cut into laboratory samples from the mature wood section located at a distance of at least 1 cm from the circumference of the trunk, as suggested by Tomczak and Jelonek [25]. Assessment of the physical and mechanical properties of pine wood was conducted according to proper standards:

- bending strength (MOR) and modulus of elasticity (MOE) based on EN 408 [26];
- compressive strength based on PN-D-04102 [27];
- density based on PN-77/D-4101 [28].

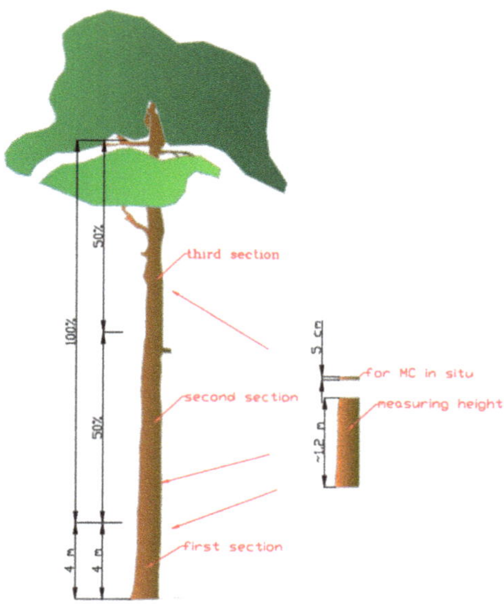

Figure 2. Scheme for obtaining samples for research.

The determination of each property was carried out on not less than 20 samples from each variant.

The obtained results were analyzed with the use of Statistica 13.0 package by performing an analysis of variance (ANOVA) at a significance level of $\alpha = 0.05$.

3. Results and Discussion

Despite that only a few investigations have been made, the obtained research results are very interesting in the context of both literature reports and planned research (Figure 3). For the analyzed mechanical properties determined in the bending test, quite clear statistical differences between the mean values were observed. What is particularly interesting is that the group of trees characterized by the highest defoliation has the highest values of the investigated features. However, despite the fact that samples from group A had a significantly lower density when compared with samples from group B, both MOR and MOE of wood obtained from the trees belonging to group A (so-called healthy trees) and B (with a defoliation of 30%) were very similar. At this stage of stand dieback, spectacular changes in the mechanical properties of wood, which, according to the literature, can be observed in the later period of degradation, have not yet been confirmed [29,30]. At present, based on the preliminary results, too few samples were collected to make unambiguous conclusions on the impact of tree dieback on the quality changes in the raw material. The observed relationships cannot be fully explained by the difference in the density of wood obtained from the selected model trees. The results of the compressive strength test performed so far indicate a significant change in strength properties. Healthy trees with a lower density showed a similar level of strength as the trees characterized by major crown damage above 30%, which can indicate the deterioration in the quality of wood harvested from dying stands. Currently, the selected model trees show a relatively good condition, which translates into the observed mechanical properties. Although these values are high, the dependencies on the degree of defoliation are no longer unequivocal. Therefore, a broader look at the problem of tree dieback and wood quality is required as numerous studies confirm that it undergoes significant decomposition in the long term [29,31,32].

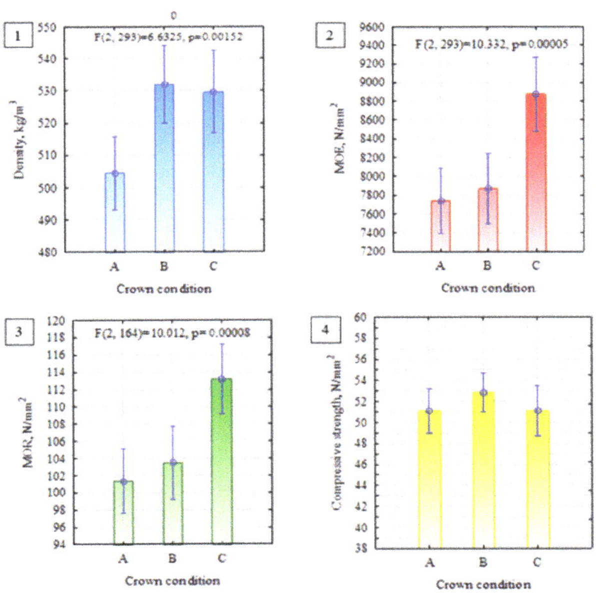

Figure 3. Properties of wood: 1—density; 2—modulus of elasticity; 3—bending strength; 4—compressive strength.

4. Conclusions

Based on the conducted preliminary research, it can be suggested that statistically significant differences in the wood properties of trees characterized by different conditions were observed. The model trees selected based on the observation of crown defoliation showed differences in both density and strength characteristics. It is especially interesting that sound, healthy trees with a lower density are characterized by a similar compressive strength compared with trees with major crown damage and much higher density. However, taking into account the preliminary nature of the study and the number of analyzed trees, more research is needed to draw clear conclusions.

Author Contributions: Conceptualization, Z.M. and R.M.; methodology, Z.M. and R.M.; validation, J.K., J.W., Z.M. and D.D.; formal analysis, Z.M., J.K. and R.M.; investigation, Z.M., J.K. and D.D.; resources, J.W.; data curation, D.D. and R.M.; writing—original draft preparation, Z.M. and R.M.; writing—review and editing, Z.M., J.K. and D.D.; visualization, J.K., J.W. and D.D.; supervision, Z.M.; project administration, R.M.; funding acquisition, R.M. All authors have read and agreed to the published version of the manuscript.

Funding: This research received no external funding.

Data Availability Statement: The data presented in this study are available upon request from the corresponding author.

Conflicts of Interest: The authors declare no conflict of interest.

References

1. Kartawik, N.; Behnke-Borowczyk, J. Fungal Communities in Pine Deadwood. *Environ. Sci. Proc.* **2022**, *22*, 41.
2. Jaworski, A. *Charakterystyka Hodowlana Drzew i Krzewów Leśnych*; PWRiL: Warszawa, Poland, 2011.
3. Białobok, S.; Boratyński, A.; Bugała, W. *Biologia Sosny Zwyczajnej*; Sorus: Kharkov, Ukraine, 1993; ISBN 83-85599-21-5.
4. Rigling, A.; Bigler, C.; Eilmann, B.; Feldmeyer-Christe, E.; Gimmi, U.; Ginzler, C.; Graf, U.; Mayer, P.; Vacchiano, G.; Weber, P.; et al. Driving factors of a vegetation shift from Scots pine to pubescent oak in dry Alpine forests. *Glob. Chang. Biol.* **2013**, *19*, 229–240. [CrossRef]
5. Camarero, J.J.; Gazol, A.; Sancho-Benages, S.; Sangüesa-Barreda, G. Know your limits? Climate extremes impact the range of Scots pine in unexpected places. *Ann. Bot.* **2015**, *116*, 917–927. [CrossRef] [PubMed]
6. Buras, A.; Schunk, C.; Zeiträg, C.; Herrmann, C.; Kaiser, L.; Lemme, H.; Straub, C.; Taeger, S.; Gößwein, S.; Klemmt, H.-J. Are Scots Pine Forest Edges Particularly Prone to Drought-Induced Mortality? *Environ. Res. Lett.* **2018**, *13*, 025001. [CrossRef]
7. Bose, A.K.; Gessler, A.; Bolte, A.; Bottero, A.; Buras, A.; Cailleret, M.; Camarero, J.J.; Haeni, M.; Hereş, A.; Hevia, A.; et al. Growth and resilience responses of Scots pine to extreme droughts across Europe depend on predrought growth conditions. *Glob. Chang. Biol.* **2020**, *26*, 4521–4537. [CrossRef] [PubMed]
8. Grabska, E.; Hawryło, P.; Socha, J. Continuous Detection of Small-Scale Changes in Scots Pine Dominated Stands Using Dense Sentinel-2 Time Series. *Remote Sens.* **2020**, *12*, 1298. [CrossRef]
9. Kunert, N. Preliminary indications for diverging heat and drought sensitivities in Norway spruce and Scots pine in Central Europe. *iForest Biogeosci. For.* **2020**, *13*, 89–91. [CrossRef]
10. Valeriano, C.; Gazol, A.; Colangelo, M.; Camarero, J.J. Drought Drives Growth and Mortality Rates in Three Pine Species under Mediterranean Conditions. *Forests* **2021**, *12*, 1700. [CrossRef]
11. Nechita, C.; Iordache, A.M.; Costinel, D.; Botoran, O.R.; Dănilă, G.; Ionete, R.E.; Varlam, M. A Tree Ring Proxy Evaluation of Declining Causes in *Pinus sylvestris* L. and *Pinus Nigra* JF Arnold in Northeastern Romania. *Forests* **2022**, *13*, 336. [CrossRef]
12. Ski, O. *Poradnik Ochrony Lasu*; Wydawnictwo Świat: Warszawa, Poland, 2001.
13. Vandekerkhove, K.; De Keersmaeker, L.; Menke, N.; Meyer, P.; Verschelde, P. When nature takes over from man: Dead wood accumulation in previously managed oak and beech woodlands in North-western and Central Europe. *For. Ecol. Manag.* **2009**, *258*, 425–435. [CrossRef]
14. Byrne, T.; Stonestreet, C.; Peter, B. Characteristics and Utilization of Post-Mountain Pine Beetle Wood in Solid Wood Products. In *The Mountain Pine Beetle: A Synthesis of Biology, Management and Impacts on Lodgepole Pine*; Canadian Forest Service: Ottawa, ON, Canada, 2006; pp. 233–253.
15. Kawalerczyk, J.; Kuliński, M.; Dziurka, D.; Mirski, R. The possibility to use a side-timber in glulam beams manufacturing for structural applications. In *Annals of Warsaw University of Life Sciences-SGGW. Forestry and Wood Technology*; Warsaw University of Life Sciences Press: Warsaw, Poland, 2021.
16. Barron, E.H. Deterioration of Southern Pine Beetle-Killed Trees. *For. Prod. J.* **1971**, *1971*, 57–59.

17. Woo, K.L.; Watson, P.; Mansfield, S.D. The Effects of Mountain Pine Beetle Attack on Lodgepole Pine Wood Morphology and Chemistry: Implications for Wood and Fiber Quality. In *Wood and Fiber Science*; The Sheridan Press: Sheridan, WY, USA, 2005; pp. 112–126.
18. Bekhta, P.; Kozak, R.; Gryc, V.; Sebera, V.; Tippner, J. Effects of Wood Particles from Deadwood on the Properties and Formaldehyde Emission of Particleboards. *Polymers* **2022**, *14*, 3535. [CrossRef] [PubMed]
19. Domański, S. *Badania Nad Przyczynami Powstawania Posuszu w Starszych Drzewostanach Sosnowych w Wielkopolskim Parku Narodowym w Ludikowie*; Państwowe Wydawnictwo Rolnicze i Leśne: Warsaw, Poland, 1953.
20. Behnke-Borowczyk, J.; Wołowska, D. The identification of fungal species in dead wood of oak. *Acta Sci. Pol. Silvarum Colendarum Ratio Ind. Lignaria* **2018**, *17*, 17–23. [CrossRef]
21. Kowalski, T.; Sowa, J.; Łakomy, P. Mycobiota in Trunks of Dying Spruce Trees in the "Puszcza Białowieska" Promotional Forest Complex and Its Ecological Function. *Sylwan* **2019**, *163*, 496–507.
22. Mazur, A.; Przybysz, K.; Kuźmiński, R.; Adamowicz, K.; Jaszczak, R.; Łakomy, P.; Kwaśna, H.; Szewczyk, W.; Turski, M.; Zientarski, J. Sukcesja Owadów Podkorowych w Sztucznie Inicjowanym Procesie Zamierania Sosny. *Acta Sci. Pol. Silvarum Colendarum Ratio Ind. Lignaria* **2015**, *14*, 241–259. [CrossRef]
23. Holeksa, J.; Kapusta, P.; Budziakowska-Kubik, E.; Izworska, K.; Kurek, P.; Piechnik, Ł.; Szarek-Łukaszewska, G.; Zielonka, T.; Żywiec, M. Stan Martwego Drewna Na Terenie Puszczy Niepołomickiej Jako Wynik Długotrwałego Użytkowania Lasu i Krótkotrwałej Ochrony w Ramach Sieci Natura 2000. *Fragm. Florist. Geobot. Pol.* **2020**, *27*, 119–139. [CrossRef]
24. Kuzminski, R.; Lakomy, P.; Mazur, A. Zamieranie Dębów-Historia, Przyczyny i Objawy. *Zarz. Ochr. Przyr. Lasach* **2007**, *1*, 194–208.
25. Tomczak, A.; Jelonek, T. Parametry Techniczne Młodocianego i Dojrzałego Drewna Sosny Zwyczajnej (*Pinus sylvestris* L.). *Sylwan* **2012**, *156*, 695–702.
26. *EN 408*; Timber Structures—Structural Timber and Glued Laminated Timber—Determination of Some Physical and Mechanical Properties. European Committee for Standardization: Brussels, Belgium, 2012.
27. *PN-D-04102*; Wood. Determination of Compressive Strength Along Fibers. Polski Komitet Normalizacyjny: Warszawa, Poland, 1979.
28. *PN-D-04101*; Wood. Determination of Density. Polski Komitet Normalizacyjny: Warszawa, Poland, 1977.
29. Oberle, B.; Dunham, K.; Milo, A.M.; Walton, M.; Young, D.F.; Zanne, A.E. Progressive, idiosyncratic changes in wood hardness during decay: Implications for dead wood inventory and cycling. *For. Ecol. Manag.* **2014**, *323*, 1–9. [CrossRef]
30. Jelonek, T.; Klimek, K.; Kopaczyk, J.; Wieruszewski, M.; Arasimowicz-Jelonek, M.; Tomczak, A.; Grzywiński, W. Influence of the Tree Decay Duration on Mechanical Stability of Norway Spruce Wood (*Picea Abies* (L.) Karst.). *Forests* **2020**, *11*, 980. [CrossRef]
31. Přívětivý, T.; Adam, D.; Vrška, T. Decay dynamics of Abies alba and Picea abies deadwood in relation to environmental conditions. *For. Ecol. Manag.* **2018**, *427*, 250–259. [CrossRef]
32. Petrillo, M.; Cherubini, P.; Fravolini, G.; Marchetti, M.; Ascher-Jenull, J.; Schärer, M.; Synal, H.-A.; Bertoldi, D.; Camin, F.; Larcher, R.; et al. Time since death and decay rate constants of Norway spruce and European larch deadwood in subalpine forests determined using dendrochronology and radiocarbon dating. *Biogeosciences* **2016**, *13*, 1537–1552. [CrossRef]

MDPI AG
Grosspeteranlage 5
4052 Basel
Switzerland
Tel.: +41 61 683 77 34

Forests Editorial Office
E-mail: forests@mdpi.com
www.mdpi.com/journal/forests

www.ingramcontent.com/pod-product-compliance
Lightning Source LLC
Chambersburg PA
CBHW061502210926
51850CB00008B/1798
* 9 7 8 3 7 2 5 8 5 2 0 1 7 *